APPLIED CATEGORICAL STRUCTURES

A Journal Devoted to Applications of Categorical Methods in Algebra, Analysis, Order, Topology and Computer Science

VOLUME 4 – 1996

KLUWER ACADEMIC PUBLISHERS
DORDRECHT / BOSTON / LONDON

ISSN 0927–2852

ISBN-13: 978-94-010-6602-0 e-ISBN-13: 978-94-009-0263-3
DOI: 10.1007/978-94-009-0263-3

APPLIED CATEGORICAL STRUCTURES

Vol. 4 No. 1 March 1996

SPECIAL ISSUE

CATEGORICAL TOPOLOGY

Guest Editor: Eraldo Giuli

Scope. The primary goal of the journal is to promote communication and increase dissemination of new results and ideas among mathematicians and computer scientists who use categorical methods in their research.

The journal focuses on applications of results, techniques and ideas from Category Theory to Mathematics, in particular Algebra, Analysis, Order and Topology and to Computer Science.

These include the study of specific topological, algebraic and algebraic-topological constructs, categorical investigations in functional analysis, in continuous order theory, in algebraic and logical type theory, in automata theory, in data bases and in languages. Furthermore the journal intends to follow the development of emerging fields in which the application of categorical methods proves to be relevant.

The journal will publish both carefully refereed research papers and survey papers. In all cases it strives for significance, originality, good exposition and the highest scientific quality in its publications.

These articles, and others, have also been published in the book: Categorical Topology by Eraldo Giuli,
ISBN 978-94-010-6602-0

Applied Categorical Structures is published quarterly.
1996 Prices: Institutional: NLG 411,–/USD 263.00 including postage and handling. Individuals: NLG 210,–/USD 124.00 including postage and handling.
Published by Kluwer Academic Publishers, Spuiboulevard 50, P.O. Box 17, 3300 AA Dordrecht, The Netherlands, and 101 Philip Drive, Norwell, MA 02061, U.S.A.

Printed on acid-free paper

Preface

This issue contains a selected group of papers presented during the *International Workshop on Categorical Topology* held from August 31 – September 4, 1994 at the University of L'Aquila, L'Aquila, Italy.

The workshop was organized by the Department of Pure and Applied Mathematics of the University of L'Aquila and was made possible by grants from the Department, the University, the Regione Abruzzo and the Cassa Di Risparmio della Provincia de l'Aquila. We wish to thank the officers of the University and public local administrations who facilitated the financial supports.

We would also like to thank the INFN (Istituto Nazionale di Fisica Nucleare) for giving the opportunity to all participants to visit the impressive laboratories under Gran Sasso and the major of Roseto who organized a nice party in the garden of the Villa Comunale.

The meeting was organized by E. Giuli, H. Herrlich and A. Tozzi and was attended by 45 mathematicians from 11 countries.

The program consisted of 34 talks on various aspects of categorical topology.

We would like to express our gratitude to the Dean of the Faculty of Science Umberto Villante for the friendly welcome he extended to the participants on behalf of the University of L'Aquila and to the Head of the Department Josef Myjak for his encouragement.

Thanks are due to the many referees of the papers presented for publication in this issue for their indispensable assistance in selecting and improving the papers.

Finally, we would like to thank the staff of the "Aula Magna" who were always ready to give announcements on the big screen, for the invaluable technical assistance, and the students S. Chiappini, M. Giuli, C. Lattanzio, G. Ippoliti and E. Lucrezi, for their help in the organization.

ERALDO GIULI

Applied Categorical Structures **4:** 1–14, 1996.

Compactness and the Axiom of Choice

Dedicated to My Friend Louis D. Nel on His Sixtieth Birthday

HORST HERRLICH
University of Bremen, Fachbereich 3, Postfach 33 04 40, 28334 Bremen, Germany

(Received: 10 October 1994; accepted: 9 February 1995)

Abstract. In the absence of the axiom of choice four versions of compactness (A-, B-, C-, and D-compactness) are investigated.

Typical results:

1. C-compact spaces form the epireflective hull in **Haus** of A-compact completely regular spaces.
2. Equivalent are:
   (a) the axiom of choice,
   (b) A-compactness = D-compactness,
   (c) B-compactness = D-compactness,
   (d) C-compactness = D-compactness and complete regularity,
   (e) products of spaces with finite topologies are A-compact,
   (f) products of A-compact spaces are A-compact,
   (g) products of D-compact spaces are D-compact,
   (h) powers X^k of 2-point discrete spaces are D-compact,
   (i) finite products of D-compact spaces are D-compact,
   (j) finite coproducts of D-compact spaces are D-compact,
   (k) D-compact Hausdorff spaces form an epireflective subcategory of **Haus**,
   (l) spaces with finite topologies are D-compact.
3. Equivalent are:
   (a) the Boolean prime ideal theorem,
   (b) A-compactness = B-compactness,
   (c) A-compactness and complete regularity = C-compactness,
   (d) products of spaces with finite underlying sets are A-compact,
   (e) products of A-compact Hausdorff spaces are A-compact,
   (f) powers X^k of 2-point discrete spaces are A-compact,
   (g) A-compact Hausdorff spaces form an epireflective subcategory of **Haus**.
4. Equivalent are:
   (a) either the axiom of choice holds or every ultrafilter is fixed,
   (b) products of B-compact spaces are B-compact.
5. Equivalent are:
   (a) Dedekind-finite sets are finite,
   (b) every set carries some D-compact Hausdorff topology,
   (c) every T_1-space has a $T_1 - D$-compactification,
   (d) Alexandroff-compactifications of discrete spaces are D-compact.

Mathematics Subject Classifications (1991). 03E25, 04A25, 54D30, 54B10, 18A40, 18B30, 54B30.

Key words: compact space, topological product, epireflective subcategory, free ultrafilter, axiom of choice, Boolean prime ideal theorem.

0. Introduction

Consider the following basic results about compact* spaces:

THEOREM 0.1 (Tychonoff 1930, 1935). *Products of families of compact spaces are compact.*

THEOREM 0.2 (Čech 1937, Stone 1937). *Compact Hausdorff spaces form an epireflective subcategory in the category* **Haus** *of Hausdorff spaces.*

Unfortunately the above results are not indisputably true. In fact, it is well known that none of them is valid in ZF set theory. Rather each requires some additional assumptions. In particular the following hold:

THEOREM 0.3 (Kelley 1950). *Theorem* 0.1 *is equivalent to the axiom of choice.*‡

THEOREM 0.4 (Rubin and Scott 1954). *Theorem* 0.2 *is equivalent to the Boolean prime ideal theorem*‡.

Unfortunately, the last two statements are not indisputably true either – since, in the absence of the axiom of choice, it is no longer clear what *compactness* should mean. Various familiar descriptions of compactness, equivalent in the presence of the axiom of choice, separate in its absence into different concepts. For some of these, Statements 0.3 and 0.4 are true, for others they are false. In particular Comfort (1968) has demonstrated that for a suitable version of compactness Theorems 0.1 and 0.2 (restricted to the realm of complete regularity) hold without any additional set theoretical assumptions in ZF.

The purpose of this note is to discuss in ZF four versions of compactness, their relations to each other, and the validity of Theorems 0.1, 0.2, and related results. Due to the following theorem, that holds in ZF, theorems of type 0.1 and 0.2 are closely related:

THEOREM 0.5 (Kennison 1965). *For isomorphism-closed full subcategories* **A** *of* **Haus** *the following conditions are equivalent:*
(1) **A** *is epireflective in* **Haus**,
(2) **A** *is closed under the formation of products and of closed subspaces.*

---

* In this paper compactness does not imply Hausdorffness, but regularity, complete regularity, normality, and zerodimensionality do.

‡ The *axiom of choice* (Zermelo 1904) states that for each family $(X_i)_{i\in I}$ of nonempty sets X_i there exists a *choice function*, i.e., a function $f\colon I \to \cup\{X_i \mid i \in I\}$ with $f(i) \in X_i$ for each $i \in I$ (in other words: $\prod_{i\in I} X_i \neq \emptyset$); equivalently: for each family $(X_i)_{i\in I}$ of pairwise disjoint nonempty sets X_i there exists a *choice set* C, i.e., a set C that contains exactly one element from each X_i.

The *Boolean prime ideal theorem* (Krull 1929, Stone 1936) states that every non-trivial Boolean algebra B has a prime ideal (in other words: $\hom(B, \mathbf{2}) \neq \emptyset$); equivalently: in every Boolean algebra, each ideal is contained in some prime ideal (= maximal ideal); equivalently: for every set X every filter on X is contained in some ultrafilter on X. The Boolean prime ideal theorem is properly weaker than the axiom of choice (Halpern 1964, Halpern and Lévy 1971).

1. A-Compactness

DEFINITION 1.1. A topological space X is called *A-compact* provided that every open cover of X contains a finite cover.

PROPOSITION 1.2. *For topological spaces X the following conditions are equivalent:*

(a) *X is A-compact,*

(b) *in X every filter has a cluster point,*

(c) *for each space Y the projection π_Y: $X \times Y \to Y$ along X is a closed map.*

Proof. (a) $\Leftrightarrow$ (b) straightforward.

(b) $\Rightarrow$ (c): Let A be a closed subset of $X \times Y$. Assume that $B = \pi_Y[A]$ is not closed in Y. Select* an element y of $\text{cl}_Y B \backslash B$. Let $\mathcal{U}$ be the neighbourhood filter of y in Y. Then $\{A \cap \pi_Y^{-1}[U] \mid U \in \mathfrak{U}\}$ is a base of a filter $\mathfrak{F}$ in $X \times Y$. Thus $\mathfrak{G} = \pi_X[\mathfrak{F}]$ is a filter in X. By (b) $\mathfrak{G}$ has a cluster point x in X. Consequently (x, y) is a cluster point of A in $X \times Y$ which does not belong to A. Contradiction.

(c) $\Rightarrow$ (b): Let $\mathfrak{F}$ be a filter in X without cluster point. Consider the space Y whose underlying set is $\mathfrak{F} \cup \{\emptyset\}$ and such that a subset C of Y is closed in Y provided that it contains the point $\emptyset$ whenever for each $F \in \mathfrak{F}$ the set $\{G \in \mathfrak{F} \mid G \subset F\}$ meets C. Then $A = \bigcup\{\text{cl}_X F \times \{F\} \mid F \in \mathfrak{F}\}$ is closed in $X \times Y$, but $\pi_Y A = \mathfrak{F}$ is not closed in Y. Thus X does not satisfy (c).

REMARKS 1.3.

(1) A proof of the Kuratowski–Mrówka theorem (that states the equivalence of (a) and (c) in 1.2) has been included, since its familiar proof employs the axiom of choice (see Mrówka (1959)).

(2) A-compact Hausdorff spaces are normal but may fail to be completely regular. Läuchli (1962/63) has constructed – in a suitable model of set theory – an infinite, locally compact Hausdorff space X such that every continuous map f: $X \to \mathbb{R}$ is constant. The one-point compactification of X has the desired properties.

(3) A-compactness is preserved under the formations of continuous images and of closed subspaces.

PROPOSITION 1.4 (Alexandroff and Urysohn 1929). *Equivalent are:*

(a) *X is A-compact and Hausdorff,*

(b) *X is H-closed and reglar.*

---

* That a subset B of Y that contains all of its accumulation points is closed in Y follows from the fact that whenever for each point y of $Y \backslash B$ there exists *some* open set U_y with $y \in U_y$ and $U_y \cap B = \emptyset$, then there exists a largest, hence *distinguished*, such U_y. This observation implies too that any set which is a neighbourhood of each of its points must be open. Consequently the familiar descriptions of continuity by means of open sets resp. by means of neighbourhoods remain equivalent.

Proof. (a) $\Rightarrow$ (b): H-closedness is obvious. The familiar proof of regularity uses the axiom of choice. Here is an alternative: Let x be an element of X. Consider the filter $\mathfrak{F}$ generated by all closed neighbourhoods of x. Since X is Hausdorff, x is the only cluster point of $\mathfrak{F}$. Let U be an open neighbourhood of x. Then either there exists $F \in \mathfrak{F}$ with $F \subset U$ or each $F \in \mathfrak{F}$ meets $X \setminus U$. However, the latter case cannot occur since it would imply that $\{F \setminus U \mid F \in \mathfrak{F}\}$ generates a filter on X without a cluster point.

(b) $\Rightarrow$ (a): Let $\mathfrak{F}$ be a filter on X and let $\mathfrak{G}$ be the filter of all neighbourhoods of members of $\mathfrak{F}$. By H-closedness $\mathfrak{G}$, being an open filter, has a cluster point x in X. By regularity x is a cluster point of $\mathfrak{F}$.

PROPOSITION 1.5. *For completely regular topological spaces X the following conditions are equivalent:*

(a) *X is A-compact,*

(b) *in X every z-filter is fixed,*

(c) *in the ring $C^\star(X)$ every ideal is fixed,*

(d) *in the ring $C(X)$ every ideal is fixed.*

Proof. Let X be a completely regular topological space. For $f \in C(X)$ and $\varepsilon \in \mathbb{R}$ define $Z_\varepsilon f = \{x \in X \mid |f(x)| \leqslant \varepsilon\}$.

(a) $\Rightarrow$ (b) immediate from Proposition 1.2.

(b) $\Rightarrow$ (c): If I is an ideal in $C^\star(X)$, then the set $\{Z_\varepsilon f \mid f \in I \text{ and } \varepsilon > 0\}$ generates a z-filter $\mathfrak{F}$ on X. If $x \in \bigcap \mathfrak{F}$, then $f(x) = 0$ for each $f \in I$.

(c) $\Rightarrow$ (d): If I is an ideal in $C(X)$, then $J = I \cap C^\star(X)$ is an ideal in $C^\star(X)$ with $\bigcap\{Z_0 f \mid f \in J\} = \bigcap\{Z_0 f \mid f \in I\}$. Thus if J is fixed, then so is I.

(d) $\Rightarrow$ (b): If $\mathfrak{F}$ is a z-filter on X, then $I = \{f \in C(X) \mid Z_0 f \in \mathfrak{F}\}$ is an ideal in $C(X)$. If $f(x) = 0$ for all $f \in I$, then $X \in \bigcap \mathfrak{F}$.

(b) $\Rightarrow$ (a): If $\mathfrak{F}$ is a filter on X, then the collection of all zero-sets in $\mathfrak{F}$ generates a z-filter $\mathfrak{G}$ on X. By complete regularity every $x \in \bigcap \mathfrak{G}$ is a cluster point of $\mathfrak{F}$. By Proposition 1.2 this implies (a).

PROPOSITION 1.6. *If X and Y are A-compact, then so is $X \times Y$.*

Proof. Immediate by Proposition 1.2.

The following result is well known:

THEOREM 1.7. *Equivalent are:*

(a) *Products of A-compact spaces are A-compact,*

(b) *products of spaces with finite topologies are A-compact,*

(c) *products of pairwise homeomorphic spaces with 3-element topologies are A-compact,*

(d) *powers X^k of A-compact (T_1)-spaces X are A-compact,*

(e) *the axiom of choice.*

Proof. The implications (a) ⇒ (b) ⇒ (c) and (a) ⇒ (d) are obvious.
(e) ⇒ (a) was established by Tychonoff (1935).
(a) ⇒ (e) is due to Kelley (1950).
(d) ⇒ (e) is due to Ward (1962) in the non-T_1 version. A simple modification of Ward's proof provides the T_1-version.
(c) ⇒ (e) is due to Alas (1969).

The following result is essentially known. The equivalence of the conditions (a), (e) and (f) has been stated without proof by Rubin and Scott (1954). Łoś and Ryll-Nardzewski (1955) demonstrated the equivalence of (f) to the following modification of (a):
(a*) products of non-empty A-compact Hausdorff spaces are non-empty and A-compact.
Mycielski (1964) established the equivalence of (f) and (c). Jech (1973) stated in form of exercises (with hints) the equivalence of the conditions (a), (c) and (f) as well as the implication (f) ⇒ (e). Banaschewski (1979) proved the equivalence of the conditions (a), (b) restricted to Hausdorff spaces, (e) and (f) in an unpublished manuscript. For the reader's convenience I include a complete (and new) proof.

THEOREM 1.8. *Equivalent are:*
(a) *Products of A-compact Hausdorff spaces are A-compact,*
(b) *products of spaces with finite underlying sets are A-compact,*
(c) *powers X^k of 2-point (discrete) spaces X are A-compact,*
(d) *A-compact zerodimensional spaces form an epireflective subcategory of* **Haus**,
(e) *A-compact Hausdorff spaces form an epireflective subcategory of* **Haus**,
(f) *the Boolean prime ideal theorem.*

Proof. (a) ⇒ (b): Let $((X_i, \tau_i))_{i\in I}$ be a family of topological spaces with finite underlying sets X_i. Let σ_i be the discrete topology on X_i. Then (a) implies that $\prod_{i\in I}(X_i, \sigma_i)$ is A-compact. As a continuous image of $\prod_{i\in I}(X_i, \sigma_i)$, the space $\prod_{i\in I}(X_i, \tau_i)$ is A-compact as well.

(b) ⇒ (c) trivial.

(c) ⇒ (d) follows from the standard construction of the A-compact zerodimensional reflection of a space X as a closed subspace of $X^{C(X,D_2)}$, where D_2 is the discrete space with underlying set $\{0, 1\}$.

(d) ⇒ (f): Let **A** be the category of A-compact Hausdorff spaces. Let $\mathfrak{F}$ be a filter on a set X. It suffices to show that $\mathfrak{F}$ can be extended to an ultrafilter (see Jech (1973)). If $\mathfrak{F}$ is fixed this is obvious. Assume that $\mathfrak{F}$ is not fixed. Consider X as a discrete space and let $\beta_X\colon X \to \beta Y$ be an **A**-reflection of X. Since the source that consists of all continuous maps from X into a 2-point discrete space is point separating and initial, β_X must be an embedding. Assume for simplicity that X is a subspace of βX and that β_X is the inclusion map. Next, embed X as an open subspace of a space Y by adding one point p whose neighbourhood filter is $\{F \cup \{p\} \mid F \in \mathfrak{F}\}$. Let $j\colon X \to Y$ be the inclusion map and let $\beta_Y\colon Y \to \beta Y$ be an **A**-reflection of Y. As before we may assume that Y is a subspace of βY and β_Y is the inclusion map. The inclusion map $j\colon X \hookrightarrow Y$ has a unique continuous extension $\tilde{j}\colon \beta X \to \beta Y$. Since $\tilde{j}[X]$ is dense in βY, so is $\tilde{j}[\beta X]$. Since $\tilde{j}[\beta X]$ is A-compact and thus closed in βY, this implies $\tilde{j}[\beta X] = \beta Y$. Let q be an element of βX with $\tilde{j}(q) = p$. Then $q \in (\beta X \setminus X)$. Let $\mathfrak{U}$ be the filter consisting of all traces of neighbourhoods of q on X. Then $\mathfrak{F} \subset \mathfrak{U}$ by continuity of $\tilde{j}$. Thus it remains to show that $\mathfrak{U}$ is an ultrafilter on X. Consider a subset A of X. The characteristic map $\chi_A\colon X \to D_2$ of X into $\{0, 1\}$, considered as a discrete space D_2, has a continuous extension $\tilde{\chi}_A\colon \beta X \to D_2$. Thus the set $B = \tilde{\chi}_A^{-1}[\tilde{\chi}_A(q)] \cap X$ belongs to $\mathfrak{U}$. Since $B = A$ or $B = (X \setminus A)$, this implies that $\mathfrak{U}$ is an ultrafilter.

(f) $\Rightarrow$ (a): Obviously (f) implies that A-compactness = B-compactness (see the next section). Thus (a) follows immediately from Proposition 2.2.

(a) $\leftrightarrow$ (e): Immediate via Theorem 0.5.

REMARKS 1.9.
(1) If X is a non-discrete 2-point space, then all powers X^k of X are A-compact.
(2) It is known that the Boolean prime ideal theorem properly implies the following weak version of the axiom of choice:

products of non-empty finite sets are non-empty.

2. B-Compactness

DEFINITION 2.1. A topological space X is called *B-compact* provided that in X every ultrafilter converges.

PROPOSITION 2.2. (1) *Finite products of B-compact spaces are B-compact.*
(2) *Products of finite spaces are B-compact.*
(3) *Products of B-compact Hausdorff spaces are B-compact.*

Proof. (1) and (3) are obvious. (2) follows from (3) as in Theorem 1.8.

THEOREM 2.3. *Equivalent are:*
(a) *Arbitrary products of B-compact (T_1)-spaces are B-compact,*
(b) *Either every ultrafilter is fixed or the axiom of choice holds.*

Proof. (a) $\Rightarrow$ (b): Assume that there exists a free ultrafilter $\mathfrak{U}$ on a set X. To show that the axiom of choice holds, let $(X_i)_{i \in I}$ be a family of non-empty sets. Let ωX be the Wallman compactification of the discrete space with underlying set X, and let p be the point to which $\mathfrak{U}$ converges in ωX. Assume for simplicity that the underlying set of ωX is disjoint from each X_i. For each $i \in I$ consider the space Y_i obtained from ωX via replacement of p by the set X_i considered as an indiscrete subspace of Y_i. (For the T_1-case enlarge the topology by adding all cofinite sets as open sets.) Then each Y_i is a B-compact (T_1)-space, and in Y_i

the filter $\mathfrak{U}_i$ generated by $\mathfrak{U}$ has X_i as its set of limit points. By (a) the product space ΠY_i is B-compact. Let $\Delta\colon X \to \Pi Y_i$ be the diagonal embedding. Then in ΠY_i the filter $\mathfrak{V}$, generated by $\Delta[\mathfrak{U}]$, is an ultrafilter, and thus converges to some point y. Consequently, for each $i \in I$, the filter $\mathfrak{U}_i = \pi_i[\mathfrak{V}]$ converges in Y_i to $\pi_i(y)$. This implies $\pi_i(y) \in X_i$, thus $y \in \Pi X_i$. So the axiom of choice holds.

(b) $\Rightarrow$ (a): If every ultrafilter is fixed, every space is B-compact, hence (a) is true. If the axiom of choice holds, (a) is trivially true.

REMARK 2.4. There are models of set theory in which B-compactness fails to be productive. In fact, Feferman (1965) and Solovay (1970) have constructed models, in which free ultrafilters exist on some sets, but all ultrafilters on countable sets are fixed.

PROPOSITION 2.5. *Every A-compact space is B-compact.*

PROPOSITION 2.6. *Equivalent are:*
(a) *A-compactness = B-compactness,*
(b) *the Boolean prime ideal theorem.*

Proof. By Proposition 2.2 and Theorem 1.8, (a) implies (b). The converse implication is obvious.

REMARKS 2.7.
(1) B-compact spaces need not to be A-compact. In fact, Blass (1977) has constructed a model of ZF in which every ultrafilter is fixed. In this model the following hold:
  (a) Every space is B-compact.
  (b) There exist B-compact zerodimensional spaces (e.g., infinite discrete spaces) that fail to be A-compact.
  (c) There exist B-compact Hausdorff spaces that fail to be regular.
  (d) There exist B-compact Hausdorff spaces that fail to be H-closed.
(2) A natural analog to Proposition 1.4 would state the equivalence of the following conditions:
  (a) X is Hausdorff and B-compact,
  (b) X is regular and in X every maximal open filter converges.

  The above observations show however that (a) does not imply (b). Moreover, the following example shows that (b) does not imply (a) as well:
  Consider a model of ZF for which each ultrafilter on the set X_0 of natural numbers is fixed, but for which there exists a free ultrafilter $\mathfrak{U}$ on some set X_1. [Such models exist: see Feferman (1965) or Solovay (1970)]. For $i \in \{0,1\}$ consider X_i as a discrete space, let $\alpha X_i = X_i \cup \{\infty_i\}$ be the Alexandroff-compactification of X_i, and let X be the subspace of the product space $X_0 \times X_1$ obtained by removal of the point (∞_0, ∞_1). Then X is zerodimensional. Let $\mathfrak{F}$ be a maximal open filter on X. The set $Y = X_0 \times X_1$

is an open, dense subset of X, thus it belongs to $\mathfrak{F}$. Hence $\mathfrak{F}_Y = \{F \cap Y \mid F \in \mathfrak{F}\}$ is an ultrafilter on (the discrete space) Y. This implies that its projection $\mathfrak{F}_0 = \pi_0[\mathfrak{F}_Y]$ is an ultrafilter on X_0. By assumption there exists $n \in X_0$ with $\{n\} \in \mathfrak{F}_0$. Thus $(\{n\} \times X_1) \in \mathfrak{F}$, which immediately implies that $\mathfrak{F}$ converges in X. Hence X satisfies (b). However the ultrafilter generated by $\{\{\infty_0\} \times U \mid U \in \mathfrak{U}\}$ does not converge in X. Thus X is not B-compact, and hence fails to satisfy condition (a).

PROPOSITION 2.8. (1) *B-compact Hausdorff spaces are epireflective in* **Haus**.
(2) *B-compact regular spaces are epireflective in* **Haus**.
(3) *B-compact completely regular spaces are epireflective in* **Haus**.

REMARKS AND QUESTIONS 2.9.
(1) By 2.7 the above three classes need not coincide.
(2) The epireflective hull in **Haus** of all A-compact Hausdorff spaces is contained in the class of all B-compact regular spaces. Under which conditions is this inclusion an equality?
(3) Whether B-compact Hausdorff (or regular) spaces form a simple epireflective subcategory of **Haus** or not may depend on set theory. In case the axiom of choice holds, the answer is positive; in Blass's (1937) model, the answer is probably negative (cf., e.g., Herrlich (1965)). Are B-compact completely regular spaces always simple in **Haus**?

3. C-Compactness

DEFINITION 3.1. A topological space X is called *C-compact* provided that X is homeomorphic to a closed subspace of some power $[0, 1]^K$ of the closed unit interval.

REMARK 3.2. Obviously C-compactness is closed-hereditary. That it is productive is not as immediate, since for a given family $(X_i)_{i \in I}$ of C-compact spaces one needs to select for each $i \in I$ some closed embedding $X_i \hookrightarrow [0, 1]^{K_i}$. However the next result shows that there is a *distinguished* such representation (cf., e.g., Herrlich (1967)).

PROPOSITION 3.3. *For topological spaces X the following conditions are equivalent:*
(a) *X is C-compact,*
(b) *the canonical map $X \to [0, 1]^{C(X,[0,1])}$ is a closed embedding.*

Proof. Obviously (b) implies (a). For the converse implication there is a purely categorical proof:
Let $m: X \hookrightarrow [0, 1]^K$ be a closed embedding and let $f: X \to [0, 1]^{C(X,[0,1])}$ be the canonical map for X. Denote the projections by $p_k: [0, 1]^K \to [0, 1]$, respectively

$\pi_g\colon [0,1]^{C(X,[0,1])} \to [0,1]$. Define $m_k = p_k \circ m$ and let $g\colon [0,1]^{C(X,[0,1])} \to [0,1]^K$ be the unique morphism such that $p_k \circ g = \pi_{m_k}$ for each $k \in K$.

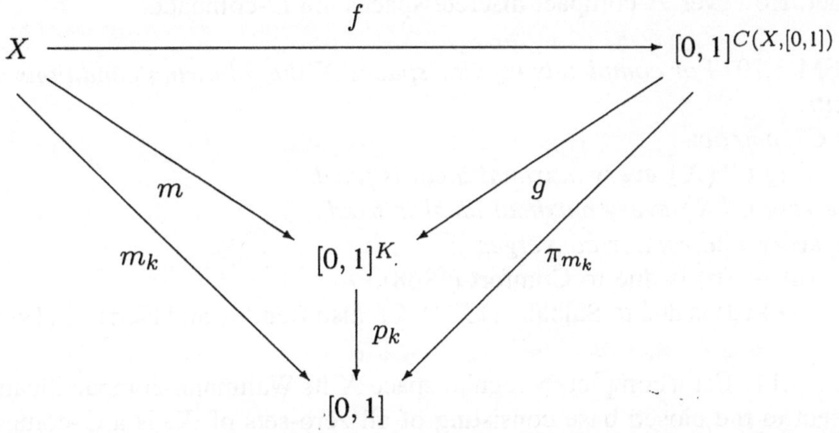

Then $p_k \circ m = m_k = \pi_{m_k} \circ f = p_k \circ g \circ f$ for each $k \in K$. Thus $m = g \circ f$. In view of the facts that in **Haus** (a) the extremal monomorphisms are precisely the closed embeddings and that (b) (Epi, Extr. Mono) is a factorization structure for morphisms, the equation $m = g \circ f$ immediately implies that f is a closed embedding.

COROLLARY 3.4. (1) *Products of C-compact spaces are C-compact.*
(2) *C-compact spaces form an epireflective subcategory of* **Haus**.

THEOREM 3.5. *C-compact spaces form the epireflective hull of all completely regular A-compact spaces.*

Proof. By Proposition 1.4 every completely regular A-compact space is C-compact. Thus the result follows via Theorem 0.5 immediately from the fact that the closed unit interval is A-compact.

REMARK 3.6. A-compact Hausdorff spaces can fail to be C-compact: In Läuchli's (1962/63) model there exist A-compact Hausdorff spaces that fail to be completely regular. Thus the class of C-compact spaces is properly contained in the epireflective hull of A-compact Hausdorff spaces in **Haus** (cf. 2.9(2)).

PROPOSITION 3.7. *Equivalent are:*
(a) *C-compactness = A-compactness and complete regularity,*
(b) *the Boolean prime ideal theorem.*

Proof. Immediate via Theorems 1.8 and 3.5.

PROPOSITION 3.8. *Every C-compact space is B-compact.*

REMARK 3.9. B-compact regular spaces need not to be completely regular, thus not C-compact. Even zerodimensional B-compact spaces may fail to be C-compact. However B-compact discrete spaces are C-compact.

THEOREM 3.10. *For completely regular spaces X the following conditions are equivalent:*
(a) *X is C-compact,*
(b) *in the ring $C^*(X)$ every maximal ideal is fixed,*
(c) *in the ring $C(X)$ every maximal ideal is fixed,*
(d) *in X every z-ultrafilter converges.*

Proof. (a) $\leftrightarrow$ (b) is due to Comfort (1968).
(b) $\leftrightarrow$ (c) $\leftrightarrow$ (d) is due to Salbany (1974). Cf. also Bentley and Herrlich (1995+).

REMARK 3.11. For a completely regular space X its Wallmann-compactification, with respect to the closed base consisting of all zero-sets of X, is a C-compact reflection of X (Salbany, 1974). An alternative description of the C-compact reflection has been provided by Chandler (1972).

REMARK 3.12. Theorems 3.5 and 3.10 above as well as the results of Bentley and Herrlich (1995+), Salbany (1974) and Comfort (1968) himself support Comfort's claim that the "correct" concept of compactness for the the theory of rings of continuous real-valued* functions is that of *C-compactness*. The more sweeping claim that *C-compactness* is the "correct" compactness-concept for topological analysis, is however unfounded. In any model of ZF without free ultrafilters on $\mathbb{N}$ the rather strange equality

$$C\text{-}compact = realcompact$$

holds. Thus C-compact subspaces of $\mathbb{R}$ maybe neither bounded nor closed. Consequently the useful *maximum principle* for continuous real-valued functions with compact domain fails for C-compactness (but holds for A-compactness).

4. D-Compactness

DEFINITION 4.1. A topological space is called *D-compact* provided that each of its infinite subsets has a complete accumulation point.

THEOREM 4.2. *Equivalent are:*
(a) D-compactness $=$ *A-compactness,*
(b) D-compactness $=$ *B-compactness,*

* For $\mathbb{Z}_2$-valued functions the "correct" concept of compactness seems to be *Boolean*. It is not known whether for zerodimensional spaces the equality *Boolean* = *C-compact* holds. See Bentley and Herrlich (1995+).

(c) *D-compactness and complete* regularity = *C-compactness,*
(d) *products of D-compact spaces are D-compact,*
(e) *products of D-compact Hausdorff spaces are D-compact,*
(f) *powers X^k of 2-point discrete spaces are D-compact,*
(g) *finite products of D-compact spaces are D-compact,*
(h) *finite coproducts of D-compact spaces are D-compact,*
(i) *products of an indiscrete space with a 2-point discrete space are D-compact,*
(j) *spaces with finite topologies are D-compact,*
(k) *D-compact Hausdorff spaces form an epireflective subcategory of* **Haus**,
(l) *D-compact zerodimensional spaces form an epireflective subcategory of* **Haus**,
(m) *the trichotomy of cardinals principle**,
(n) *the axiom of choice.*

Proof. Since (n) implies each of the above conditions, since each of the above conditions implies (f) or (i), and since (m) implies (n) (Hartogs 1915), it suffices to show the implications (i) ⇒ (m) and (f) ⇒ (m).

It suffices in each case to show that for disjoint infinite sets A and B the cardinals $\operatorname{Card} A$ and $\operatorname{Card} B$ are comparable. Let X be a discrete space with underlying set $\{0,1\}$.

(i) ⇒ (m): Let Y be the indiscrete space with underlying set $A \cup B$. Then $X \times Y$ is D-compact. Thus the set $C = (\{0\} \times A) \cup (\{1\} \times B)$ has a complete accumulation point (x, y) in $X \times Y$. Assume, without loss of generality that $x = 0$. Then $\{0\} \times (A \cup B)$ is a neighbourhood of (x, y) that meets C in $\{0\} \times A$. This implies $\operatorname{Card} A = \operatorname{Card}(\{0\} \times A) = \operatorname{Card} C = \operatorname{Card} A + \operatorname{Card} B$. Thus $\operatorname{Card} B \leqslant (\operatorname{Card} A + \operatorname{Card} B)$ implies $\operatorname{Card} B \leqslant \operatorname{Card} A$.

(f) ⇒ (m): Let 0 be not contained in $A \cup B$. Form the set $K = \{0\} \cup A \cup B$. For each subset S of K let $\chi_S \colon K \to \{0,1\}$ be the characteristic function of S. Then the set $C = \{\chi_{\{a\}} \mid a \in A\} \cup \{\chi_{\{0,b\}} \mid b \in B\}$ has a complete accumulation point χ in the D-compact space X^K. Assume, without loss of generality that the 0-th coordinate $\chi(0)$ of χ equals 0. Then $\{0\} \times X^{A \cup B}$ is a neighbourhood of χ that meets C in $\{\chi_{\{a\}} \mid a \in A\}$. As before this implies $\operatorname{Card} B \leqslant \operatorname{Card} A + \operatorname{Card} B = \operatorname{Card} C = \operatorname{Card} A$, hence $\operatorname{Card} B \leqslant \operatorname{Card} A$.

REMARK 4.3. In the above Theorem the conditions (a), (b), respectively (c) may be replaced by the following ones:
(a′) every A-compact space is D-compact,
(b′) every B-compact Hausdorff space is D-compact,
(c′) every C-compact space is D-compact.

Thus, even though it is not known whether D-compactness implies A-compactness,

* The *trichotomy of cardinals principle* states that for each pair (a, b) of cardinals exactly one of the relations $a < b$, $a = b$, $b < a$ holds; equivalently: for each pair (A, B) of sets $\operatorname{Card} A \leqslant \operatorname{Card} B$ or $\operatorname{Card} B \leqslant \operatorname{Card} A$.

the following somewhat strange result holds in ZF: if all A-compact spaces are D-compact, then all D-compact spaces are A-compact. Shortly:

(1) (A-compact $\Rightarrow$ D-compact) $\Rightarrow$ (D-compact $\Rightarrow$ A-compact).
Similarly:

(2) For Hausdorff spaces:
(B-compact $\Rightarrow$ D-compact) $\Rightarrow$ (D-compact $\Rightarrow$ B-compact).

(3) For completely regular spaces:
(C-compact $\Rightarrow$ D-compact) $\Rightarrow$ (D-compact $\Rightarrow$ C-compact).

PROPOSITION 4.4. *Equivalent are:*

(a) *every set carries a D-compact Hausdorff topology,*

(b) *every set carries a D-compact T_1-topology,*

(c) *Alexandroff-compactifications of discrete spaces are D-compact,*

(d) *every T_1-space has a $T_1 - D$-compactification,*

(e) *every Dedekind-finite set is finite,*

(f) *every cardinal is comparable with $\aleph_0$.*

Proof. (e) $\Rightarrow$ (c) $\Rightarrow$ (a) $\Rightarrow$ (b): Immediate.

(b) $\Rightarrow$ (e): Let X be an infinite set. Supply X with a D-compact T_1-topology. In the resulting space the set X has a complete accumulation point x. Then for $y \in (X\backslash\{x\})$, the set $X\backslash\{y\}$ is a neighbourhood of x. Thus $\text{Card}(X\backslash\{y\}) = \text{Card}\, X$. Thus X is Dedekind-infinite.

(e) $\leftrightarrow$ (f): Immediate and known.

(e) $\Rightarrow$ (d): Let $(X, \mathfrak{X})$ be an arbitrary T_1-space. Take $Y = X \cup \{\infty\}$, where ∞ is not an element of X. If $\mathfrak{Y}$ concists of all subsets B of Y that satisfy

(i) $(B \cap X) \in \mathfrak{X}$ and

(ii) $\infty \in B \Rightarrow Y\backslash B$ finite,

then $(Y, \mathfrak{Y})$ is a $T_1 - D$-compactification of $(X, \mathfrak{X})$.

(d) $\Rightarrow$ (e): Let X be an infinite set. Let Y be a $T_1 - D$-compactification of the discrete space with underlying set X. Then X has a complete accumulation point y in Y. Let x be an element of $X\backslash\{y\}$. Then $Y\backslash\{x\}$ is a neighbourhood of y in Y. Thus $\text{Card}(X\backslash\{x\}) = \text{Card}\, X$. Hence X is Dedekind-infinite.

REMARK 4.5. Since the axiom of choice implies the comparability of cardinals, each of the conditions in Theorem 4.2 implies each of the conditions in Proposition 4.4. The implication [powers X^k of a discrete 2-point space X are D-compact] $\Rightarrow$ [Alexandroff-compactifications of discrete spaces are D-compact] follows more directly from the observation that the Alexandroff-compactification of a discrete space with underlying set Y is homeomorphic to a closed subspace of X^Y. For the next result compare Sierpiński (1918):

PROPOSITION 4.6. *Equivalent are:*

(a) *The unit interval* $[0, 1]$ *is D-compact,*

(b) *for infinite cardinals a and b less or equal to card* $(\mathbb{R})$ *the equality* $a + b = \text{Max}\{a, b\}$ *holds.*

Proof. (a) $\Rightarrow$ (b): If a and b are infinite cardinals less or equal to Card($\mathbb{R}$), then there exist infinite subsets A of $[0, \frac{1}{3}]$ and B of $[\frac{2}{3}, 1]$ with $\mathrm{Card}(A) = a$ and $\mathrm{Card}(B) = b$. By (a), the set $A \cup B$ has a complete accumulation point in $[0, 1]$. This implies $a + b = \mathrm{Max}\{a, b\}$.

(b) $\Rightarrow$ (a): Let A be an infinite subset of $[0, 1]$. By induction, define intervals $[a_n, b_n]$ as follows: $[a_0, b_0] = [0, 1]$.

$$[a_{n+1}, b_{n+1}] = \begin{cases} [a_n, \frac{a_n+b_n}{2}], & \text{if} \quad \mathrm{Card}\, A = \mathrm{Card}(A \cap [a_n, \frac{a_n+b_n}{2}]) \\ [\frac{a_n+b_n}{2}, b_n], & \text{otherwise.} \end{cases}$$

By (b), the equality $\mathrm{Card}\, A = \mathrm{Card}(A \cap [a_n, b_n])$ holds for each n. Thus the unique element x of $\bigcap\limits_n [a_n, b_n]$ is a complete accumulation point of A. [Equivalently: x is the greatest lower bound of the set $\{y \in [0, 1] \mid \mathrm{Card}\, A = \mathrm{Card}(A \cap [0, y])\}$.]

COROLLARY 4.7. $[0, 1]$ *is D-compact provided that it is well orderable.*

References

Alas, O. T.: The axiom of choice and two particular forms of Tychonoff theorem, *Portugal. Math.* **28** (1969), 75–76.

Alexandroff, P. and Urysohn, P.: Mémoire sur les espaces topologiques compacts, *Verh. Nederl. Akad. Wetensch. Aft. Naturk. Sect. 1* **14** (1929), 1–96.

Banaschewski, B.: Compactification and the axiom of choice, Unpublished manuscript, 1979.

Bentley, H. L. and Herrlich, H.: Compactness and rings of continuous functions – without the axiom of choice, To appear (1995+).

Blass, A.: A model without ulrafilters, *Bull. Acad. Sci. Polon., Sér. Sci. Math. Astr. Phys.* **25** (1977), 329–331.

Čech, E.: On bicompact spaces, *Ann. Math.* **38** (1937), 823–844.

Chandler, R. E.: An alternative construction of βX and νX, *Proc. Amer. Math. Soc.* **32** (1972), 315–318.

Comfort, W. W.: A theorem of Stone–Čech type, and a theorem of Tychonoff type, without the axiom of choice; and their realcompact analogues, *Fund. Math.* **63** (1968), 97–110.

Feferman, S.: Some applications of the notion of forcing and generic sets, *Fund. Math.* **56** (1965), 325–345.

Gillman, L. and Jerison, M.: *Rings of Continuous Functions,* Van Nostrand, 1960.

Halpern, J. D.: The independence of the axiom of choice from the Boolean prime ideal theorem, *Fund. Math.* **55** (1964), 57–66.

Halpern, J. D. and Lévy, A.: The Boolean prime ideal theorem does not imply the axiom of choice, *Proc. of Symposium Pure Math. of the AMS* **13** (1971), Part I, 83–134.

Hartogs, F.: Über das Problem der Wohlordnung, *Math. Annalen* **76** (1915), 138–143.

Herrlich, H.: Wann sind alle stetigen Abbildungen in Y konstant? *Math. Zeitschr.* **90** (1965), 152–154.

Herrlich, H.: $\mathfrak{E}$-kompakte Räume, *Math. Zeitschr.* **96** (1967), 228–255.

Jech, T. J.: *The Axiom of Choice,* North-Holland, Amsterdam, 1973.

Kelley, J. L.: The Tychonoff product theorem implies the axiom of choise, *Fund. Math.* **37** (1950), 75–76.

Kennison, J. F.: Reflective functors in general topology and elsewhere, *Trans. Amer. Math. Soc.* **118** (1965), 303–315.

Krull, W.: Die Idealtheorie in Ringen ohne Endlichkeitsbedingungen, *Mathem. Annalen* **101** (1929), 729–744.
Läuchli, H.: Auswahlaxiom in der Algebra, *Commentarii Math. Helvetici* **37** (1962–63), 1–18.
Łoś, J. and Ryll-Nardzewski, C.: On the application of Tychonoff's theorem in mathematical proofs, *Fund. Math.* **38** (1951), 233–237.
Łoś, J. and Ryll-Nardzewski, C.: Effectiveness of the representation theory for Boolean algebras, *Fund. Math.* **41** (1955), 49–56.
Moore, G. H.: *Zermelo's Axiom of Choice. Its Origins, Developments and Influence,* Springer, New York, 1982.
Mrówka, S.: Compactness and product spaces, *Colloq. Math.* **7** (1959), 19–22.
Mycielski, J.: Two remarks on Tychonoff's product theorem, *Bull. Acad. Polon. Sci. Sér. Sci. Math., Astr. Phys.* **12** (1964), 439–441.
Rubin, H. and Rubin, J. E.: *Equivalents of the Axiom of Choice II,* North Holland, Amsterdam, 1985.
Rubin, H. and Scott, D.: Some topological theorems equivalent to the Boolean prime ideal theorem, *Bull. Amer. Math. Soc.* **60** (1954), 389.
Salbany, S.: On compact* spaces and compactifications, *Proc. Amer. Math. Soc.* **45** (1974), 274–280.
Sierpiński, W.: L'axiome de M. Zermelo et son rôle dans la théorie des ensembles et l'analyse, *Bull. l'Acad. Sci. Cracovie Cl. Sci. Math. Sér. A* (1918), 97–152.
Solovay, R. M.: A model of set theory in which all sets of reals are Lebesgue measurable, *Ann. Math.* **92** (1970), 1–56.
Stone, M. H.: The theory of representations for Boolean algebras, *Trans. Amer. Math. Soc.* **40** (1936), 37–111.
Stone, M. H.: Applications of the theory of Boolean rings to general topology, *Trans. Amer. Math. Soc.* **41** (1937), 375–481.
Tarski, A.: Ein Überdeckungssatz für endliche Mengen, *Fund. Math.* **30** (1938), 156–163.
Tychonoff, S.: Über die topologische Erweiterung von Räumen, *Mathem. Annalen* **105** (1930), 544–561.
Tychonoff, S.: Ein Fixpunktsatz, *Mathem. Annalen* **111** (1935), 767–776.
Ward, L. E.: A weak Tychonoff theorem and the axiom of choice, *Proc. Amer. Math. Soc.* **13** (1962), 757–758.
Zermelo, E.: Beweis, daß jede Menge wohlgeordnet werden kann, *Mathem. Annalen* **59** (1904), 514–516.

Applied Categorical Structures **4:** 15–29, 1996.

On Categorical Notions of Compact Objects

MARIA MANUEL CLEMENTINO*
Departamento de Matemática, Universidade de Coimbra, 3000 Coimbra, Portugal

(Received: 3 November 1994; accepted: 6 July 1995)

Abstract. Due to the nature of compactness, there are several interesting ways of defining compact objects in a category. In this paper we introduce and study an internal notion of *compact objects relative to a closure operator* (following the Borel–Lebesgue definition of compact spaces) and a notion of *compact objects with respect to a class of morphisms* (following Áhn and Wiegandt [2]). Although these concepts seem very different in essence, we show that, in convenient settings, compactness with respect to a class of morphisms can be viewed as Borel–Lebesgue compactness for a suitable closure operator. Finally, we use the results obtained to study compact objects relative to a class of morphisms in some special settings.

Mathematics Subject Classifications (1991). 18A30, 54D30, 18B30, 54B30.

Key words: factorization system, closure operator.

Introduction

The study of compactness in a general categorical setting is widely justified by its importance in mathematics. The concept of closure operators (introduced by Dikranjan and Giuli in [7]), being a precious tool to develop topological concepts and techniques in categories, also plays a rôle in the study of compactness. In fact, in a category equipped with a closure operator, different notions of compactness – according to different facets of the concept of compact spaces – have been investigated. The compact objects arising this way have interesting properties, which in general depend on the way the closure operator behaves (see, e.g., [13, 3, 8, 6] and [5]). But there are also notions of compactness, namely in Algebra, that do not depend on a closure operator. Some of the latter were unified by Áhn and Wiegandt [2] in the concept of *F-compact objects*.

The goal of this paper is to study and relate two distinct notions of compactness: an internal notion, that depends on a closure operator (following the Borel–Lebesgue condition for compact spaces), and an external notion, depending on a fixed class of morphisms (which is essentially the notion of F-compact objects in [2]). Namely, we will show that, in a convenient setting, the latter

* Partial financial assistance by Centro de Matemática da Universidade de Coimbra and by a NATO Collaborative Grant (CRG 940847) is gratefully acknowledged.

notion, apparently not depending on a closure operator, coincides with the internal notion of compact objects relative to a special closure operator.

To develop this study we first introduce the above mentioned notions of compact objects (Section 1). Then, in Section 2, considering an orthogonal closure operator in the sense of Sousa [14], we establish our main result – Theorem 2.2 – that identifies the two notions. Finally, in Section 3, the results obtained lead us to interesting examples that illustrate the behaviour of the compact objects relative to a class of morphisms, mainly in topological contexts.

I would like to thank Eraldo Giuli and Walter Tholen for useful discussions on the subject of this paper.

1. Compact Objects

In this section we study two concepts, seemingly very different in nature, of compactness in categories.

First we present the background on closure operators we need throughout. To define closure operators, one considers a category $\mathcal{X}$ and a class $\mathcal{M}$ of morphisms of $\mathcal{X}$, closed under composition and containing the isomorphisms, such that $\mathcal{X}$ is $\mathcal{M}$-complete (that is, the pullback of a morphism in $\mathcal{M}$ along any morphism and multiple pullbacks of (arbitrary) families of morphisms in $\mathcal{M}$ exist and belong to $\mathcal{M}$). We recall that, under these assumptions, $\mathcal{M}$ is a class of monomorphisms and there exists a class $\mathcal{E}$ of morphisms such that $(\mathcal{E},\mathcal{M})$ is a factorization system for morphisms (see [10]). For each object X of $\mathcal{X}$, in the class $\mathcal{M}/X$ of morphisms in $\mathcal{M}$ with codomain X – which we call *subobjects* of X – one can consider a pre-order $\leqslant$ defined by $m \leqslant n :\Leftrightarrow \exists k\colon n\cdot k = m$; if $m \leqslant n$ and $n \leqslant m$ one says that $m \cong n$, defining this way an equivalence relation on $\mathcal{M}/X$. Every X has a largest subobject, $1_X\colon X \to X$, and a least one, $0_X\colon O_X \to X$. For each morphism $f\colon X \to Y$, there are functors $f(\ )\colon \mathcal{M}/X \to \mathcal{M}/Y$, where $f(m)$ is the $\mathcal{M}$-part of the $(\mathcal{E},\mathcal{M})$-factorization of $f\cdot m$, and $f^{-1}(\ )\colon \mathcal{M}/Y \to \mathcal{M}/X$, where $f^{-1}(n)$ is the pullback of n along f, which are adjoints: $f(\ )$ is left adjoint to $f^{-1}(\ )$.

A *closure operator* c *on* $\mathcal{X}$ *with respect to* $\mathcal{M}$ is a family $c = (c_X\colon \mathcal{M}/X \to \mathcal{M}/X)_{X\in\mathcal{X}}$ of extensive and monotone maps c_X such that $f(c_X(m)) \leqslant c_Y(f(m))$ for each $\mathcal{X}$-morphism $f\colon X \to Y$ and subobject m of X. For each $m \in \mathcal{M}$, $m \leqslant c_X(m)$ (for simplicity we in general write $c(m)$ instead of $c_X(m)$), hence there is a morphism $m_{c(m)}$ such that $m \cong c(m)\cdot m_{c(m)}$. A subobject m is *closed* if $m \cong c(m)$, and a morphism $f\colon X \to Y$ is *dense* if $c(f(1_X)) \cong 1_Y$. A closure operator c is called *idempotent* (resp. *weakly hereditary*) if, for each $m \in \mathcal{M}$, $c(m)$ is closed (resp. $m_{c(m)}$ is dense). We say that c is *grounded* if, for each $X \in \mathcal{X}$, $c(0_X) \cong 0_X$. For more information on closure operators see [7] and [9].

1.1. Borel–Lebesgue Compact Objects

Let $\mathcal{X}$ be an $\mathcal{M}$-complete category, for a given class $\mathcal{M}$ of morphisms of $\mathcal{X}$ closed under composition and containing the isomorphisms, and c a closure operator in $\mathcal{X}$, with respect to $\mathcal{M}$. As usually, we denote by $\mathcal{E}$ the class of morphisms such that $(\mathcal{E}, \mathcal{M})$ is a factorization system for morphisms in $\mathcal{X}$.

DEFINITION 1.1. An object X of $\mathcal{X}$ is said to be *Borel–Lebesgue compact relative to* c (or simply BL-*compact relative to* c) whenever, for each family $(m_i)_{i\in I}$ of subobjects of X such that $\bigwedge_I c(m_i) \cong c(0_X)$, there exists a finite subset J of I such that $\bigwedge_J m_j \leqslant c(0_X)$.

We denote by BL*Comp* the full subcategory of BL-compact objects of $\mathcal{X}$; BL*Comp*(c) is also used whenever it is necessary to indicate which is the closure operator we are using.

EXAMPLES 1.2. It is easily seen that:

1. the BL-compact objects in the category $\mathcal{T}op$ of topological spaces and continuous maps relative to the usual (Kuratowski) closure operator k are the *compact spaces*;
2. in the category $\mathcal{P}r\mathcal{T}op$ of pretopological spaces, the BL-compact objects relative to the usual closure operator are the *compact pretopological spaces* in the sense of Čech [4];
3. the BL-compact objects in the category $\mathcal{L}oc$ of locales relative to the natural closure operator (which assigns to each sublocale the least closed sublocale containing it – see [11, 2.4]) are the *compact locales*.

Next we analyse the properties of the subcategory BL*Comp*. In the results we establish the properties of the factorization system and of the closure operator play an important rôle.

PROPOSITION 1.3. *If the closure operator c is weakly hereditary, then the subcategory* BL*Comp is closed under closed subobjects.*

Proof. Let X be BL-compact and $m\colon M \to X$ be a closed subobject of X. If $(m_i)_I$ is a family of subobjects of M such that $\bigwedge_I c(m_i) \cong c(0_M)$, then $\bigwedge_I c(m \cdot m_i) \cong \bigwedge_I m \cdot c(m_i) \cong m \cdot c(0_M) \cong c(0_X)$ because c is weakly hereditary. Since X is BL-compact, there exists a finite subset J of I such that $\bigwedge_J m \cdot m_j \leqslant c(0_X)$. Therefore $\bigwedge_J m_j \leqslant c(0_M)$ and so M is BL-compact. □

PROPOSITION 1.4. *If the morphism $f\colon X \to Y$ satisfies the condition*

$$\forall m \in \mathcal{M}/Y \quad f^{-1}(m) \cong 0_X \Leftrightarrow m \cong 0_Y, \tag{1.1}$$

X is BL-*compact and $c(0_Y) \cong 0_Y$, then Y is also* BL-*compact.*

Proof. Let X be BL-compact, $f: X \to Y$ satisfy condition (1.1) and $c(0_Y) \cong 0_Y$. If $(m_i)_I$ is a family of subobjects of Y such that $\bigwedge_I c(m_i) \cong 0_Y$, then we have that

$$\bigwedge_I c_X(f^{-1}(m_i)) \leqslant \bigwedge_I f^{-1}(c_Y(m_i)) \cong f^{-1}\Big(\bigwedge_I c_Y(m_i)\Big) \cong f^{-1}(0_Y) \cong 0_X.$$

Therefore there exists a finite subset J of I such that $\bigwedge_J f^{-1}(m_j) \cong 0_X$, since X is BL-compact. Finally we conclude that $\bigwedge_J m_j \cong 0_Y$ because $f^{-1}(\bigwedge_J m_j) \cong \bigwedge_J f^{-1}(m_j) \cong 0_X$ and condition (1.1) is fulfilled by f. □

COROLLARY 1.5. *If the closure operator c is grounded and every element of $\mathcal{E}$ satisfies condition* (1.1), *then the subcategory* $\mathsf{BL}\mathit{Comp}$ *is closed under images.*

1.2. Compact Objects with Respect to a Class of Morphisms

We recall that a *directed set* is a pair $(I, \leqslant)$ where I is a set and $\leqslant$ is a reflexive and transitive relation on I such that, for each pair i, j of elements of I, there exists a $k \in I$ satisfying $i \leqslant k$ and $j \leqslant k$. A *codirected* (or *inverse*) *system in* $\mathcal{X}$ is a pair $((X_i)_I, (\phi^j_k: X_j \to X_k)_{k \leqslant j})$, where $(X_i)_I$ is a family of objects of $\mathcal{X}$ indexed by a directed set $(I, \leqslant)$ and $(\phi^j_k: X_j \to X_k)_{k \leqslant j}$ is a family of morphisms of $\mathcal{X}$ such that $\phi^i_i = 1_{X_i}$ for each $i \in I$, and, if i, j, k are elements of I and $k \leqslant j \leqslant i$, then $\phi^j_k \cdot \phi^i_j = \phi^i_k$.

DEFINITION 1.6. Let $\mathcal{S}$ be a class of morphisms of $\mathcal{X}$. We say that an object X of $\mathcal{X}$ is *compact with respect to* $\mathcal{S}$ (or simply $\mathcal{S}$-*compact*) if, whenever $((X_i)_I, (\phi^j_k: X_j \to X_k)_{k \leqslant j})$ is an inverse system with limit $\varprojlim X_i$ and $(f_i: X \to X_i)_I$ is a compatible cone pointwise in $\mathcal{S}$ (i.e., f_i belongs to $\mathcal{S}$ for all $i \in I$), then the morphism $f: X \to \varprojlim X_i$ induced by $(f_i)_I$ also belongs to $\mathcal{S}$.

We will denote by $Comp(\mathcal{S})$ the full subcategory of $\mathcal{S}$-compact objects of $\mathcal{X}$.

We remark that, given a functor $F: \mathcal{X} \to \mathcal{Y}$, an $\mathcal{X}$-object is F-compact in the sense of Áhn and Wiegandt [2] if and only if it is $\mathcal{S}$-compact when $\mathcal{S} = \{f \in Mor\mathcal{X}\,;\, Ff \in Epi\mathcal{Y}\}$.

The proofs of the following propositions are straightforward.

PROPOSITION 1.7. *If the class of morphisms $\mathcal{S}$ is part of a factorization system for sources $(\mathcal{S}, \mathcal{N})$ in $\mathcal{X}$, then, for each object X of $\mathcal{X}$, the following conditions are equivalent:*

(i) *X is $\mathcal{S}$-compact;*

(ii) *if* $((X_i)_I, (\phi_k^j)_{k \leqslant j})$ *is an inverse system with limit and* $(f_i\colon X \to X_i)_I$ *is a compatible cone pointwise in* $\mathcal{S}$*, then the source in* $\mathcal{N}$ *of the* $(\mathcal{S},\mathcal{N})$*-factorization of* $(f_i)_I$ *is a limit cone;*

(iii) *if* $((X_i)_I, (\phi_k^j)_{k \leqslant j})$ *is an inverse system with limit and* $(f_i\colon X \to X_i)_I$ *is a compatible cone which belongs to* $\mathcal{N}$ *and is pointwise in* $\mathcal{S}$*, then* $(f_i)_I$ *is a limit cone.*

PROPOSITION 1.8. *If the class of morphisms* $\mathcal{S}$ *is part of a factorization system for sources* $(\mathcal{S},\mathcal{N})$ *in* $\mathcal{X}$*, then* $Comp(\mathcal{S})$ *is closed under morphisms in* $\mathcal{S}$ *(i.e., if* $f\colon X \to Y$ *belongs to* $\mathcal{S}$ *and* X *belongs to* $Comp(\mathcal{S})$ *then* Y *also belongs to* $Comp(\mathcal{S})$*.)*

EXAMPLES 1.9.

1. In the paper [2] Áhn and Wiegandt present several algebraic examples which illustrate the importance of this notion in Algebra. Namely,
   (a) if $\mathcal{X}$ is the category R-Mod of R-modules and $\mathcal{E}$ is the class of surjective homomorphisms, then the $\mathcal{E}$-compact objects are the *discrete linearly compact modules*;
   (b) if $\mathcal{X}$ is the category of linearly topological Hausdorff R-modules and continuous homomorphisms, and $\mathcal{E}$ is the class of surjective continuous homomorphisms, then the $\mathcal{E}$-compact objects are the *linearly compact modules*;
   (c) in the previous category, if $\mathcal{E}$ is the class of dense homomorphisms, then *every object of* $\mathcal{X}$ *is* $\mathcal{E}$*-compact.*

   These facts can be generalized to complete abelian categories, as was developed in [2].
2. In topological settings there are also interesting examples. We mention below some topological situations that have remarkable similarities with the above ones.
   (a) If $\mathcal{X}$ is a topological category over $\mathcal{S}et$ and $\mathcal{E}$ is the class of morphisms whose underlying map is surjective, then the $\mathcal{E}$-compact objects are just the *objects of* $\mathcal{X}$ *whose underlying set is finite* (see Theorem 3.4); so, in particular, compact objects in $\mathcal{S}et$ with respect to the class of surjective maps are the *finite sets* (we remark that the proof of this assertion that appears in [2] is not correct).
   (b) In $\mathcal{T}op_1$ the compact objects with respect to the class of surjective continuous maps are the *compact* T_1*-spaces* (see Theorem 3.8).
   (c) If in $\mathcal{T}op$ ($\mathcal{H}aus$) we consider the class $\mathcal{E}$ of dense continuous maps, then *every topological (Hausdorff) space is* $\mathcal{E}$*-compact* (see [2] and [5]).

2. $\mathcal{E}$-Compact Objects Are BL-Compact

Next we will show that, in convenient settings, the compact objects with respect to a class of morphisms are exactly the Borel–Lebesgue compact objects relative to a special closure operator.

From now on we consider a non-thin and $\mathcal{M}$-complete category $\mathcal{X}$ with pushouts and equalizers, where $\mathcal{M}$ is a class of morphisms closed under composition and containing the isomorphisms; $\mathcal{X}$ is therefore equipped with a factorization system $(\mathcal{E}, \mathcal{M})$. Furthermore we assume that $\mathcal{X}$ has a terminal object T and that $\mathcal{E}$ is a class of epimorphisms determined by T in the following way:

$$\mathcal{E} = \{f \in Mor\mathcal{X} \ ; \ T \text{ is projective with respect to } f\}.$$

These assumptions have the following interesting consequences.

- *The class $\mathcal{E}$ is stable under pullbacks.*
- *The terminal object is a generator*: if $u, v\colon X \to Y$ are such that $u \cdot x = v \cdot x$ for every $x\colon T \to X$, then their equalizer belongs to $\mathcal{E}$ by the way $\mathcal{E}$ is determined by T, hence it is an isomorphism, that is, $u = v$.
- *Every morphism with domain T belongs to $\mathcal{M}$.*
- *In the copower $(T \coprod T, (\tau_1, \tau_2))$, $\tau_1 \neq \tau_2$.*
- *For each object X of $\mathcal{X}$, the morphism $t_X\colon X \to T$ belongs to $\mathcal{E}$ if and only if X is not pre-initial.*

In this setting we consider the orthogonal closure operator *ort*, with respect to $\mathcal{M}$, induced by $\{T\}$ in the sense of Sousa [14], which we describe in the sequel.

For each $m\colon M \to X$ in $\mathcal{M}$, form the pushout of m along $t_M\colon M \to T$:

$$\begin{array}{ccc} M & \xrightarrow{\ m\ } & X \\ {\scriptstyle t_M}\downarrow & & \downarrow{\scriptstyle f_M} \\ T & \xrightarrow[\ x_M\ ]{} & X/M; \end{array} \qquad (2.1)$$

the closure $ort(m)$ is the pullback of x_M along f_M.

The closure operator *ort* is idempotent (since the pullback of a pushout is already a pushout) and, under our assumptions, it is also grounded. In fact, if we form the pullback (p, t_P) of the pushout (x_{O_X}, f_{O_X}) of $0_X\colon O_X \to X$ and $t_{O_X}\colon O_X \to T$, the following diagram commutes

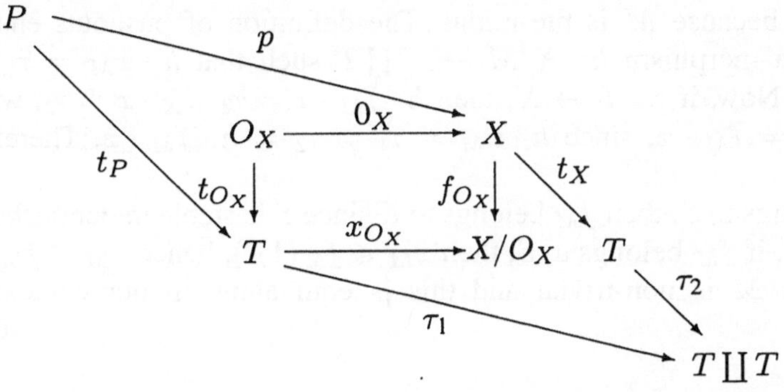

since O_X is pre-initial. Hence there exists a morphism $h\colon X/O_X \to T\coprod T$ such that $h \cdot x_{O_X} = \tau_1$ and $h \cdot f_{O_X} = \tau_2 \cdot t_X$. If there were a morphism $x\colon T \to P$, then

$$\tau_1 = \tau_1 \cdot t_P \cdot x = h \cdot x_{O_X} \cdot t_P \cdot x = h \cdot f_{O_X} \cdot p \cdot x = \tau_2 \cdot t_X \cdot p \cdot x = \tau_2,$$

which contradicts our assumptions. Therefore P is pre-initial, which implies that $P \cong 0_X$ and so c is grounded as claimed.

This closure operator has another important feature: it is the largest closure operator that keeps fixed the subobjects with domain T (i.e., every subobject with domain T is closed), when one considers the partial order $\leqslant$, in the conglomerate of closure operators on $\mathcal{X}$ with respect to $\mathcal{M}$, defined by: $d \leqslant e :\Leftrightarrow (\forall m \in \mathcal{M})\ d(m) \leqslant e(m)$.

The behaviour of the pushout diagram (2.1) will be very important in our study. Next we state some of its properties.

LEMMA 2.1. *For $m \in \mathcal{M}$, consider the pushout diagram* (2.1). *Then:*

(a) $x_M \leqslant f_M(m) \;\Leftrightarrow\; x_M \leqslant f_M(1_X)$;

(b) $f_M \in \mathcal{E} \;\Leftrightarrow\; t_M \in \mathcal{E}$.

Proof. To prove the non-trivial part of (a) assume that $x_M \nleqslant f_M(m)$, or, equivalently, that M is pre-initial, since, if $x\colon T \to M$, then $x_M \cong f_M(m \cdot x)$. Consider the copower $(T\coprod T, (\tau_1, \tau_2))$ and the following diagram

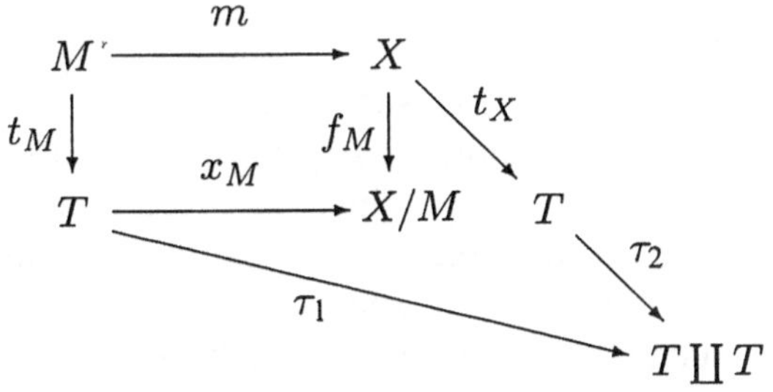

which commutes because M is pre-initial. The definition of pushouts ensures the existence of a morphism $h\colon X/M \to T\coprod T$ such that $h \cdot x_M = \tau_1$ and $h \cdot f_M = \tau_2 \cdot t_X$. Now, if $x\colon T \to X$, then $h \cdot f_M \cdot x = \tau_2 \cdot t_X \cdot x = \tau_2$, which implies that $x_M \neq f_M \cdot x$, since $h \cdot x_M = \tau_1 \neq \tau_2 = h \cdot f_M \cdot x$. Therefore, $x_M \nleqslant f_M(1_X)$.

(b): If t_M belongs to $\mathcal{E}$, then f_M belongs to $\mathcal{E}$ since $\mathcal{E}$ is stable under pushouts. On the other hand, if f_M belongs to $\mathcal{E}$, then $x_M \leqslant f_M(1_X)$, hence $x_M \leqslant f_M(m)$ by (a). Therefore M is non-trivial and this is equivalent, in our context, to $t_M \in \mathcal{E}$. □

THEOREM 2.2. *For an object X of $\mathcal{X}$, the following conditions are equivalent:*

(i) *X is Borel–Lebesgue compact relative to ort;*

(ii) *X is compact with respect to $\mathcal{E}$.*

Proof. (i) ⇒ (ii): Let X be **BL**-compact and consider an inverse system $((X_i)_I, (\phi_k^j\colon X_j \to X_k)_{k\leqslant j})$ with limit $(Y, (\lambda_i)_I)$, a compatible cone $(f_i\colon X \to X_i)_I$ pointwise in $\mathcal{E}$ and the morphism $f\colon X \to Y$ it induces. We want to show that T is projective with respect to f.

Let $y\colon T \to Y$ and $m_i \cong f_i^{-1}(\lambda_i \cdot y)$, for $i \in I$. Each $m_i\colon M_i \to X$ is closed, because it is the pullback of $\lambda_i \cdot y$, and non-trivial since $t_i\colon M_i \to T$, as the pullback of a morphism in $\mathcal{E}$, belongs to $\mathcal{E}$. Moreover, it is easily seen that, if $j, k \in I$ and $k \leqslant j$, then $m_j \leqslant m_k$. In fact, as $\phi_k^j \cdot \lambda_j \cdot y = \lambda_k \cdot y$, we obtain the following commutative diagram

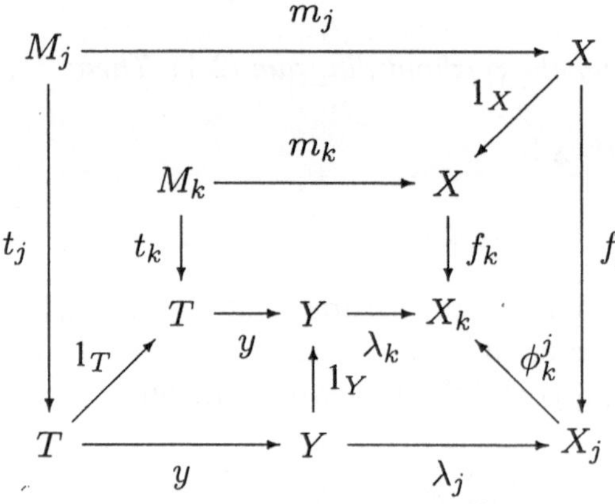

and then there exists a morphism $h_k^j\colon M_j \to M_k$ such that $m_k \cdot h_k^j = m_j$ by definition of pullbacks. We thus have that the family $(m_i)_I$ consists of non-trivial closed subobjects of the **BL**-compact object X such that, for every finite subset J of I, $\bigwedge_J m_j \ncong 0_X$. Hence, since *ort* is grounded, $\bigwedge m_i \ncong 0_X$ by the definition of **BL**-compact objects. So there is an $x\colon T \to X$ such that $x \leqslant \bigwedge m_i$, that is there are morphisms $x_i\colon T \to M_i$ such that $m_i \cdot x_i = x$ for each $i \in I$. Therefore

$\lambda_i \cdot f \cdot x = f_i \cdot m_i \cdot x_i = \lambda_i \cdot y \cdot t_i \cdot x_i = \lambda_i \cdot y$, which implies that $f \cdot x = y$ because (λ_i) is a monosource. Hence f belongs to $\mathcal{E}$ as claimed.

(ii) $\Rightarrow$ (i): Let X be an $\mathcal{E}$-compact object and $(m_i)_I$ a family of subobjects of X such that, for each finite subset J of I, $\bigwedge_J m_j \not\cong 0_X$. We want to show that $\bigwedge_I ort(m_i)\colon M \to X \not\cong 0_X$ which is equivalent to the existence of a subobject $x\colon T \to X$ such that $x \leqslant \bigwedge_I ort(m_i)$.

The set

$$\mathcal{F} := \Big\{ \bigwedge_J ort(m_j);\ J \text{ is a finite subset of } I \Big\}$$

is a directed set when ordered by the relation $\leqslant$ on subobjects of X. Consider, for each $n\colon N \to X$ in $\mathcal{F}$, the morphism f_N obtained by pushout of $t_N\colon N \to T$ along n. By our assumptions f_N belongs to $\mathcal{E}$ since N is non-trivial. For each pair $v\colon V \to X$, $w\colon W \to X$ of elements of $\mathcal{F}$ such that $v \leqslant w$, by definition of pushouts there exists a morphism $\phi^V_W\colon X/V \to X/W$ making the following diagram

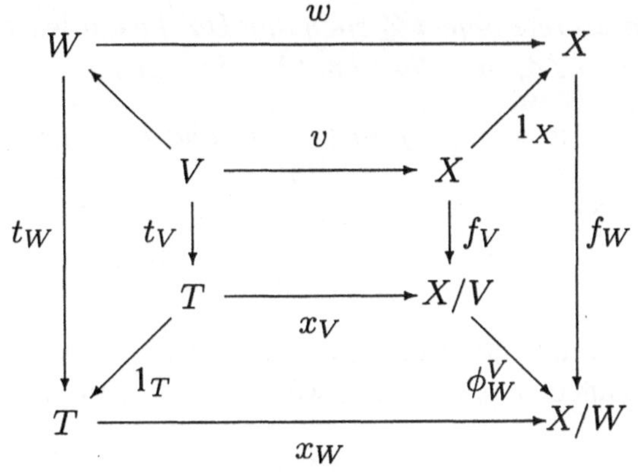

commute. Therefore $((X/N)_{n\colon N \to X \in \mathcal{F}}, (\phi^V_W)_{v \leqslant w})$ is a codirected system and (f_N), which is pointwise in $\mathcal{E}$, is a compatible cone. We thus have that the morphism $f\colon X \to \varprojlim X/N$ it induces also belongs to $\mathcal{E}$.

Now consider the morphism $y\colon T \to \varprojlim X/N$ induced by (x_N). Since $f \in \mathcal{E}$, there exists $x\colon T \to X$ such that $f \cdot x = y$. Hence $f_N \cdot x = x_N$, which implies, under our assumptions, that $x \leqslant n$ for each $n \in \mathcal{F}$. Since in particular every subobject of the form $ort(m_i)$ belongs to $\mathcal{F}$, we may conclude that $x \leqslant ort(m_i)$ for every $i \in I$, and therefore $x \leqslant \bigwedge_I ort(m_i)$. □

REMARK 2.3. We point out that the assumption that $\mathcal{X}$ is not thin is not essential in Theorem 2.2. In fact, although our proof depends on that fact, it is easily seen that, if $\mathcal{X}$ is thin and it satisfies all the other assumptions we considered, then:

– $Comp(\mathcal{E}) = \mathcal{X}$;

– the closure operator ort is trivial (i.e., for every subobject $m\colon M \to X$, $ort_X(m) \cong 1_X$), hence BL-$Comp(ort) = \mathcal{X}$.

Therefore the equality $Comp(\mathcal{E}) =$ BL-$Comp(ort)$ still holds.

3. Relations between the Two Notions

Now our goal is to investigate the relations between $\mathcal{E}$-compact and BL-compact objects relative to a given closure operator.

First we describe the closure operator ort induced by T in some special contexts. This closure operator is easily determined when the category $\mathcal{X}$ has an indiscrete object Z relative to the functor $\mathcal{X}(T, -)$ – which we will call U for simplicity – such that $\mathcal{X}(T, Z)$ has at least two elements. (We recall that, given a functor $F\colon \mathcal{X} \to \mathcal{Y}$, an object Z of $\mathcal{X}$ is *indiscrete relative to* F – or simply F-*indiscrete* – if, for each morphism $f\colon FX \to FZ$ of $\mathcal{Y}$, there exists a morphism $f'\colon X \to Z$ of $\mathcal{X}$ such that $Ff' = f$.)

LEMMA 3.1. *Let $\mathcal{X}$ have a U-indiscrete object Z such that UZ has at least two elements. If $m\colon M \to X$ belongs to $\mathcal{M}$, and diagram* (2.1) *is a pushout, then:*

(a) $(\forall x, y\colon T \to X)\ f_M \cdot x = f_M \cdot y \Leftrightarrow x = y$ *or* $(x \leqslant m$ *and* $y \leqslant m)$;

(b) $(\forall x\colon T \to X)\ f_M \cdot x = x_M \Leftrightarrow x \leqslant m$;

(c) *diagram* (2.1) *is a pullback.*

Proof. To prove the non-trivial implication of (a) consider $x, y\colon T \to X$ such that $f_M \cdot x = f_M \cdot y$ and, for instance, $x \nleqslant m$. Then we may define the map

$$g\colon UX \longrightarrow UZ$$
$$z \longmapsto \begin{cases} z_0 & \text{if } z = x, \\ z_1 & \text{otherwise,} \end{cases}$$

where $z_0 \neq z_1$. Since Z is U-indiscrete there exists an $\mathcal{X}$-morphism $h\colon X \to Z$ fulfilling the equality $Uh = g$. From the fact that $x \nleqslant m$ it follows that $h \cdot m = z_1 \cdot t_M$, since T is a generator, and therefore by the definition of pushouts there is a morphism k making the following diagram

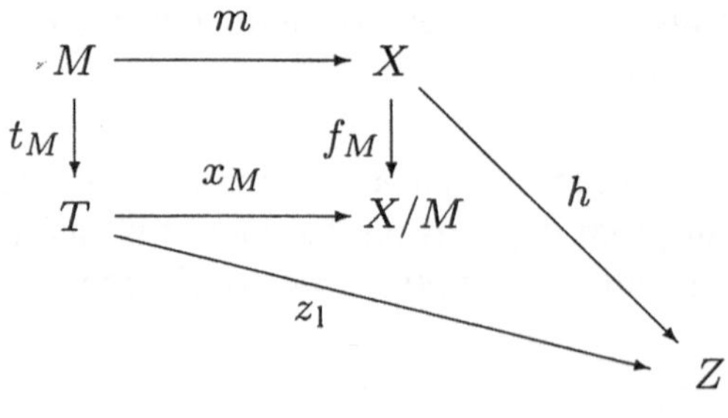

commute. Hence $z_0 = g(x) = h \cdot x = k \cdot f_M \cdot x = k \cdot f_M \cdot y = h \cdot y = g(y)$, and so $x = y$ as claimed.

Now, assertion (b) follows from (a) and Lemma 2.1. Indeed, if $f_M \cdot x = x_M$, then $x_M \leqslant f_M(m)$ by Lemma 2.1, and therefore the way $\mathcal{E}$ is determined by T implies that there exists y: $T \to X$ such that $y \leqslant m$ and $f_M \cdot y = x_M (= f_M \cdot x)$. Hence from (a) it follows that $x = y \leqslant m$ as claimed. The other implication is obvious.

Finally (c) follows from (b) taking again into account the way T determines $\mathcal{E}$. □

Assertion (c) says that every pushout diagram (2.1) is a pullback, and so every subobject is *ort*-closed. That is:

PROPOSITION 3.2. *If $\mathcal{X}$ has a U-indiscrete object Z such that UZ has at least two elements, then the closure operator ort induced by T is the discrete closure operator (i.e., for every subobject m, $m \cong c(m)$).*

COROLLARY 3.3. *If $\mathcal{X}$ has a U-indiscrete object Z such that UZ has at least two elements, then the $\mathcal{E}$-compact objects of $\mathcal{X}$ are* BL*-compact relative to every grounded closure operator on $\mathcal{X}$ with respect to $\mathcal{M}$.*

Proof. One has to notice only that, if c is a grounded closure operator, then every BL-compact object relative to *ort* is BL-compact relative to c. □

THEOREM 3.4. *If $\mathcal{X}$ is a topological category and $\mathcal{E}$ is the class of surjective morphisms, then the $\mathcal{E}$-compact objects are those objects of $\mathcal{X}$ whose underlying set is finite.*

Now let us fix a closure operator c in $\mathcal{X}$ with respect to $\mathcal{M}$. We already know how to relate $\mathcal{E}$-compact and BL-compact objects in the whole category $\mathcal{X}$ whenever $\mathcal{X}$ satisfies the conditions we considered above. Next we will be interested in the behaviour of $\mathcal{E}$-compact objects in some subcategories of $\mathcal{X}$.

First we consider the full subcategory $\mathcal{T}_1$ of $\mathcal{X}$ having as objects those $\mathcal{X}$-objects whose subobjects with domain T are c-closed. If we denote also by c the restriction of the closure operator c to $\mathcal{T}_1$, the condition $c \leqslant ort$ holds, since *ort* is the largest closure operator that keeps fixed the subobjects with domain T. Therefore, if $\mathcal{C}$ is a subcategory of $\mathcal{X}$ contained in $\mathcal{T}_1$, then every BL-compact object in $\mathcal{C}$ relative to c is BL-compact relative to *ort*, that is, it is $\mathcal{E}$-compact in $\mathcal{C}$ (by $\mathcal{E}$-compact in $\mathcal{C}$ we mean $\mathcal{E} \cap Mor\mathcal{C}$-compact in $\mathcal{C}$).

REMARK 3.5. As it is well-known, the *regular closure operator $reg_{\mathcal{D}}$ induced by any subcategory $\mathcal{D}$* of $\mathcal{X}$ (see [7] and [9]) also keeps fixed the subobjects in $\mathcal{D}$ with domain T. Therefore we always have that $reg_{\mathcal{D}} \leqslant ort$ (when restricted to $\mathcal{D}$).

In the case of $\mathcal{C} = \mathcal{T}_1$ the remaining inclusion, $ort \leqslant c$, can be assured under some suitable conditions. For that we make use of the concept of c-quotients we introduce below.

DEFINITION 3.6. We say that an $\mathcal{X}$-morphism $f\colon X \to Y$ is a *c-quotient* if c_Y is completely determined by c_X via f in the following sense

$$(\forall m \in \mathcal{M}/Y) \quad c_Y(m) \cong f(c_X(f^{-1}(m))).$$

PROPOSITION 3.7. *Let $\mathcal{X}$ have a U-indiscrete object Z such that UZ has at least two elements, and assume that c is a closure operator in $\mathcal{X}$, with respect to $\mathcal{M}$, such that the extremal epimorphisms of $\mathcal{X}$ are c-quotients. Then $c =$ ort when restricted to the subcategory $\mathcal{T}_1$ if and only if c is grounded and idempotent.*

Proof. To prove the non-trivial implication it suffices to show that, if $m\colon M \to X$ is a c-closed non-trivial subobject of an object X in $\mathcal{T}_1$, then the pushouts of (m, t_M) in $\mathcal{X}$ and in $\mathcal{T}_1$ coincide (and therefore m is ort-closed), which is equivalent for X/M to be an object of $\mathcal{T}_1$. First we remark that f_M is an extremal epimorphism, since it is the pushout of the split epimorphism t_M, and so it is a c-quotient. Consider $z\colon T \to X/M$. If $z = x_M$, then z is c-closed, since $c_{X/M}(z) \cong f_M(c_X(f_M^{-1}(z))) \cong f_M(c_X(m)) \cong f_M(m) \cong z$. If $z \neq x_M$ then, since $f_M \in \mathcal{E}$, there exists $x\colon T \to X$ such that $f \cdot x = z$. By Lemma 3.1 this morphism x is unique, therefore, by the way T determines $\mathcal{E}$, it follows that $f_M^{-1}(z) = x$, and therefore, since x is c-closed, also z is c-closed. Hence X/M belongs to $\mathcal{T}_1$, and the conclusion follows. □

THEOREM 3.8. *Let $\mathcal{X}$ have a U-indiscrete object Z such that UZ has at least two elements. If we have defined an idempotent and grounded closure operator c such that extremal epimorphisms in $\mathcal{X}$ are c-quotients, then an object in $\mathcal{T}_1$ is* **BL***-compact relative to c if and only if it is $\mathcal{E}$-compact in $\mathcal{T}_1$.*

Rather than $\mathcal{T}_1$ there is a larger (full) subcategory of $\mathcal{X}$,

$$\mathcal{T}_0 := \{X \in \mathcal{X};\ \text{if } x, y\colon T \to X \text{ and } c(x) \cong c(y), \text{ then } x \cong y\},$$

where $\mathcal{E}$-compact objects can also be related to **BL**-compact objects.

PROPOSITION 3.9. *Let $\mathcal{X}$ have a U-indiscrete object Z such that UZ has at least two elements, and assume that c is an idempotent closure operator such that extremal epimorphisms in $\mathcal{X}$ are c-quotients. Then, when restricted to $\mathcal{T}_0$, $ort \leqslant c$.*

Proof. As in the proof of Proposition 3.7, we have to check only that, if $m\colon M \to X$ is a c-closed non-trivial subobject of an object in $\mathcal{T}_0$, then also X/M belongs to $\mathcal{T}_0$.

Let $x, y\colon T \to X/M$ be such that $c_{X/M}(x) \cong c_{X/M}(y)$. Since $c_{X/M}(x_M) \cong f_M(c_X(m)) \cong f_M(m) \cong x_M$, either $x = y = x_M$ or $x \neq x_M$ and $y \neq x_M$. In the latter case $f_M^{-1}(x) \cong x'\colon T \to X$ and $f_M^{-1}(y) \cong y'\colon T \to X$ by Lemma 3.1. From the inequalities $x \leqslant c_{X/M}(y) \cong f_M(c_X(y'))$ and $y \leqslant c_{X/M}(x) \cong f_M(c_X(x'))$ it follows that $x' \leqslant c_X(y')$ and $y' \leqslant c_X(x')$ because f_M belongs to $\mathcal{E}$. Hence $x' = y'$ since X belongs to $\mathcal{T}_0$. Finally we may conclude that $x = y$, and, consequently, that X/M belongs to $\mathcal{T}_0$. □

THEOREM 3.10. *If $\mathcal{X}$ has a U-indiscrete object Z such that UZ has at least two elements, the closure operator c is idempotent and grounded, and the extremal epimorphisms in $\mathcal{X}$ are c-quotients, then an $\mathcal{E}$-compact object in $\mathcal{T}_0$ is* BL-*compact relative to c.*

Next we will give a brief description of what is known about $\mathcal{E}$-compact objects in subcategories of $\mathcal{T}op$.

EXAMPLES 3.11. Consider the class $\mathcal{E}$ of surjective continuous maps in $\mathcal{T}op$.

1. In $\mathcal{T}op$, by Proposition 3.2, *ort is the discrete closure operator* (which is also the regular closure operator $reg_{\mathcal{T}op}$ induced by $\mathcal{T}op$) and therefore, since the BL-compact objects relative to the discrete closure operator are obviously the finite spaces, the $\mathcal{E}$-compact objects are the finite spaces. Hence,

$$\text{BL-}\mathcal{C}omp(reg_{\mathcal{T}op}) = \mathcal{C}omp(\mathcal{E}) \subset \text{BL-}\mathcal{C}omp(k).$$

2. In the subcategory $\mathcal{T}op_0$ of T_0-spaces, consider the usual closure operator k and the *b-closure* (which is the regular closure operator $reg_{\mathcal{T}op_0}$ induced by $\mathcal{T}op_0$), whose definition follows: for each T_0-space X and each subset M of X

$$b_X(M) = \{x \in X;\ \forall U \in \mathcal{V}_x\ \ U \cap M \cap \overline{\{x\}} \neq \emptyset\}.$$

The closure operator *ort*, defined by for each T_0-space X and each subset M of X

$$ort_X(M) = \overline{M} \cap \Big(\bigcap\{O \subseteq X;\ M \subseteq O \text{ and } O \text{ is open}\}\Big)$$

(see [15]), is different from b and k, and it satisfies $b < ort < k$ (cf. Remark 3 and Proposition 3.9). Also, it is easily seen that there are compact T_0-spaces which are not $\mathcal{E}$-compact in $\mathcal{T}op_0$ (take, e.g., the Sierpiński space $\mathsf{S} = \{0, 1\}$, where $\{1\}$ is the non-trivial open subset, and $X = \prod_{n\in\mathbb{N}} \mathsf{S}_n \setminus \{(1)\}$). In conclusion,

$$\text{BL-}\mathcal{C}omp(reg_{\mathcal{T}op_0}) \subseteq \mathcal{C}omp(\mathcal{E}) \subset \text{BL-}\mathcal{C}omp(k).$$

(The first inclusion is probably strict, but we do not have an example of a BL-compact object relative to *ort* which is not BL-compact relative to b.)

3. From Proposition 3.7 it follows immediately that, in the subcategory $\mathcal{T}op_1$ of T_1-spaces, $ort = k$, and therefore the $\mathcal{E}$-compact objects in $\mathcal{T}op_1$ are exactly the compact T_1-spaces.

   We remark that, if we consider the category $\mathcal{T}op$ endowed with the regular closure operator $reg_{\mathcal{T}op_1}$, then all the assumptions of the Theorem 3.8 but the one on extremal epimorphisms hold. Moreover, it is easily seen that the T_1-spaces which are BL-compact relative to $reg_{\mathcal{T}op_1}$ are the finite ones (and therefore the condition on extremal epimorphisms in $\mathcal{X}$ of Theorem 3.8 is essential). We thus have

$$\mathsf{BL}\text{-}\mathcal{C}omp(reg_{\mathcal{T}op_1}) \subset \mathcal{C}omp(\mathcal{E}) = \mathsf{BL}\text{-}\mathcal{C}omp(k).$$

4. Consider now the subcategory $\mathcal{T}op_2$ of Hausdorff spaces. There it is easy to check that $k < ort$, and therefore compact Hausdorff spaces (which are the BL-compact objects relative to k and to $reg_{\mathcal{T}op_2}$ since these two closure operators coincide in $\mathcal{T}op_2$) are $\mathcal{E}$-compact, but there are non-compact Hausdorff spaces which are $\mathcal{E}$-compact (consider, for instance, $X = [0, 1]$, where all points have their natural neighbourhoods except 0, whose neighbourhood system is $\mathcal{V}_0 = \{U\backslash\{1/n;\ n \in \mathbb{N}\};\ U \text{ contains } [0, \delta) \text{ for some } \delta > 0\}$). Therefore, we have that:

$$\mathsf{BL}\text{-}\mathcal{C}omp(reg_{\mathcal{T}op_2}) = \mathsf{BL}\text{-}\mathcal{C}omp(k) \subset \mathcal{C}omp(\mathcal{E}).$$

References

1. Adámek, J., Herrlich, H., and Strecker, G. E.: *Abstract and Concrete Categories*, Wiley, New York-Chichester-Brisbane-Toronto-Singapore, 1990.
2. Áhn, P. N. and Wiegandt, R.: Compactness in categories and interpretations, Preprint, 1990.
3. Castellini, G.: Compact objects, surjectivity of epimorphisms and compactifications, *Cahiers Topologie Geom. Differentielle Categoriques* **31** (1990), 53–65.
4. Čech, E.: *Topological Spaces*, Revised by Z. Frolík and M. Katětov, Academia, Praha, 1966.
5. Clementino, M. M.: Separação e Compacidade em Categorias, PhD Thesis, Universidade de Coimbra, 1992.
6. Clementino, M. M., Giuli, E., and Tholen, W.: Topology in a category: compactness, Preprint.
7. Dikranjan, D. and Giuli, E.: Closure operators I, *Topology Appl.* **27** (1987), 129–143.
8. Dikranjan, D. and Giuli, E.: Compactness, minimality and closedness with respect to a closure operator, in: *Categorical Topology and Its Relations to Analysis, Algebra and Combinatorics*, Proc. Int. Conf. Prague, World Scientific, Singapore-New Jersey-London-Hong Kong, 1988, pp. 284–296.
9. Dikranjan, D., Giuli, E., and Tholen, W.: Closure operators II, in: *Categorical Topology and Its Relations to Analysis, Algebra and Combinatorics*, Proc. Int. Conf. Prague, World Scientific, Singapore-New Jersey-London-Hong Kong, 1988, pp. 297–335.
10. Freyd, P. J. and Kelly, G. M.: Categories of continuous functors, I, *J. Pure Appl. Algebra* **2** (1972), 169–191. Erratum *ibid.* **4** (1974), 121.
11. Johnstone, P. T.: *Stone Spaces*, Cambridge Univ. Press, Cambridge, 1982.
12. MacLane, S.: *Categories for the Working Mathematician*, Springer-Verlag, Berlin-Heidelberg-New York, 1971.
13. Manes, E. G.: Compact Hausdorff objects, *Topology Appl.* **4** (1974), 341–360.

14. Sousa, L.: Orthogonality and closure operators, *Cahiers Topologie Geom. Differentielle Categoriques*, to appear.
15. Sousa, L.: α-sober spaces via the orthogonal closure operator, *Applied Categ. Structures* **4** (1996), 87–95 (this issue).

Applied Categorical Structures **4:** 31–41, 1996.

Reflective Relatives of Adjunctions

G. RICHTER
Universität Bielefeld, Fakultät für Mathematik, Universitätsstraße 25, D-33615 Bielefeld, Germany

(Received: 28 October 1994; accepted: 4 May 1995)

Abstract. Every map $T \to UX$ from a set T to the underlying set UX of a compact Hausdorff space X admits a unique continuous extension $\beta T \to X$ from the Čech–Stone-compactification βT of T to X. Is it true for an arbitrary space X with this unique extension property to be already compact Hausdorff? No, there is a sophisticated counterexample [8]. Consequently, it makes sense to investigate the full subcategory of all such spaces in **Top**, say $\mathbf{Comp}_\beta$, which turns out to be reflective, containing compact Hausdorff spaces as reflective and bicoreflective subcategory. This paper deals with a new topological description of the spaces in $\mathbf{Comp}_\beta$, which yields more natural examples up to a finally dense class. Moreover, it turns out that there are very abstract categorical reasons for the concrete topological observations above.

Mathematics Subject Classifications (1991). 54B30, 18B30, 18A40, 54A20, 54D30, 54D35.

Key words: Čech–Stone-compactification, continuous images of compact Hausdorff spaces, convergence structures, (almost) reflective and coreflective subcategories, factorization-systems for morphisms and sources, topological functors.

1. Introduction

Consider the following abstract setting of categories and functors

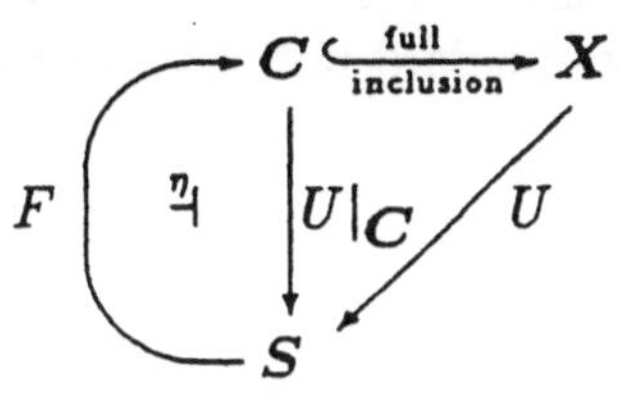

There is always a largest subcategory of $\mathbf{X}$, say $\mathbf{C}_F$, such that F acts as a left adjoint for the respective restriction of U using the same unit η, namely

$$X \in \mathbf{C}_F :\iff \forall f: S \longrightarrow UX\ \exists!\bar{f}: FS \longrightarrow X: U\bar{f}\eta S = f.$$

This category is easily seen to be closed under the formation of limits in $\mathbf{X}$, if U maps limit-sources to mono-sources. Moreover, $\mathbf{C}_F$ is closed with respect to (U-retract, mono-source)-factorizations, in general [8].

There is a 'weakly' counterpart of $\mathbf{C}_F$, say $w\mathbf{C}_F$, if one omits the uniqueness of the extension $\bar{f}$ above. This category is even closed under U-retracts, in general, and products, if the U-images of the projections form a mono-source in $\mathbf{S}$.

For the concrete example $\mathbf{C} = \mathbf{Comp}_2$, the category of compact Hausdorff spaces, U the underlying set functor of $\mathbf{Top}$, the category of all topological spaces and continuous maps between them, and $F = \beta$, the Čech–Stone-compactification, much more is known about $\mathbf{C}_F = \mathbf{Comp}_\beta$. This category is even reflective in $\mathbf{Top}$ and $X \in \mathbf{Comp}_\beta$ iff the *Kelleyfication* KeX, i.e. X equipped with the final topology with respect to all continuous maps $C \to X$, $C \in \mathbf{Comp}_2$, is compact Hausdorff. Consequently, $\mathbf{Comp}_2$ is bicoreflective and, for different reasons, reflective in $\mathbf{Comp}_\beta$. Moreover, $\mathbf{Comp}_\beta$ is closed hereditary and every of its spaces has unique sequential limits [8].

In Section 2, $w\mathbf{Comp}_\beta$ will be shown to be almost reflective in $\mathbf{Top}$ in the sense of H. Herrlich [5] and $X \in w\mathbf{Comp}_\beta$ iff X is a continuous image of some compact Hausdorff space. These spaces are much better behaved than arbitrary compact ones. They turn out to be characterized by the existence of a compatible compact Hausdorff convergence structure [3] (i.e. a 'Limitierung' in the sense of H. R. Fischer [4]) on themselves.

Just as any other finitely coproductive class of spaces containing all singletons, continuous images of compact Hausdorff spaces can be used to introduce a convergence structure. Say that a filter *converges compactly regular* to a point x if it contains some filterbasis of continuous images of compact Hausdorff spaces, converging to x with respect to the given topology. This enables to characterize the spaces in $\mathbf{Comp}_\beta$ by compactly regular convergence being compact Hausdorff, which yields new and very natural examples up to a finally dense class.

The third section deals with the abstract categorical background of the concrete topological observations above, using suitable factorization structures for sources and morphisms on $\mathbf{X}$ in order to get $\mathbf{C}$, $\mathbf{C}_F$ reflective and a counterpart for the bicoreflectivity of $\mathbf{Comp}_2$ in $\mathbf{Comp}_\beta$. Moreover $w\mathbf{C}_F$ turns out to be almost reflective in $\mathbf{X}$, especially, if U is topological. For standard notions and results we refer to [1].

2. Concrete Topological Observations

Recall, that a *convergence structure* $\varkappa$ on a set X is given by sets $\varkappa(x)$ of filters on X for each $x \in X$(the ones 'converging to x') containing the respective principal filter $\dot{x}$ and being closed under the formation of refinements and finite intersections. There is a closure operator defined by

$$X \supseteq S \longmapsto S^{-\varkappa} := \{y \in X \mid S \in \mathfrak{H} \in \varkappa(y)\},$$

which fails to be idempotent, in general. Just as for topological spaces, a convergence structure $\varkappa$ is called *regular*, if

$$\mathfrak{F} \in \varkappa(x) \Longrightarrow \mathfrak{F}^{-\varkappa} \in \varkappa(x),$$

where $\mathfrak{F}^{-\varkappa}$ denotes the filter generated by the sets $F^{-\varkappa}$, $F \in \mathfrak{F}$. The following is a relative notion of regularity.

DEFINITION 1. Let $\varkappa$, λ be convergence structures on X. Then $\varkappa$ is called *λ-regular*, if

$$\mathfrak{F} \in \varkappa(x) \Longrightarrow \mathfrak{F}^{-\varkappa} \in \lambda(x)$$

for all $x \in X$, $\mathfrak{F} \in \varkappa(x)$.

Obviously, the latter implies $\mathfrak{F} \in \lambda(x)$, hence $\varkappa$ is finer than λ.

PROPOSITION 2. *The following are equivalent for a topological space X and its convergence structure λ:*
(i) *X is a continuous image of some compact Hausdorff space;*
(ii) $X \in w\mathbf{Comp}_\beta$;
(iii) *X admits a λ-regular convergence structure $\varkappa$, which is compact (i.e. every ultrafilter converges) and Hausdorff (i.e. $\varkappa$-convergence is unique).*

Proof. With respect to the underlying set functor U of **Top**, continuous images are U-retracts. Now $w\mathbf{Comp}_\beta$ is easily seen to be closed under the formation of U-retracts, just as $w\mathbf{Comp}_F$ in the general setting mentioned in the introduction. Therefore, (i) implies (ii). Assuming (ii), one gets a continuous map e: $\beta UX \to X$ with $Ue\eta UX = \mathrm{id}\, UX$. Using one of the usual constructions of βT as set of all ultrafilters on T and $\eta T(x) = \dot{x}$ [7], this means $e(\dot{x}) = x$ for all $x \in X$. Now define for $x \in X$

$$\mathfrak{F} \in \varkappa(x) :\Longleftrightarrow \mathfrak{F} = \bigcap_{i=1}^{n} \mathfrak{U}_i, \quad \mathfrak{U}_i \in \beta UX, \quad e(\mathfrak{U}_i) = x, \quad i = 1, \ldots, n.$$

Then $\dot{x} \in \varkappa(x)$ and $\varkappa(x)$ is closed under the formation of finite intersections.

Moreover, any refinement $\mathfrak{G}$ of some $\mathfrak{F}$ as above can be written as

$$\mathfrak{G} = \bigcap_{j=J} \mathfrak{U}_j \quad \text{for some} \quad \emptyset \neq J \subseteq \{1, \ldots, n\},$$

because every ultrafilter refinement of $\mathfrak{G}$ is easily seen to coincide with some $\mathfrak{U}_j$, $j \in \{1, \ldots, n\}$ and $\mathfrak{G}$ is the intersection of its ultrafilter refinements.

Altogether, $\varkappa$ turns out to be a convergence structure on X, which is compact and Hausdorff by construction. The corresponding closure for $S \subseteq X$ turns out to be

$$\begin{aligned} S^{-\varkappa} &= \{y \in X \mid S \in \mathfrak{H} \in \varkappa(y)\} = \{y \in X \mid S \in \mathfrak{V} \in \beta UX,\ e(\mathfrak{V}) = y\} \\ &= e(\{\mathfrak{V} \in \beta UX \mid S \in \mathfrak{V}\}). \end{aligned}$$

The neighborhoods of $\mathfrak{V} \in \beta UX$ are given by the elements $V \in \mathfrak{V}$ as follows

$$\{\mathfrak{W} \in \beta UX \mid V \in \mathfrak{W}\} =: \mathcal{V}.$$

By continuity of e at $\mathfrak{U}_i$, $i = 1, \dots, n$, as above one gets for any neighborhood U of x elements $V_i \in \mathfrak{U}_i$ with $e(\mathcal{V}_i) \subseteq U$, hence for $F := \bigcup_{i=1}^n V_i \in \mathfrak{F}$

$$\begin{aligned} F^{-\varkappa} &= e\left(\left\{\mathfrak{W} \in \beta UX \,\middle|\, \bigcup_{i=1}^n V_i \in \mathfrak{W}\right\}\right) = e\left(\bigcup_{i=1}^n \{\mathfrak{W} \in \beta UX \mid V_i \in \mathfrak{W}\}\right) \\ &= \bigcup_{i=1}^n e(\mathcal{V}_i) \subseteq U. \end{aligned}$$

This shows that $\mathfrak{F}^{-\varkappa}$ converges to x in the given topology on X, which completes the proof of (iii).

Vice versa, $\varkappa$ as presumed in (iii) defines a map e: $\beta UX \to X$ by $e(\mathfrak{U}) = x$ if $\mathfrak{U} \in \varkappa(x)$, especially $e(\dot{x}) = x$, hence e is surjective. Moreover, e is continuous, because $\mathfrak{U}^{-\varkappa}$ converges to x in the given topology on X. This yields for any neighborhood U of x some $V \in \mathfrak{U}$ with $e(\mathcal{V}) = V^{-\varkappa} \subseteq U$. □

PROPOSITION 3. *w**Comp**$_\beta$ is almost reflective in* **Top**.

Proof. Since w**Comp**$_\beta$ is closed with respect to images, it remains to construct a weak reflection w: $X \to \hat{X}$ for any space X. Consider the following pushout in **Top**, where UX is interpreted as discrete space:

$$\begin{array}{ccc} UX & \xrightarrow{i=\text{identity}} & X \\ {\scriptstyle \eta X}\downarrow & & \downarrow{\scriptstyle w} \\ \beta UX & \xrightarrow{\quad j \quad} & \hat{X} \end{array}$$

Obviously, $\hat{X}$ is a continuous image of βUX via j, hence in w**Comp**$_\beta$. For every continuous f: $X \to X' \in w$**Comp**$_\beta$ there is an extension $\bar{f}$: $\beta UX \to X'$ of fi, i.e. $\bar{f}\eta X = fi$, and, consequently, a map $\hat{f}$: $\hat{X} \to X'$ with $\hat{f}w = f$ (and $\hat{f}j = \bar{f}$). □

REMARK 4. The category **Comp** of all compact spaces is known to be not almost reflective in **Top** [6]. This shows again, that not every compact space appears as continuous image of a compact Hausdorff one [2, 11].

The following is immediate from the definition of compactly regular convergence as mentioned in the introduction.

LEMMA 5. *Compactly regular convergence is preserved by continuous maps between topological spaces.*

Since every ultrafilter on the image of an arbitrary map is the image of some ultrafilter one gets

COROLLARY 6. *Every ultrafilter on a continuous image of a compact Hausdorff space converges compactly regular to some point.*

REMARK 7. In other words, compactly regular convergence is compact on $X \in w\mathbf{Comp}_\beta$. This observation yields an alternative proof for the following extension

$$X = (\mathbb{N} \times \mathbb{N}) \cup \{(\infty, 0),\ (0, \infty)\}$$

of the discrete space $\mathbb{N} \times \mathbb{N}$ to be not in $w\mathbf{Comp}_\beta$ (but in $\mathbf{Comp}$) [2, 11]. Take as neighborhoodbasis for $(\infty, 0)$ the sets

$$U_n = \{(i,j) \mid i \geqslant n\}, \quad n \in \mathbb{N},$$

of all points on the right of some vertical line, and for $(0, \infty)$

$$V_f = \{(i,j) \mid j \geqslant f(i)\}, \quad f\colon \mathbb{N} \to \mathbb{N},$$

the sets of all points above the graph of some map $f\colon \mathbb{N} \to \mathbb{N}$.
Then for any compact $C \subseteq U_n$ the sets

$$C \setminus U_k \subseteq \mathbb{N} \times \mathbb{N}, \quad k \in \mathbb{N},$$

are compact, hence finite. This proves C to be in the complement of some V_f. Likewise, any compact $D \subseteq V_f$ turns out to be in the complement of some U_n. Otherwise there would be a sequence (i_n, j_n) in D, with (i_n) strictly increasing, hence converging to $(\infty, 0) \notin D$. But this sequence is in the complement of some V_g with $g(i_n) > j_n$. Consequently, the infinite discrete subset $\{(i_n, j_n) \mid n \in \mathbb{N}\}$ is closed in D, a contradiction. This shows, that every ultrafilter converging to $(\infty, 0)$ and $(0, \infty)$ fails to converge compactly regular.

THEOREM 8. *A space X belongs to $\mathbf{Comp}_\beta$ iff compactly regular convergence on X is compact Hausdorff.*

Proof. For $X \in \mathbf{Comp}_\beta \subseteq w\mathbf{Comp}_\beta$, compactly regular convergence is already known to be a compact convergence structure, by Corollary 6. In order to prove uniqueness, it suffices to consider some ultrafilter $\mathfrak{U}$ on X compactly regular converging to $x \in X$ and to show $x = \varepsilon X(\mathfrak{U})$ for the unique continuous extension $\varepsilon X\colon \beta UX \to X$ of $\mathrm{id}\, UX$. Therefore, define $e\colon \beta UX \to X$ by

$$e(\mathfrak{U}) := x \quad \text{and} \quad e(\mathfrak{V}) := \varepsilon X(\mathfrak{V})$$

otherwise. Then e is continuous at $\mathfrak{U}$, because $\mathfrak{U}$ contains a filterbase $\mathfrak{B}$ of continuous images of compact Hausdorff spaces converging to x, hence every neighborhood V of x contains some $B \in \mathfrak{B}$ which defines a neighborhood

$$\mathcal{B} := \{\mathfrak{V} \in \beta UX \mid B \in \mathfrak{V}\}$$

of $\mathfrak{U}$. Now $e(\mathfrak{U}) = x \in V$ and for the other $\mathfrak{V} \in \mathcal{B}$

$$e(\mathfrak{V}) = \varepsilon X(\mathfrak{V}) = \lim \mathfrak{V} \in B \subseteq V$$

in $\mathrm{Ke}\, X \in \mathbf{Comp}_2$, where B is closed, consequently $e(\mathcal{B}) \subseteq V$.

For $\beta UX \ni \mathfrak{V} \neq \mathfrak{U}$ there is a neighborhood $\mathcal{V}$ of $\mathfrak{V}$ not containing $\mathfrak{U}$ and $e|\mathcal{V} = \varepsilon X|\mathcal{V}$ shows continuity of e at $\mathfrak{V}$. Moreover, convergence of principal filters $\dot{y}$ is unique in X being a T_1-space, thus $e(\dot{y}) = \varepsilon X(\dot{y}) = y$. Therefore, e is a continuous extension of $\mathrm{id}\, UX$ as well and uniqueness forces $e = \varepsilon X$, especially $x = e(\mathfrak{U}) = \varepsilon X(\mathfrak{U})$.

For the converse observe that images of continuous maps $f\colon C \to X$, $C \in \mathbf{Comp}_2$, are closed under the formation of compactly regular limits, simply by uniqueness using Lemma 5. This proves compactly regular convergence to be regular as defined in the introduction, hence $X \in w\mathbf{Comp}_\beta$ by Proposition 2.

For any continuous extension $\bar{f}\colon \beta T \to X$ of a given map $f\colon T \to UX$ and $\mathfrak{U} \in \beta T$, the ultrafilter $\langle \mathfrak{U} \rangle$ generated by $\mathfrak{U}$ on βT converges to $\mathfrak{U}$; thus its image $\bar{f}(\langle \mathfrak{U} \rangle)$ to $\bar{f}(\mathfrak{U})$ even compactly regular. But $\bar{f}(\langle \mathfrak{U} \rangle)$ coincides with $f(\mathfrak{U})$, hence $\bar{f}$ is uniquely determined by f, provided that compactly regular convergence is unique. □

REMARK 9. A look at the proof above shows that $\mathfrak{U} \in \beta UX$ converges compactly regular to x in X iff $\lim \mathfrak{U} = x$ in $\mathrm{Ke}\, X$. Consequently, compactly regular convergence is even topological and coincides with the Kelleyfication, if it is compact Hausdorff.

EXAMPLE 10. The first example of a non Hausdorff space in $\mathbf{Comp}_\beta$ was a sophisticated closed subspace of the unit interval $[0, 1]$ equipped with an additional point $*$ and a neighborhood basis for $*$ given by all sets

$$V_{\varepsilon,(x_n)} := \{*\} \cup]0, \varepsilon[\setminus \{x_n \mid n \in \mathbb{N}\}$$

with $\varepsilon > 0$ and $(x_n) \to 0$ in $[0, 1]$ [8]. Now it can be shown, that the whole space and, therefore, all of its closed subspaces belong to $\mathbf{Comp}_\beta$.

Consider an infinite member B of a filterbasis $\mathfrak{B}$ converging to $*$ with $0 \notin B$. Then $B \cap V_{1/n,0} \neq \emptyset$, hence $B \cap]0, 1/n[\neq \emptyset$ for all $n \in \mathbb{N}^\bullet$. This yields a strictly decreasing sequence (x_n) in $B \cap]0, 1[$ converging to 0. Now $S := \{x_n \mid n \in \mathbb{N}\}$ is infinite, discrete and closed in B, because $0 \notin B$ and $S \cap V_{1,(x_n)} = \emptyset$. Therefore, B fails to be compact. This shows, that $\dot{*}$ is the only ultrafilter converging to $*$ compactly regular and that compactly regular convergence is the usual convergence on the compact Hausdorff sum $\{*\} \coprod [0, 1]$.

Next consider the topological spaces $T_{\mathfrak{U},t}$ given by a set T, a nonprincipal ultrafilter $\mathfrak{U}$ on T, and $t \in T$ such that $\mathfrak{U} \cap \dot{t}$ is the neighborhoodfilter of t and $T \setminus \{t\}$ discrete. These spaces are finally dense in $\mathbf{Top}_1$ the full subcategory of all T_1-spaces in $\mathbf{Top}$. Consequently, their $\mathbf{Comp}_\beta$-reflections are finally dense in $\mathbf{Comp}_\beta$. These reflections are easily seen to be the weak reflections as constructed in the proof of Proposition 3 if their codomain belongs to $\mathbf{Comp}_\beta$. Therefore, look at $\widehat{T}_{\mathfrak{U},t} =$

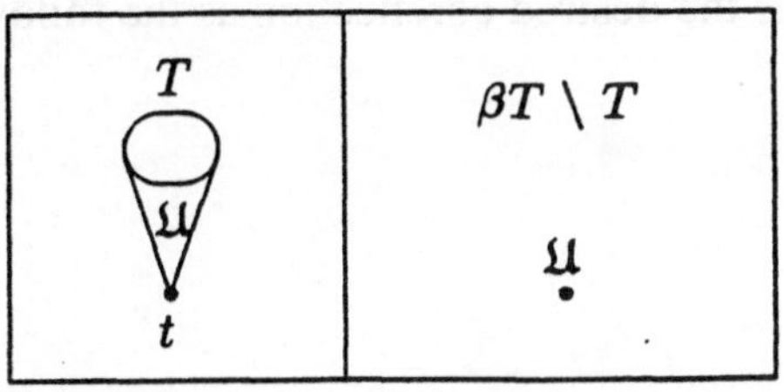

,

containing $T_{\mathfrak{U},t}$ as an open subspace, in which compactly regular convergence is trivial. In fact, every infinite subset $B \subseteq T$ has an infinite subset C in the complement of some $U \cup \{t\}$, $U \in \mathfrak{U}$. Consequently, C is discrete and closed in B, thus B fails to be compact. This shows, that compactly regular convergence on $\widehat{T}_{\mathfrak{U},t}$ is the usual convergence on βT, hence compact Hausdorff. □

REMARK 11. The convergence of principal filters as well as Fréchet filters is always compactly regular. But there is no type of nonprincipal ultrafilters with that property, as the previous example shows.

3. Abstract Categorical Background

This section deals with the abstract setting as mentioned in the introduction. To begin with, there is a counterpart for $\mathbf{Comp}_2$ to be bicoreflective in $\mathbf{Comp}_\beta$ using the counit ε of the adjunction $F \dashv U|_{\mathbf{C}_F}$.

PROPOSITION 12. *Any* $(\mathbf{E}, \mathbf{M})$*-factorization structure for morphisms in* $\mathbf{X}$ *yields two full subcategories* $\check{\mathbf{C}}$, $\hat{\mathbf{C}} \subseteq \mathbf{C}_F$ *by*

$$X \in \check{\mathbf{C}} :\iff \varepsilon X \in \mathbf{E},$$
$$X \in \hat{\mathbf{C}} :\iff (\varepsilon X = me,\ e \in \mathbf{E},\ m \in \mathbf{M} \Rightarrow m,\ Um\ mono).$$

$\check{\mathbf{C}}$ *is coreflective in* $\hat{\mathbf{C}}$ *and* U *maps the coreflections to isomorphisms.*

Proof. Consider the $(\mathbf{E}, \mathbf{M})$-factorization $\varepsilon X = me$, $X \in \hat{\mathbf{C}}$, e: $FUX \to \check{X}$. Then $\check{X} \in \check{\mathbf{C}}$, because for any $\mathbf{S}$-morphism f: $S \to U\check{X}$ there exists $\bar{f} := eFUmFf$: $FS \to \check{X}$ in $\mathbf{X}$ with

$$UmU\bar{f}\eta S = UmUeUF(Umf)\eta S = U\varepsilon X \eta UXUmf = Umf,$$

hence $U\bar{f}\eta S = f$ by assumption on Um. Moreover, for any $\underline{f}$: $FS \to \check{X}$

$$U\underline{f}\eta S = f \implies U(m\underline{f})\eta S = Umf = U(m\bar{f})\eta S$$
$$\underset{X \in \mathbf{C}_F}{\implies} m\underline{f} = m\bar{f} \underset{m\ \text{mono}}{\implies} \underline{f} = \bar{f}.$$

This proves $\check{X} \in \mathbf{C}_F$ and for $S = U\check{X}$, $f = \mathrm{id}\, U\check{X}$

$$\varepsilon\check{X} = \bar{f} = eFUm \in \mathbf{E},$$

because $U\varepsilon X = UmUe$ being a retract and Um a monomorphism forces Um to be an isomorphism. Now $m\colon\ \check{X} \to X$ is the desired coreflection as the following commutative diagram shows

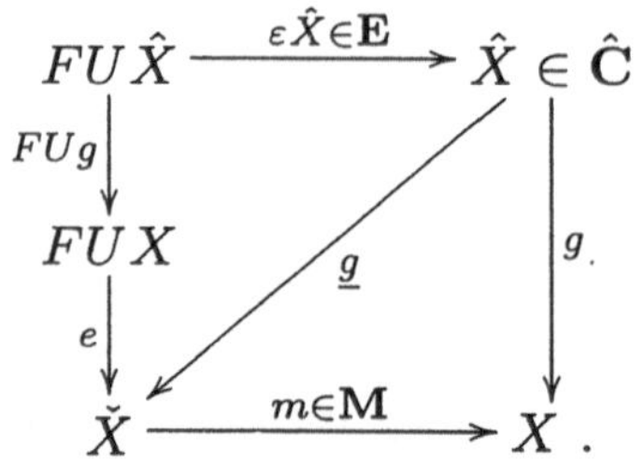

□.

EXAMPLE 13. Let U be the underlying set functor of **Top**, **E** the class of all quotient maps and, consequently, **M** the class of all continuous injections. For $F = \beta$, one gets $\check{\mathbf{C}} = \mathbf{Comp}_2$ and $\hat{\mathbf{C}} = \mathbf{Comp}_\beta$. For an arbitrary full subcategory $\mathbf{C} \subseteq \mathbf{Top}$ one gets $\hat{\mathbf{C}} = \mathbf{C}_F$ as well and $\check{\mathbf{C}}$ is the algebraic subcategory mentioned in [9; 2.2]. This generalizes to an arbitrary **M**-topological functor $U\colon\ \mathbf{X} \to \mathbf{S}$ with respect to a given $(\mathbf{E}, \mathbf{M})$-factorization structure for sources in **S**, where **M** consists of mono-sources and $UF(\mathbf{E}) \subseteq \mathbf{E}$ [10; 2.4(6)].

In order to get a counterpart for the reflectivity of $\mathbf{Comp}_2 \hookrightarrow \mathbf{Comp}_\beta \hookrightarrow \mathbf{Top}$ some more restrictive assumptions on **X** and U are required.

LEMMA 14. *If* **X** *has an* $(\mathbf{E}, \mathbf{M})$*-factorization structure for sources, where* **M** *consists of mono-sources and* $FU(\mathbf{E}) \subseteq \mathbf{E}$, *then* $\mathbf{C}_F$ *is closed under the formation of* $(\mathbf{E}, \mathbf{M})$*-factorizations. Especially,* **C** *can be replaced by its* **E***-reflective hull in* $\mathbf{C}_F$.

Proof. Consider the $(\mathbf{E}, \mathbf{M})$-factorization of some source in $\mathbf{C}_F$

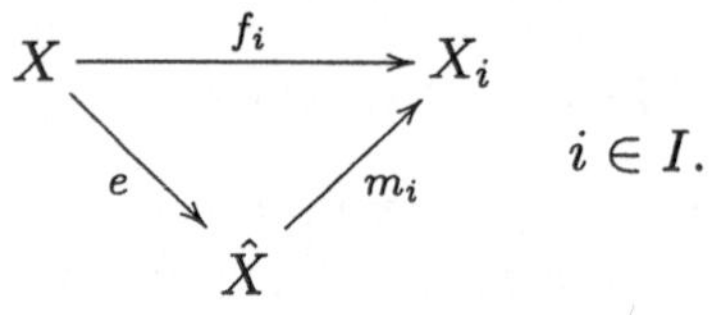

Using $FUe \in \mathbf{E}$, there is a morphism $\hat{\varepsilon}\colon\ FU\hat{X} \to \hat{X}$ as indicated

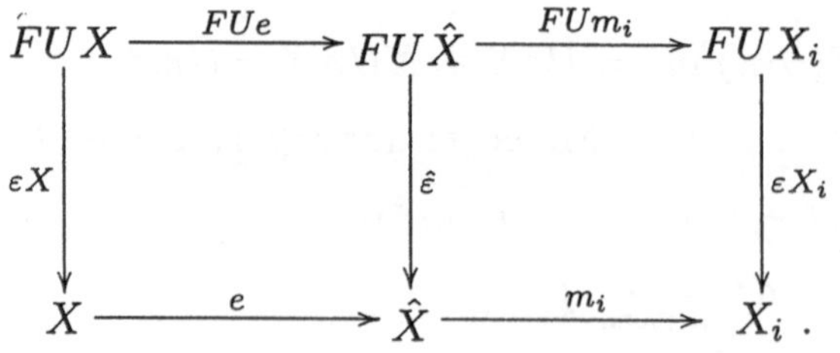

An easy calculation shows $\hat{\varepsilon}\eta U\hat{X} = \mathrm{id}\, U\hat{X}$. Moreover, any **S**-morphism $f\colon S \to U\hat{X}$ admits an extension $\hat{f} := \hat{\varepsilon}FUf$, the uniqueness of which follows from $(m_i)_I$ being a mono-source. □

The following is straight forward.

LEMMA 15. *If $G \dashv U$, then there exists a canonical natural transformation $\nu\colon GU \to FU$, which is a pointwise (weak) reflection into $\mathbf{C}_F(w\mathbf{C}_F)$.*

THEOREM 16. *With the same assumptions as in Lemma 14, let the counit δ of the adjunction $G \dashv U$ be pointwise in* **E**. *Then $\mathbf{C}_F$ is reflective in* **X**.

Proof. Consider some object X in **X**, the source $(X, f\colon X \to Y)_{Y \in \mathbf{C}_F}$, its composition with δX, the extensions along νX as in Lemma 15, and its $(\mathbf{E}, \mathbf{M})$-factorization $m_f e$. Using $\delta X \in \mathbf{E}$, there is rX as indicated

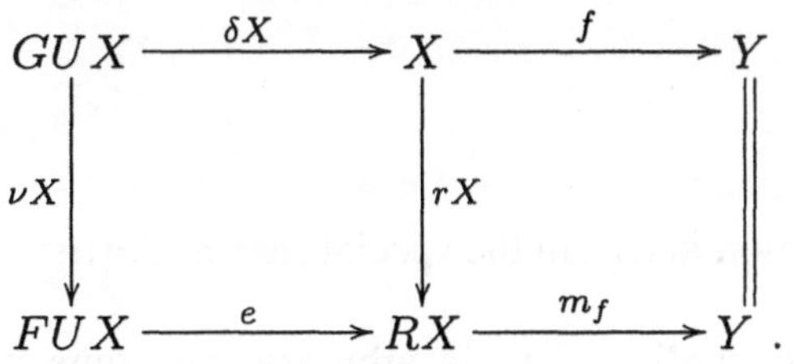

Moreover, $RX \in \mathbf{C}_F$ by Lemma 14, and the extension m_f of f along rX turns out to be unique by Lemma 15, because e is an epimorphism. □

REMARK 17. There is a slightly different construction for rX, using the extensions of all Uf along ηUX and the $(\mathbf{E}, \mathbf{M})$-factorization of the latter. This approach works without assuming U to be right adjoint. It requires U to be faithful, U-images of sources in **M** to be mono-sources and U-initial, instead. These properties follow from what is presumed in Theorem 16.

Although rX as constructed above has the weak extension property with respect to all $f\colon X \to Y \in w\mathbf{C}_F$, it is by no means obvious, whether RX belongs to $w\mathbf{C}_F$ or not, in general. However, this holds if e is a U-retract or $w\mathbf{C}_F$ is closed or closed up with respect to $(\mathbf{E}, \mathbf{M})$-factorizations.

For the special case $\mathbf{C}_F = \mathbf{Comp}_\beta$, $w\mathbf{C}_F = w\mathbf{Comp}_\beta$ consider for **E** all continuous surjections and **M** the conglomerate of continuous initial mono-sources.

PROPOSITION 18. *Let $U\colon \mathbf{X} \to \mathbf{S}$ be topological. Then $w\mathbf{C}_F$ is almost reflective in* **X**.

Proof. Consider $G \overset{\gamma}{\underset{\delta}{\dashv}} U$ given by the discrete objects in $\mathbf{X}$, especially, $\gamma S = \mathrm{id}\, S$, $U\delta X = \mathrm{id}\, UX$, and $U\nu X = \eta UX$, $\nu\colon\ GU \to FU$ as in Lemma 15. For every object X in $\mathbf{X}$,

$$\begin{array}{ccc} UX = UGUX & \xrightarrow{U\delta X = \mathrm{id}\, UX} & UX \\ {\scriptstyle \eta UX =}\Big\downarrow{\scriptstyle U\nu X} & & \Big\downarrow{\scriptstyle \eta UX} \\ UFUX & \xrightarrow{\mathrm{id}\, UFUX} & UFUX \end{array}$$

is a pushout in $\mathbf{S}$, the final lift of which along U is a pushout in $\mathbf{X}$

$$\begin{array}{ccc} GUX & \xrightarrow{\delta X} & X \\ \Big\downarrow{\scriptstyle \nu X} & & \Big\downarrow{\scriptstyle w} \\ FUX & \xrightarrow{j} & \hat{X}\,. \end{array}$$

Now, w turns out to be a weak reflection just as in the special case $w\mathbf{Comp}_\beta$. □

REMARK 19. The same technique applies if U is arbitrary, pushouts of U-retracts along $\mathbf{X}$-morphisms exist, and are U-retracts. Another possible generalization is to consider full subcategories $\mathbf{Y}$ between $\mathbf{C}$ and $\mathbf{X}$ which are closed under the formation of the particular pushouts used above. Then $w\mathbf{C}_F \cap \mathbf{Y}$ is almost reflective in $\mathbf{Y}$. Take for instance $\mathbf{Y} = \mathbf{Top}_i$, $i = 0, 1$, the category of all T_i-spaces etc.

References

1. Adámek, J., Herrlich, H., and Strecker, G. E.: *Abstract and Concrete Categories*, John Wiley, 1990.
2. Balachandran, V. V.: Minimal bicompact spaces, *J. Indian Math. Soc. (N. S.)* **12** (1948), 47–48.
3. Binz, E.: *Continuous Convergence on $C(X)$*, Lecture Notes in Mathematics 469, Springer-Verlag, Berlin–Heidelberg–New York, 1975.
4. Fischer, H. R.: Limesräume. *Math. Ann.* **137** (1959), 169–303.
5. Herrlich, H.: Almost reflective subcategories of **Top**, *Topology Appl.* **49** (1993), 251–264.
6. Hušek, M.: Čech–Stone-like compactifications for general topological spaces, *Comment. Math. Univ. Carolinae* **33** (1992), 159–163.
7. Manes, E.: *A Triple Miscellany: Some Aspects of the Theory of Algebras over a Triple*, Thesis, Weslyan University, 1967.
8. Richter, G.: A characterization of the Stone–Čech-compactification, in *Categorical Topology*, World Scientific, Singapore, 1989, pp. 462–476.
9. Richter, G.: Characterizations of algebraic and varietal categories of topological spaces, *Topology Appl.* **42** (1991), 109–125.

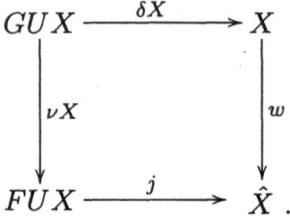

10. Richter, G.: Algebra $\subset$ Topology?!, in *Category Theory at Work*, Heldermann, Berlin, 1991, pp. 261–273.
11. Stone, A. H.: Compact and compact Hausdorff, in *Aspects of Topology*, London Math. Soc. Lecture Note Series: 93, Cambridge, 1985, pp. 315–324.

Applied Categorical Structures **4:** 43–55, 1996.

The Katětov Dimension of Proximity Spaces

An internal approach

H. L. BENTLEY
Department of Mathematics, University of Toledo, Toledo, Ohio 43606 U.S.A.

M. HUŠEK
Department of Mathematics, Charles University, Sokolovská 83, 18600 Prague, Czech Republic

and

R. G. ORI
Department of Mathematics, University of Durban-Westville, Private Bag X54001, 5400 Durban, South Africa

(Received: 7 February 1995; accepted: 4 May 1995)

Abstract. An analogue of Katětov's theorem on the equality between the dimension of a Tychonov space and the analytic dimension of its ring of bounded real-valued continuous maps is established for proximity spaces and proximally continuous maps by an internal method of proof. A new kind of filter, called proximally prime filter, arises naturally as a tool in this theory.

Mathematics Subject Classifications (1991). 54F45, 54E05, 54C40.

Key words: proximity space, analytic subalgebra, proximally prime filter, dimension.

1. Introduction

Katětov's theorem on the coincidence of the dimension of a completely regular space and the Katětov dimension of the space's algebra of bounded real-valued continuous maps is one of the great theorems of general topology [7]. One can find an exposition of Katětov's theory in [3]. (In honor of Katětov who founded this theory, we shall use the term *Katětov dimension* instead of *analytic dimension*, the term by which it has been previously referred.) Hejcman [4] proved that that equality holds also for the dimension of a proximity space (covering dimension based on finite uniform covers) and the Katětov dimension of the proximity space's algebra of bounded proximally continuous real-valued maps. Hejcman's proof was by means of the Smirnov compactification and the fact that Katětov's result is valid for compact Hausdorff spaces.

The present work is a presentation of an analogous theory for a proximity space carried out by internal means, i.e., without reference to the Smirnov compactification. The possibility of such an internal approach naturally may be aesthetically appealing, but there are also other reasons for this work. One is that the methods developed are hoped to be applicable for extending Katětov's theory to even more general spaces than proximity spaces. The second, and perhaps the strongest, reason for this work is that the main tool which enables this internal approach is a new kind of filter which we call *proximally prime* and which is of great interest for its own sake.

Our exposition follows the pattern of the one in Chapter 16 of the book by Gillman and Jerison, and our references to those results, which shall be frequent, will be made in the form, e.g., [3; Lemma 16.2]. Our proximity theorems are analogues of the topological theorems that appear there. For these analogues, there are two points to be made: First, by comparing the topological theorem in Gillman–Jerison with our proximity theorem, a greater understanding of both should result. Second, the topological theorem is an easy consequence of the proximity theorem (by considering a compact Hausdorff space to have its unique compatible proximity) and vice versa (by way of the Smirnov compactification).

Because of limitations in space, the proofs we give for the theorems about dimension concern only the finite dimensional, non-zero case; at any rate, the proofs for the zerodimensional case are easy. Omitted proofs are either easy or are known.

We assume familiarity with Chapter 16 *Dimension* of [3] and with the basic properties of proximity spaces as in, e.g., [8] or [5]. Every proximity space becomes a totally bounded uniform space by means of the definitions: A cover $\mathcal{U}$ of a proximity space X is said to *strongly refine* a cover $\mathcal{V}$ provided that for every $U \in \mathcal{U}$ there exists $V \in \mathcal{V}$ such that $U < V$ (by $U < V$ we mean, as usual, that V is a proximal neighborhood of U). A cover $\mathcal{U}$ of X is called a *uniform cover* of X provided it is strongly refined by some finite cover of X. We shall not distinguish between a proximity space and the totally bounded uniform space that arises by means of the preceding definition. In particular, functions are proximally continuous iff they are uniformly continuous with respect to the uniform structure arising as above (for domain and range).

Throughout this paper, we assume that the real numbers carry the usual proximity, i.e., $A < B$ iff with respect to the usual metric, some ε-neighborhood of A is a subset of B. We let $\mathrm{P}^*(X)$ denote the algebra of all bounded proximally continuous real-valued maps on X. An algebra over the reals is assumed always to contain a unit element 1 different from zero, and the same for subalgebras. $\mathrm{P}^*(X)$ is assumed to carry the supremum metric and all topological references on $\mathrm{P}^*(X)$ are to the topology determined by this structure.

Occasionally, as a means of indicating an intuitive basis for certain results or definitions, we shall make reference to the Smirnov compactification of X, which we shall denote by σX. In such cases, we shall make the (unnecessary, but simplifying) implicit assumption that X is a separated proximity space. However, for all our results which don't involve the Smirnov compactification: If $G \subset X$

then we define op $G = \sigma X \backslash \mathrm{cl}_{\sigma X}(X \backslash G)$; it is the largest open subset of σX such that $X \cap \mathrm{op}\, G \subset G$, and if G is open in X then $G = X \cap \mathrm{op}\, G$. Recall that $\mathcal{U}$ is a uniform cover of X iff op $\mathcal{U}$ is a uniform cover (= refined by a finite open cover) of σX. For subsets A and B of X, we have $A < B$ iff $\mathrm{cl}_{\sigma X} A \subset \mathrm{op}\, B$. Every bounded real-valued proximally continuous map g on X can be uniquely extended to a continuous map σg on σX.

2. The Stone–Weierstrass Theorem

There is a proximity version of the Stone–Weierstrass Theorem proved in the book by E. Čech [1; Theorem 25.E.2] which is stated as follows: *If B is a subalgebra of* $\mathrm{P}^*(X)$ *which initially generates X then B is dense in* $\mathrm{P}^*(X)$. The following theorem is in a more convenient form for our purposes.

THEOREM 1 (Stone–Weierstrass theorem for proximity spaces, proximity analogue of [3; 16.4]). *Let X be a proximity space and let B be a subalgebra of* $\mathrm{P}^*(X)$. *Then the closure of B is precisely the set of all maps $g \in \mathrm{P}^*(X)$ such that $g\mathcal{H}$ converges whenever $\mathcal{H}$ is a filter on X with $f\mathcal{H}$ convergent for all $f \in B$.*

Proof. We leave the proof of the forward implication as an exercise. For the backwards implication, let $g \in \mathrm{P}^*(X)$ with g not in the closure of B. Then g is not proximally continuous with respect to the proximity p initially generated by B. Thus there are two p-proximal sets E and F such that the distance between $g[E]$ and $g[F]$ is positive, say r. There is a p-cluster (German: Büschel) having E and F as members (see [5; 6.4.5]). The dual of this p-cluster is a p-round p-Cauchy filter $\mathcal{H}$ (see [5; chart on p. 180 and 6.4.21(6) (b) $\Longleftrightarrow$ (c)]) satisfying the condition that each member of $\mathcal{H}$ meets both E and F. Then $g\mathcal{H}$ is Cauchy and hence it converges. But it cannot converge since g images of members of $\mathcal{H}$ must have diameters at least r.

3. Proximally Prime Filters

We now introduce the main tool for the internal approach to Katětov's theorem for proximity spaces. A filter $\mathcal{F}$ on a proximity space X is called *proximally prime* provided that whenever $A \cup B \in \mathcal{F}$ with $\{A, B\}$ far in X then $A \in \mathcal{F}$ or $B \in \mathcal{F}$.

The next two results show relationships between proximally prime filters and subsets of the Smirnov compactification; they are here for completeness and to show some intuitive basis for considering proximally prime filters, but we shall not use them.

PROPOSITION 2. *Let $\mathcal{F}$ be a filter on a proximity space X. Then $\mathcal{F}$ is proximally prime iff the set $\bigcap\{\mathrm{cl}_{\sigma X} F \mid F \in \mathcal{F}\}$ is a connected subspace of σX.*

PROPOSITION 3. *Let X be a proximity space and let A be a closed nonempty subspace of σX. Define $\mathcal{F} = \mathcal{F}(A)$ to be the set of all subsets F of X such that for some open subset G of σX, we have $A \subset G$ and $G \cap X \subset F$. Then*
(a) *A is connected iff the filter $\mathcal{F}$ is proximally prime.*
(b) $A = \bigcap\{\mathrm{cl}_{\sigma X} F \mid F \in \mathcal{F}\}$.

We say that a collection $\mathcal{P}$ of subsets of a proximity space X is a *proximal partition* provided $\mathcal{P}$ is finite and any two distinct members of $\mathcal{P}$ are far. A proximal partition $\mathcal{P}$ is called a *division* of a filter $\mathcal{H}$ provided $\bigcup \mathcal{P} \in \mathcal{H}$.

A remark: If $\mathcal{F}$ is a filter on a proximity space X then $\mathcal{F}$ is proximally prime iff every division of $\mathcal{F}$ has a member in common with $\mathcal{F}$.

Another remark: If $\mathcal{P}$ is a proximal partition on a proximity space X, then $\mathcal{P}$ is a uniform cover of $\bigcup \mathcal{P}$.

The following theorem is a characterization of proximally prime filters by means of proximally continuous maps; we shall have need of this result in the sequel.

PROPOSITION 4. *Let $\mathcal{F}$ be a filter on a proximity space X. Then the following are equivalent:*
(a) *$\mathcal{F}$ is proximally prime.*
(b) *For every $f \in \mathrm{P}^*(X)$, the set of adherence points of $f\mathcal{F}$ is connected.*
(c) *For every $f \in \mathrm{P}^*(X)$, if $f\mathcal{F}$ has only a finite number of adherence points then it converges.*

Proof. (a) $\to$ (b): Let $f \in \mathrm{P}^*(X)$. If the set E of adherence points of $f\mathcal{F}$ is not connected then there are two open sets G and H meeting and covering E that have a positive distance from each other. The sets $f^{-1}[G]$ and $f^{-1}[H]$ are far and their union belongs to $\mathcal{F}$ so one of them must belong to $\mathcal{F}$. But that is not possible since the image of the other contains at least one element of E. The implication from (b) to (c) is trivial. Now for the remaining implication, assume that (c) is true. Let $A, B \subset X$ with $\{A, B\}$ far and $A \cup B \in \mathcal{F}$. By Efremovič's Theorem [2], there exists a proximally continuous map $f\colon X \to [0,1]$ which sends A into 0 and B into 1. Since $A \cup B \in \mathcal{F}$, 0 and 1 are the only possible adherence points of $f\mathcal{F}$. If it converges to 0 then $A = (A \cup B) \cap f^{-1}[0, 1/2] \in \mathcal{F}$, while if it converges to 1 then $B = (A \cup B) \cap f^{-1}[1/2, 1] \in \mathcal{F}$.

4. Envelopment

This section is devoted to establishing Theorem 10 below, a result which will be crucial in establishing our main theorem, Theorem 23. If $\mathcal{H}$ is a filter on a proximity space X then a filter $\mathcal{G}$ on X is said to be *$\mathcal{H}$-minimal* provided $\mathcal{G}$ is a minimal element of the set $\{\mathcal{B} \mid \mathcal{B}$ is a proximally prime filter on X and $\mathcal{H} \subset \mathcal{B}\}$.

LEMMA 5. *Let X be a proximity space and let $\mathcal{H}$ be a filter on X.*
(a) *$\mathcal{H}$ is contained in a proximally prime filter.*
(b) *For every proximally prime filter $\mathcal{A}$ on X with $\mathcal{H} \subset \mathcal{A}$ there exists an $\mathcal{H}$-minimal filter $\mathcal{G}$ with $\mathcal{H} \subset \mathcal{G} \subset \mathcal{A}$.*

Proof. (a): Every ultrafilter is proximally prime.
(b): A standard application of Zorn's lemma will do it.

LEMMA 6. *Let X be a proximity space. Then for any filter $\mathcal{H}$ on X, we have $\mathcal{H} = \bigcap\{\mathcal{F} \mid \mathcal{F}$ is $\mathcal{H}$-minimal$\}$.*

Proof. Let $A \subset X$ with $A \notin \mathcal{H}$. Then $\mathcal{H} \cup \{X \backslash A\}$ has the finite intersection property and hence is contained in an ultrafilter $\mathcal{U}$. By Lemma 5, there exists an $\mathcal{H}$-minimal filter $\mathcal{G}$ with $\mathcal{G} \subset \mathcal{U}$. It follows that $A \notin \mathcal{U}$.

LEMMA 7. *Let X be a proximity space, let $\mathcal{H}$ be a filter on X, and let $\mathcal{G}$ be a proximally prime filter on X with $\mathcal{H} \subset \mathcal{G}$. Then $\{G \in \mathcal{G} \cap \mathcal{P} \mid \mathcal{P}$ is a division of $\mathcal{H}\}$ is a base for an $\mathcal{H}$-minimal filter $\mathcal{D}$ on X with $\mathcal{D} \subset \mathcal{G}$.*

Proof. If $\mathcal{P}_1$ and $\mathcal{P}_2$ are divisions of $\mathcal{H}$ then so is $\mathcal{P} = \mathcal{P}_1 \wedge \mathcal{P}_2$, the collection of all possible intersections of a member of $\mathcal{P}_1$ with a member of $\mathcal{P}_2$. It follows that the given set is a filter base and so it generates a filter $\mathcal{D}$. To show that $\mathcal{D}$ is proximally prime, assume that $\{E, F\}$ is far with $E \cup F \in \mathcal{D}$. There is a division $\mathcal{P}$ of $\mathcal{H}$ such that some $A \in \mathcal{G} \cap \mathcal{P}$ satisfies $A \subset E \cup F$. Since $\mathcal{G}$ is proximally prime, $E \in \mathcal{G}$ or $F \in \mathcal{G}$. Assume $E \in \mathcal{G}$. Then $\mathcal{B} = \{A \cap E,\ A \cap F\} \cup (\mathcal{P} \backslash \{A\})$ is a division of $\mathcal{H}$ with $A \cap E \in \mathcal{G} \cap \mathcal{B}$. Therefore $A \cap E \in \mathcal{D}$. Hence $E \in \mathcal{D}$ and we have shown that $\mathcal{D}$ is proximally prime. Assume that $\mathcal{A}$ is a proximally prime filter on X with $\mathcal{H} \subset \mathcal{A} \subset \mathcal{D}$. Let $\mathcal{P}$ be a division of $\mathcal{H}$ and let $G \in \mathcal{G} \cap \mathcal{P}$. Then $\mathcal{P}$ is a division of $\mathcal{A}$ and so there exists $P \in \mathcal{A} \cap \mathcal{P}$. Thus $P, G \in \mathcal{D}$ and hence $P \cap G \neq \emptyset$. Therefore $P = G$ and so $G \in \mathcal{A}$.

LEMMA 8. *Let X be a proximity space and let $\mathcal{H}$ be a filter on X. Then for every $\mathcal{H}$-minimal filter $\mathcal{G}$ on X, the collection $\{G \in \mathcal{G} \cap \mathcal{P} \mid \mathcal{P}$ is a division of $\mathcal{H}\}$ is a base for $\mathcal{G}$.*

LEMMA 9. *Let X be a proximity space and let $\mathcal{H}$ be a filter on X. Let $(\mathcal{H}_i)_{i \in I}$ be the family of all $\mathcal{H}$-minimal filters. For each $i \in I$, let $H_i \in \mathcal{H}_i$. Then there exists a finite subset J of I with $\bigcup_{i \in I} H_i \in \mathcal{H}$.*

Proof. By Lemma 8, for each $i \in I$ there exist a division $\mathcal{P}_i$ of $\mathcal{H}$ and a set $P_i \in \mathcal{H}_i \cap \mathcal{P}_i$ with $P_i \subset H_i$. We proceed by contradiction. Suppose that for every finite subset J of I we have $\bigcup_{i \in I} H_i \notin \mathcal{H}$. Consider the collection $\{H \backslash \bigcup_{i \in I} P_i \mid H \in \mathcal{H}$ and J is a finite subset of $I\}$; it is a filter base on X so it generates a filter $\mathcal{G}$. Also, $\mathcal{H} \subset \mathcal{G}$. Let $\mathcal{D}$ be a $\mathcal{G}$-minimal filter. By Lemma 8, $\mathcal{D}$ is generated by the filter base $\{D \in \mathcal{D} \cap \mathcal{P} \mid \mathcal{P}$ is a division of $\mathcal{G}\}$. We shall show that $\mathcal{D}$ is an $\mathcal{H}$-minimal filter. To that end, Lemma 7 implies that it will be sufficient to show that $\{D \in \mathcal{D} \cap \mathcal{S} \mid \mathcal{S}$ is a division of $\mathcal{H}\}$ is a base for $\mathcal{D}$. Let $D \in \mathcal{D}$.

Then there exists a division $\mathcal{Q}$ of $\mathcal{G}$ and a set $Q \in \mathcal{Q}$ with $Q \subset D$. There exist $H \in \mathcal{H}$ and a finite subset J of I with $\bigcup \mathcal{Q} \supset H \backslash \bigcup_{i \in J} P_i$. Then $\mathcal{P} = \bigwedge_{i \in J} \mathcal{P}_i$ is a proximal partition which satisfies: for all $P \in \mathcal{P}$, either P is a subset of $X \backslash \bigcup_{i \in J} P_i$ or it is disjoint from it. Let $\mathcal{R} = (\mathcal{Q} \wedge \mathcal{P}) \cup \{P \in \mathcal{P} \mid P \subset \bigcup_{i \in J} P_i\}$. Since the set $H \cap \bigcup \mathcal{P}$ is a member of $\mathcal{H}$ and is a subset of $\bigcup \mathcal{R}$, we have that $\mathcal{R}$ is a division of $\mathcal{H}$. The restriction of $\mathcal{R}$ to $(H \cap \bigcup \mathcal{P}) \backslash \bigcup_{i \in J} P_i$ is a division of $\mathcal{G}$, so it contains a member M of $\mathcal{D}$, and refines $\mathcal{Q}$, thus $M \subset Q \subset D$. Finally, this establishes that $\mathcal{D}$ is an $\mathcal{H}$-minimal filter. Therefore, for some $i \in I$, we have $\mathcal{D} = \mathcal{H}_i$. $X \in \mathcal{H}$ so, with $K = \{i\}$, $X \backslash P_i \in \mathcal{D}$. But that is impossible since $P_i \in \mathcal{H}_i = \mathcal{D}$.

In a proximity space X, we say that a cover $\mathcal{U}$ *envelops* a filter $\mathcal{H}$ provided that for every proximally prime filter $\mathcal{F}$ on X with $\mathcal{H} \subset \mathcal{F}$ we have $\mathcal{F} \cap \mathcal{U} \neq \emptyset$.

THEOREM 10 (Proximity analogue of [3; 16.26]). *If a uniform cover $\mathcal{U}$ of a proximity space X envelope a filter $\mathcal{H}$, then there exists a division $\mathcal{P}$ of $\mathcal{H}$ such that $\mathcal{P}$ refines $\mathcal{U}$.*

Proof. Let $(\mathcal{H}_i)_{i \in I}$ be the family of all $\mathcal{H}$-minimal filters. Then by the definition of enveloping, for all $i \in I$, $\mathcal{H}_i \cap \mathcal{U} \neq \emptyset$, so select $U_i \in \mathcal{H}_i \cap \mathcal{U}$. By Lemma 8, for each $i \in I$ there exists a division $\mathcal{P}_i$ of $\mathcal{H}$ and there exists $P_i \in \mathcal{H}_i \cap \mathcal{P}_i$ with $P_i \subset U_i$. By Lemma 9, there exists a finite subset J of I with $\bigcup_{i \in J} P_i \in \mathcal{H}$. Then $H = \bigcap_{i \in J} (\bigcup \mathcal{P}_i)$ is a member of $\mathcal{H}$. It follows that the collection $\mathcal{B} = \bigwedge_{i \in J} \mathcal{P}_i$ is a division of $\mathcal{H}$, and so $\mathcal{P} = \mathcal{B} \wedge \{P_i \mid i \in I\}$ is a division of $\mathcal{H}$ that refines $\mathcal{U}$.

5. Analytic Subalgebras

If B is a normed algebra over the real number field, and A is a closed subalgebra of B then we say that A is *analytic* provided whenever $x \in B$ and $a_1, \ldots, a_n \in A$ with $x^n + a_1 x^{n-1} + \cdots + a_{n-1} x + a_n = 0$ it follows that $x \in A$. It is immediate that given any subset K of the algebra B, there is a smallest analytic subalgebra containing K: we shall denote it by $\mathrm{L}(K)$. An analytic subalgebra is said to be *finitely generated* provided it is of the form $\mathrm{L}(K)$ for some finite subset K of B. The *Katětov dimension* of the normed algebra B is defined to be the smallest cardinal m such that for every finitely generated analytic subalgebra A of B, there exists $K \subset B$ with $\mathrm{card}(K) \leqslant m$ and $A \subset \mathrm{L}(K)$. Evidently, the Katětov dimension is a nonnegative whole number or $\aleph_0$.

Remark: One has to exercise a certain amount of care: A subalgebra of B can have greater Katětov dimension than does B (see [3; 16.36]).

Katětov's theorem is the fact that the Katětov dimension of the algebra $\mathrm{C}^*(X)$ is equal to the covering dimension (based on finite cozero covers) of X for any Tychonov space X.

We need to have a simpler characterization of analytic subalgebras and we need to establish relationships between proximally prime filters and analytic subalgebras. These facts follow.

LEMMA 11. *Let X be a proximity space and let A be a closed subalgebra of* $P^*(X)$ *such that* $f \in P^*(X)$ *and* $f^2 \in A$ *implies* $f \in A$. *Let* $\mathcal{H}$ *be a filter on* X *such that* $f\mathcal{H}$ *converges for all* $f \in A$. *Then there exists a proximally prime filter* $\mathcal{F} \subset \mathcal{H}$ *on* X *such that* $f\mathcal{F}$ *converges for all* $f \in A$.

Proof. The proof that Gillman–Jerison gave for their 16.2 works as well for the algebra $P^*(X)$ in order to show that every closed subalgebra is a sublattice. Let $\mathcal{F}$ be the filter on X generated by the collection of all subsets of X of the form $f^{-1}[0,1]$ where $0 \leqslant f \in A$ and $f\mathcal{H}$ converges to 0. Observe that $\mathcal{F}$ is also generated by the collection of all subsets of X of the form $f^{-1}U$ where $f \in A$ with $f\mathcal{H}$ convergent and with U a neighborhood of the point to which $f\mathcal{H}$ converges. From that observation, it follows easily that for all $f \in A$, $f\mathcal{H}$ converges iff $f\mathcal{F}$ converges, and when they converge to the same point. Let H_1 and H_2 be far subsets of X such that $H = H_1 \cup H_2 \in \mathcal{F}$. H contains a set of the form $g^{-1}[0,1]$ for some $g \in A$ with $0 \leqslant g$ and such that $g\mathcal{H}$ converges to 0. Then $g\mathcal{F}$ converges to 0. Let $h = g \wedge 1$ and define the real-valued function f on X by $f(x) = 1 - h(x)$ if $x \in X \backslash H_1$ and $f(x) = h(x) - 1$ if $x \in X \backslash H_2$. Since for $x \in (X \backslash H_1) \cap (X \backslash H_2)$, $h(x) = g(x) = 1$, it follows that f is well defined. Since $\{X \backslash H_1, X \backslash H_2\}$ is a uniform cover of X and since the two maps $1-h$ and $h-1$ are uniformly continuous, it follows that f is uniformly continuous as well. Since $f^2 = (h-1)^2 \in A$ we have $f \in A$. Therefore $f\mathcal{H}$ converges; say it converges to the real number r. Then $f\mathcal{F}$ converges to r. Since $g\mathcal{F}$ converges to 0, also $h\mathcal{F}$ converges to 0. Since $|f| = |h-1|$, then $|f|\mathcal{F}$ converges to 1. Therefore, $r = 1$ or $r = -1$. If $r = 1$, then the set $f^{-1}[1/2, 2] \cap g^{-1}[0,1]$ is a subset of H_2 which belongs to $\mathcal{F}$. If $r = -1$, then the set $f^{-1}[-2, -1/2] \cap g^{-1}[0,1]$ is a subset of H_1 which belongs to $\mathcal{F}$.

THEOREM 12 (Proximity analogue of [3; Lemma 16.31]). *Let X be a proximity space and let A be a closed subalgebra of* $P^*(X)$ *such that* $f \in P^*(X)$ *and* $f^2 \in A$ *implies* $f \in A$. *Let* $\mathcal{F}$ *be minimal among all filters on* X *whose image under* f *converges for all* $f \in A$. *Then* $\mathcal{F}$ *is proximally prime.*

Proof. By Lemma 11, there exists a proximally prime filter $\mathcal{G} \subset \mathcal{F}$ such that $f\mathcal{G}$ converges for all $f \in A$. By the minimality property of $\mathcal{F}$, we have $\mathcal{F} = \mathcal{G}$.

THEOREM 13 (Proximity analogue of [3; Problem 16B]). *Let X be a proximity space and let A be a closed subalgebra of X such that whenever* $f \in P^*(X)$ *and* $f^2 \in A$ *then* $f \in A$. *Then A is an analytic subalgebra of* $P^*(X)$.

Proof. Let $g_1, \dots, g_n \in A$ and let $f \in P^*(X)$ such that $f^n + g_1 f^{n-1} + \cdots + g_{n-1} f + g_n = 0$. We use Theorem 1 to show that $f \in \operatorname{cl} A$. Let $\mathcal{H}$ be a filter on

X such that $h\mathcal{H}$ converges for all $h \in A$. To show that $f\mathcal{H}$ converges, we use Lemma 11. So there exists a proximally prime filter $\mathcal{F} \subset \mathcal{H}$ on X such that $h\mathcal{F}$ converges for all $h \in A$. Let $r_1, \ldots, r_n$ be such that $g_i\mathcal{F} \to r_i$ for each i. Since $f[X]$ is bounded $f\mathcal{F}$ has an adherence point. For any adherence point t of $f\mathcal{F}$, since r_i is an adherence point of $g_i\mathcal{F}$ for each i, we have $t^n + r_1 t^{n-1} + \cdots + r_{n-1}t + r_n$ is an adherence point of $(f^n + g_1 f^{n-1} + \cdots + g_{n-1} f + g_n)\mathcal{F} = 0[\mathcal{F}]$ and hence $t^n + r_1 t^{n-1} + \cdots + r_{n-1}t + r_n = 0$. This shows that $f\mathcal{F}$ has only finitely many adherence points, and by Proposition 4 we are through.

LEMMA 14. *Let X be a proximity space, let A be an analytic subalgebra of* $\mathrm{P}^*(X)$, *and let $g \in \mathrm{P}^*(X)$. Then $g \in A$ iff $g\mathcal{F}$ converges whenever $\mathcal{F}$ is a proximally prime filter on X with $f\mathcal{F}$ convergent for all $f \in A$.*

Proof. One direction is trivial. For the other direction use Theorem 1 and Lemma 11.

LEMMA 15 (Proximity analogue of [3; 16.30]). *Let $\mathcal{F}$ be a proximally prime filter on a proximity space X and let A be the set of all $f \in \mathrm{P}^*(X)$ such that $f\mathcal{F}$ converges. Then A is an analytic subalgebra of* $\mathrm{P}^*(X)$.

Proof. It is straightforward to check that A is a closed subalgebra of $\mathrm{P}^*(X)$. We use Theorem 13 to show that A is analytic. Let $f \in \mathrm{P}^*(X)$ and assume that $f^2 \in A$. Then by hypothesis, $f^2\mathcal{F}$ converges, say to r. There are two possibilities, either $r = 0$ or $r > 0$. Assume first that $r = 0$. For any $\varepsilon > 0$ we have that $(f^2)^{-1}[-\varepsilon^2, \varepsilon^2] \subset f^{-1}[-\varepsilon, \varepsilon]$. Since the set before the inclusion lies in $\mathcal{F}$, then so does the set after the inclusion. Hence $f\mathcal{F}$ converges to 0 and therefore $f \in A$. This finishes the case when $r = 0$ so now assume that $r > 0$. Let $F_1 = f^{-1}(-\infty, -\sqrt{r/2})$ and let $F_2 = f^{-1}(\sqrt{r/2}, \infty)$. Then $F_1 \cup F_2 = (f^2)^{-1}(r/2, \infty) \in \mathcal{F}$ and therefore, $F_1 \in \mathcal{F}$ or $F_2 \in \mathcal{F}$. Suppose first that $F_2 \in \mathcal{F}$. To show that $f\mathcal{F}$ converges to $\sqrt{r}$, let $\varepsilon > 0$ and let $\eta = \varepsilon(\sqrt{r/2} + \sqrt{r})$. Then $(f^2)^{-1}(r - \eta, r + \eta)$ is a member of $\mathcal{F}$ and hence its intersection with F_2 is also a member of $\mathcal{F}$. Since that intersection is contained in $f^{-1}(\sqrt{r} - \varepsilon, \sqrt{r} + \varepsilon)$, it follows that $f\mathcal{F}$ converges to $\sqrt{r}$ as claimed and therefore $f \in A$. Similarly one can show that if $F_1 \in \mathcal{F}$ then $f\mathcal{F}$ converges to $-\sqrt{r}$, and in this case also we have $f \in A$.

THEOREM 16 (Proximity analogue of [3; 16.32]). *Let X be a proximity space and let H be a subset of* $\mathrm{P}^*(X)$. *Then $L(H)$ is precisely the set of all $g \in \mathrm{P}^*(X)$ such that $g\mathcal{F}$ converges whenever $\mathcal{F}$ is a proximally prime filter on X such that $f\mathcal{F}$ converges for all $f \in H$.*

Proof. Apply the preceding two lemmas along with the fact that the intersection of analytic subalgebras is analytic.

We let $\mathrm{P}_n^*(X)$ be the set of all bounded proximally continuous maps on the proximity space X into $\mathbf{R}^n$, $\mathbf{R}^n$ having the maximum metric, and we place on $\mathrm{P}_n^*(X)$ the sup-norm metric. All topological references on $\mathrm{P}_n^*(X)$ are to this structure. In either of these metric spaces, we let $\mathrm{B}_\varepsilon(x)$ and $\mathrm{K}_\varepsilon(x)$ denote, respectively, the open and closed balls.

THEOREM 17 (Proximity analogue of [3; 16.33]). *Let X be a proximity space and let $f \in \mathrm{P}_n^*(X)$. Then the analytic subalgebra A_f of $\mathrm{P}^*(X)$ generated by the set $\{f_1, \ldots, f_n\}$ where $f = (f_1, \ldots, f_n)$, is precisely the set of all functions $g \in \mathrm{P}^*(X)$ such that $g\mathcal{F}$ converges whenever $\mathcal{F}$ is a proximally prime filter with $f\mathcal{F}$ convergent.*

Proof. Use Theorem 16.

6. Katětov Dimension Theory

If $\mathcal{W}$ is a cover of a set Y and $\mathcal{G}$ is a filter on Y, then we say that $\mathcal{G}$ is $\mathcal{W}$*-Cauchy* iff $\mathcal{W} \cap \mathcal{G} \neq \emptyset$.

The following lemma is, of course, known (maybe as folklore).

LEMMA 18. *Let X be a proximity space and let $F_1, \ldots, F_m \subset X$ such that for distinct i and j, $\{F_i, F_j\}$ are far. Then there exist $E_1, \ldots, E_m \subset X$ such that for distinct i and j, $\{E_i, E_j\}$ are far and for all i, $F_i < E_i$.*

Proof. For all distinct pairs i, j, select $P_{i,j}, Q_{i,j}$ with $F_i < P_{i,j}$, $F_j < Q_{i,j}$ and $\{P_{i,j}, Q_{i,j}\}$ far. Define E_k to be the intersection of the two sets $\bigcap_{i \neq j} P_{k,j}$ and $\bigcap_{i \neq k} Q_{i,k}$.

THEOREM 19 (Proximity analogue of [3; 16.27]). *Let X be a proximity space and let $g \in \mathrm{P}_n^*(X)$. Let $\mathcal{U}$ be a finite uniform cover of X such that whenever $\mathcal{H}$ is a filter with $g\mathcal{H}$ convergent then $\mathcal{U}$ envelops $\mathcal{H}$.*

(a) *There exists a finite uniform cover $\mathcal{W}$ of $g[X]$ such that whenever $\mathcal{H}$ is a filter with $g\mathcal{H}$ being $\mathcal{W}$-Cauchy then $\mathcal{U}$ envelops $\mathcal{H}$.*

(b) *$\mathcal{U}$ is refined by a uniform cover of X having order $\leqslant n$.*

Proof. (a): Let $(\mathcal{H}_i)_{i \in I}$ be the family of all filters on X whose image under g converges, and for each $i \in I$ select y_i in $\mathbf{R}^n$ such that $g\mathcal{H}_i$ converges to y_i. For each $i \in I$ let $\mathcal{G}_i$ denote the collection of all subsets of X which contain a set of the form $g^{-1}[K_\varepsilon(y_i)]$ with $\varepsilon > 0$. Note that for each $i \in I$, $\mathcal{G}_i$ is a filter on X such that $g\mathcal{G}_i$ converges to y_i and $\mathcal{G}_i \subset \mathcal{H}_i$. By Theorem 10 for every $i \in I$ there exists a division $\mathcal{P}_i$ of $\mathcal{G}_i$ such that $\mathcal{P}_i$ refines $\mathcal{U}$. Write $\mathcal{P}_i = \{F_{i1}, \ldots, F_{im}\}$. For all $i \in I$ and $k = 1, \ldots, m$, there exists $U_{ik} \in \mathcal{U}$ with $F_{ik} \subset U_{ik}$. For all i, there exists $\varepsilon_i > 0$ such that $g^{-1}[K_{\varepsilon_i}(y_i)] \subset \bigcup_{k=1}^m F_{ik}$. Now the closure of $g[X]$ is contained in $\bigcup_{i \in I} \mathrm{B}_{\varepsilon_i}(y_i)$. (For any point in that closure, look at the filter generated by the preimages under g of the intersection of open neighborhoods of the point with the set $g[X]$.) Since the closure of $g[X]$ is compact, it is contained in finitely many of those open balls, i.e., there exists a finite set $J \subset I$ so that the closure of $g[X]$ is contained in $\bigcup_{i \in J} \mathrm{B}_{\varepsilon_i}(y_i)$. Let $\mathcal{V}$ denote the set of those

$B_{\varepsilon_i}(y_i)$ with $i \in J$. Let $\mathcal{W}$ be the trace of $\mathcal{V}$ on $g[X]$. Let $\mathcal{D}$ be a filter with $g\mathcal{D}$ $\mathcal{W}$-Cauchy. To complete the proof of (a), we must show that $\mathcal{U}$ envelops $\mathcal{D}$. To that end, let $\mathcal{F}$ be a proximally prime filter on X with $\mathcal{D} \subset \mathcal{F}$. For some $D \in \mathcal{D}$ and some $i \in J$, we have $g[D] \subset B_{\varepsilon_i}(y_i) \subset K_{\varepsilon_i}(y_i)$. By the construction of ε_i it follows that the set D is contained in the union $\bigcup_{k=1}^{m} F_{ik}$. Therefore, that union is a member of $\mathcal{D}$ and hence also of $\mathcal{F}$. Since $\mathcal{F}$ is proximally prime, for some k, $F_{ik} \in \mathcal{F}$. Hence, $U_{ik} \in \mathcal{F}$ and it follows that $\mathcal{F} \cap \mathcal{U} \neq \emptyset$.

(b): Let $\mathcal{W}$ be as stated in (a). The uniform cover $\mathcal{W}$ is strongly refined by some uniform cover $\mathcal{A}$ of $g[X]$. The closure of $g[X]$ is a compact subspace of $\mathbf{R}^n$ and hence is of dimension $\leqslant n$. The proximity space $g[X]$ has the same dimension as its closure. Therefore, $\mathcal{A}$ is refined by a finite uniform cover $\mathcal{V} = \{V_1, \dots, V_s\}$ of $g[X]$ such that $\mathcal{V}$ has order $\leqslant n$. For each $i = 1, \dots, s$, let $\mathcal{H}_i$ be the collection of all subsets of X which contain a set of the form $g^{-1}M$ with $V_i < M$ in $g[X]$. Then for each i, $\mathcal{H}_i$ is a filter on X such that $g\mathcal{H}_i$ is $\mathcal{W}$-Cauchy, and it follows from (a) that $\mathcal{U}$ envelops $\mathcal{H}_i$. By Theorem 10 there exists a division $\mathcal{P}_i$ of $\mathcal{H}_i$ which refines $\mathcal{U}$. Let $\mathcal{D}_i$ be the trace of $\mathcal{P}_i$ on the set $g^{-1}[V_i]$. For all i, $\mathcal{D}_i$ is a uniform cover of $g^{-1}[V_i]$ while $\{g^{-1}[V_i] \mid i = 1, \dots, s\} = g^{-1}\mathcal{V}$ is a uniform cover of X. If follows that $\mathcal{D} = \bigcup_{i=1}^{s} \mathcal{D}_i$ is a uniform cover of X. Clearly $\mathcal{D}$ refines $\mathcal{U}$ and has order $\leqslant n$.

LEMMA 20 (Proximity analogue of [3; 16.24]). *Let X be a proximity space. Given $f_1, \dots, f_s \in P_n^*(X)$ and given $\varepsilon > 0$, there exists a finite uniform cover $\mathcal{V}$ of X such that* $\operatorname{diam} f_k[V] \leqslant \varepsilon$ *for each $k \leqslant s$ and each $V \in \mathcal{V}$.*

LEMMA 21. *Let $\{A_1, \dots, A_s\}$ and $\{B_1, \dots, B_s\}$ be finite uniform covers of a proximity space X such that for all k, $A_k < B_k$. Then there exists, for each $k = 1, \dots, s$, $h_k \in P^*(X)$ with $0 \leqslant h_k$, with $A_k < X \backslash Z(h_k) < B_k$, and such that $\sum_{k=1}^{s} h_k = 1$.*

THEOREM 22 (Proximity analogue of [3; 16.28]). *Let X be a proximity space, let $\mathcal{U}$ be a finite uniform cover of X, and let n be a natural number. Define $G(\mathcal{U})$ to be the set of all $g \in P_n^*(X)$ such that whenever $\mathcal{H}$ is a filter on X with $g\mathcal{H}$ convergent, then $\mathcal{U}$ envelops $\mathcal{H}$. Then*

(a) *$G(\mathcal{U})$ is open in $P_n^*(X)$.*

(b) *If X has dimension $\leqslant n$ then $G(\mathcal{U})$ is dense in $P_n^*(X)$.*

Proof. (a): Let $g \in G = G(\mathcal{U})$. By Theorem 19 there exists a finite uniform cover $\mathcal{W}$ of $g[X]$ such that for any filter $\mathcal{H}$ on X with $g\mathcal{H}$ being $\mathcal{W}$-Cauchy we have that $\mathcal{U}$ envelops $\mathcal{H}$. There exists $\varepsilon > 0$ such that the trace on $g[X]$ of the collection of all open balls about the points of $g[X]$ of radius ε refines $\mathcal{W}$. It is not difficult to show that $B_{\varepsilon/4}(g) \subset G$. Therefore G is open.

(b): Assume that X has dimension $\leqslant n$. Given $f \in P_n^*(X)$ and $\varepsilon > 0$ we are to find $g \in G$ within distance ε of f. By Lemma 20, there exists a finite uniform cover $\mathcal{V}$ of X such that $\operatorname{diam} f[V] \leqslant \varepsilon/2$ for each $V \in \mathcal{V}$. Since X

has dimension $\leqslant n$, the finite uniform cover $\mathcal{V} \wedge \mathcal{U}$ has a refinement by a finite uniform cover $\mathcal{C} = \{C_1, \ldots, C_s\}$ having order $\leqslant n$. By [6; IV.19], there exists a uniform cover $\mathcal{A} = \{A_1, \ldots, A_s\}$ with each $A_k < C_k$. By Lemma 21, there exist $h_1, \ldots, h_s \in \mathrm{P}^*(X)$ with $0 \leqslant h_k$ and $A_k < X \backslash Z(h_k) < C_k$ for all k, and with $\sum_{k=1}^{s} h_k = 1$. Define $S_k = X \backslash Z(h_k)$. Select any $x_1, \ldots, x_s$ with $x_k \in S_k$ and let $T_k = \mathrm{B}_{\varepsilon/2}(f(x_k))$. We apply [3; 16.23] to the sets T_k to obtain points y^k and proceed exactly as in the middle of page 255 of [3], with $g(x) = \sum_{k=1}^{s} h_k(x) y^k$, to conclude that g is within ε of f. It remains to show that $g \in G$. Let $\mathcal{H}$ be a filter on X such that $g\mathcal{H}$ converges, say to $y \in \mathbf{R}^n$. We must show that $\mathcal{U}$ envelops $\mathcal{H}$. To that end, let $\mathcal{F}$ be a proximally prime filter on X with $\mathcal{H} \subset \mathcal{F}$. We must show that $\mathcal{F} \cap \mathcal{U} \neq \emptyset$. By definition of g and the properties of h, $g[X]$ is contained in the union over all $J \subset \{1, \ldots, s\}$ having cardinal $\leqslant n+1$ of the convex hull of the set $\{y^j \mid j \in J\}$. The union is over a finite index set and the hulls are closed so $\mathrm{cl}(g[X])$ is also contained in that union and hence so is y. Let H be the set of all $J \subset \{1, \ldots, s\}$ with J having cardinal at most $n+1$, with y being in the convex hull of the set $\{y^j \mid j \in J\}$, and with each of the corresponding barycentric coordinates of y being strictly positive. For each $J \in H$ let $a_k^J > 0$ with $\sum_{k \in J} a_k^J = 1$ and $\sum_{k \in J} a_k^J y^k = y$. By [3; 16.23], such a_k^J are unique, and they exist by definition of H. We extend the definition of a_k^J by defining $a_k^J = 0$ for all $k \in \{1, \ldots, s\} \backslash J$. A simple exercise shows that for all $J, L \in H$ with $J \neq L$ we have $0 < \|(a_k^J)_{k=1}^s - (a_k^L)_{k=1}^s\|$, the norm being the maximum norm on $\mathbf{R}^s$. H being finite, we can select $\eta > 0$ so that $3\eta < \|(a_k^J)_{k=1}^s - (a_k^L)_{k=1}^s\|$, for all $J, L \in H$ with $J \neq L$, and further $2\eta < a_k^J$ for all $k \in J \in H$. Define D to be the set of all elements $(b_k)_{k=1}^s$ in $\mathbf{R}^s$ with each $b_k \geqslant 0$ and with $\sum_{k=1}^{s} b_k = 1$. Define q: $D \to \mathbf{R}^n$ by $q((b_k)_{k=1}^s) = \sum_{k=1}^{s} b_k y^k$. With r: $D \to \mathbf{R}^n$ being the inclusion map, we have that $q = r \circ h$. For each $J \in H$ define E_J to be the set of all $(b_k)_{k=1}^s \in C$ such that $\|(b_k)_{k=1}^s - (a_k^J)_{k=1}^s\| < \eta$. We shall show that $\mathcal{P} = \{h^{-1}[E_J] \mid J \in H\}$ is a division of $\mathcal{F}$. A simple exercise shows that for distinct J, L any element of E_J and any element of E_L are at least the distance η apart. Therefore, distinct elements of $\mathcal{P}$ are far. For each $J \subset \{1, \ldots, s\}$, q maps the convex hull of $\{e^j \mid j \in J\}$ onto the convex hull of $\{y^j \mid j \in J\}$ bijectively and continuously, and because of compactness, as a uniform isomorphism. (Here $e_k^j = 0$ if $k \neq j$ and $= 1$ if $k = j$.) Therefore, for each $J \subset \{1, \ldots, s\}$, there exists $\delta_J > 0$ such that $\|\sum_{j \in J} b_j y^j - \sum_{j \in J} b'_j y^j\| < \delta_J$ implies $\|(b_j)_{j=1}^s - (b'_j)_{j=1}^s\| < \eta$, where we define $b_j = 0$ for $j \notin J$, etc. Let $\delta = \min\{\delta_J \mid J \subset \{1, \ldots, s\}\}$. $g^{-1}\mathrm{B}_\delta(y)$ is a member of $\mathcal{F}$ and is contained in the union of all the elements of $\mathcal{P}$. Therefore, that union is also an element of $\mathcal{F}$, and hence $\mathcal{P}$ is a division of $\mathcal{F}$. To complete the proof, we need only that $\mathcal{P}$ refines $\mathcal{U}$. Let $J \in H$. Since $J \neq \emptyset$, fix any element k of J. Then one can show that $h^{-1}[E_J] \subset S_k$.

THEOREM 23 (Proximity analogue of Katětov's theorem [3; 16.35]). *The following are equivalent for any proximity space X:*
(a) *X has dimension $\leqslant n$.*
(b) *$P^*(X)$ has Katětov dimension $\leqslant n$.*

Proof. To show that (a) implies (b), let $(f_k)_{k=1}^s$ be a finite sequence of elements of $P^*(x)$. By Lemma 20, for each natural number m there exists a finite uniform cover $\mathcal{U}_m$ of X such that for every $k \leqslant s$ and $U \in \mathcal{U}_m$ the set $f_k[U]$ has diameter $\leqslant 1/m$. Since X has dimension $\leqslant n$, there exists an open dense subset $G(\mathcal{U}_m)$ of $P_n^*(X)$ as described in Theorem 22. Since $P_n^*(X)$ is a complete metric space, we have, by the Baire category theorem, that the intersection of all the $G(\mathcal{U}_m)$ is nonempty. Select g in that intersection and let $g = (g_1, \dots, g_n)$. It will finish the proof if we show that the analytic subalgebra A_g with base $\{g_1, \dots, g_n\}$ contains every f_k. We apply Theorem 17. Let $\mathcal{F}$ be a proximally prime filter on X such that $g_i\mathcal{F}$ converges (say to a_i) for every i. We must show that $f_k\mathcal{F}$ converges. Define $a = (a_1, \dots, a_n)$. Then $g\mathcal{F}$ converges to a. For every m, by definition of $G(\mathcal{U}_m)$ and since $g \in G(\mathcal{U}_m)$, it follows that $\mathcal{U}_m$ envelops $\mathcal{F}$, and, since $\mathcal{F}$ is proximally prime, we have $\mathcal{U}_m \cap \mathcal{F} \neq \emptyset$. Select $U_m \in \mathcal{U}_m \cap \mathcal{F}$ and recall that the set $f_k[U_m]$ has diameter $\leqslant 1/m$. It follows that $f_k\mathcal{F}$ is Cauchy, and hence converges.

To prove that (b) implies (a), let $\mathcal{U}$ be a finite uniform cover of X, say $\mathcal{U} = \{U_1, \dots, U_s\}$. By [6; IV.19], there exists a uniform open cover $\mathcal{W} = \{W_1, \dots, W_s\}$ with $W_k < U_k$ for all k. By Efremovič's Theorem, for all k, there exists a proximally continuous g_k: $X \to [0, 1]$ which maps W_k to 0 and $X \backslash U_k$ to 1. By hypothesis, $\{g_1, \dots, g_s\}$ is contained in an analytic subalgebra A_f for some $f \in P_n^*(X)$. We apply Theorem 19. Let $\mathcal{H}$ be a filter such that $f\mathcal{H}$ converges. We must show that $\mathcal{U}$ envelops $\mathcal{H}$. Let $\mathcal{F}$ be a proximally prime filter on X with $\mathcal{H} \subset \mathcal{F}$. We must show that $\mathcal{F} \cap \mathcal{U} \neq \emptyset$. Since $f\mathcal{F}$ converges it follows from Theorem 17 that $g_k\mathcal{F}$ converges for every k. $\bigcup \mathcal{W} = X$ and X meets every member of $\mathcal{F}$ so it follows that some W_k meets every member of $\mathcal{F}$. By construction of g_k and since W_k meets every member of $\mathcal{F}$, we have that $g_k\mathcal{F}$ converges to 0. $g_k^{-1}[-1/2, 1/2]$ is a member of $\mathcal{F}$ and is a subset of U_k. Therefore, $U_k \in \mathcal{F}$, and we have shown that $\mathcal{U}$ envelops $\mathcal{H}$. By Theorem 19(b), $\mathcal{U}$ has a refinement of order $\leqslant n$. Hence X has dimension $\leqslant n$.

References

1. Čech, E.: *Topological Spaces*, Interscience, London, 1966.
2. Efremovič, V. A.: The geometry of proximity. I., *Mat. Sbornik N. S.* **31**(73) (1952), 189–200.
3. Gillman, L. and Jerison, M.: *Rings of Continuous Functions*, Van Nostrand, Princeton, 1960.
4. Hejcman, J.: On analytical dimension of rings of bounded uniformly continuous functions, *Comment. Math. Univ. Carol.* **28** (1987), 325–335.
5. Herrlich, H.: *Topologie II: Uniforme Räume*, Heldermann-Verlag, Berlin, 1988.
6. Isbell, J. R.: *Uniform Spaces*, American Mathematical Society, Providence, 1964.

7. Katětov, M.: On rings of continuous functions and the dimension of compact spaces, *Časopis Pěst Mat. Fys.* **75** (1950), 1–16 (Russian, English and Czech summaries.)
8. Naimpally, S. A. and Warrack, B. D.: *Proximity Spaces*, Cambridge University Press, Cambridge, 1970.

Applied Categorical Structures **4**: 57–68, 1996.

Generalized Reflective cum Coreflective Classes in Top and Unif

MIROSLAV HUŠEK*
Dept. of Mathematics, Charles University, Sokolovská 83, 186 00 Prague, Czech Republic

and

ANNA TOZZI
Dept. of Mathematics, University of L'Aquila 67100, Italy

(Received: 3 November 1994; accepted: 27 September 1995)

Abstract. The Herrlich's problem from [8] whether there are nontrivial classes of topological spaces that are both almost reflective or injective and almost coreflective or projective, is investigated in a more general setting using cone and cocone modifications of the classes used in the problem. We look also at the problem for uniform spaces. Typical results: There is no nontrivial multiprojective and orthogonal class of topological spaces; There is a reflective class of uniform spaces that is almost coreflective in `Unif`.

Mathematics Subject Classification (1991). 54B30.

Key words: coreflectivity, reflectivity, projectivity, injectivity.

The paper consists of four sections and two final diagrams summarizing the state of relations between generalized coreflective and reflective classes in `Top` and `Unif`. The first section recalls the main definitions, perhaps with some small modifications. The second section sums up basic properties of the above classes that we need or that help in understanding some relations; we restate here the Herrlich's problem from [8] for our situation. The last sections contain the main results of the paper for `Top` or `Unif`, resp.

Terms and concepts we use in this paper are well-known except, perhaps, the empty object X (notation $\emptyset$) in a category $\mathcal{K}$ that is defined by the properties $\mathcal{K}(X,Y)$ is a singleton for every object Y of $\mathcal{K}$, and $\mathcal{K}(Y,X) = \emptyset$ for every object Y of $\mathcal{K}$ different from X.

All subcategories are supposed to be full and so we shall mostly use classes of objects instead of subcategories. We shall also suppose that all subcategories are closed under isomorphisms.

* Work on this paper was initiated while the first author was a C.N.R. visitor of the University of L'Aquila. Partial financial assistence by Charles University Grant 349/1994 is also acknowledged.

1. Basic Definitions

Let $\mathcal{K}$ be a category and $\mathcal{C}$ be a class of objects of $\mathcal{K}$ closed under isomorphisms.

DEFINITION 1. $\mathcal{C}$ is said to be *weakly coreflective* in $\mathcal{K}$ if every object K of $\mathcal{K}$ has a weak coreflection $c_K\colon C_K \to K$ in $\mathcal{C}$, i.e., $C_K \in \mathcal{C}$ and for any morphism $f\colon C \to K$, with $C \in \mathcal{C}$, there is some $\tilde{f}\colon C \to C_K$ such that $c_K \circ \tilde{f} = f$.

A weakly coreflective class in $\mathcal{K}$ which is closed under retracts in $\mathcal{K}$ is called *almost coreflective* in $\mathcal{K}$.

We follow the terminology of [8]. A different terminology is used, e.g., in [1].

DEFINITION 2. $\mathcal{C}$ is said to be *weakly multicoreflective* in $\mathcal{K}$ if every object of $\mathcal{K}$ has a weak multicoreflection in $\mathcal{C}$, i.e., for every $\mathcal{K}$-object K there exists a nonvoid set $\mathcal{C}_K$ of morphisms from objects of $\mathcal{C}$ into K such that any morphism from a nonempty object C of $\mathcal{C}$ into K factorizes via a morphism from $\mathcal{C}_K$.

A weakly multicoreflective class in $\mathcal{K}$ which is closed under retracts in $\mathcal{K}$ is called *almost multicoreflective* in $\mathcal{K}$.

A weakly multicoreflective class in $\mathcal{K}$ with unique factorizations (i.e., in the previous notation, for every $f\colon C \to K, C \in \mathcal{C}$, there exists a unique $c_f\colon C_f \to K$ from $\mathcal{C}_K$ and a unique $g\colon C \to C_f$ with $c_f g = f$) is called *multicoreflective* in $\mathcal{K}$.

The concept of a weakly multicoreflective class was used, e.g., in [15], that of multicoreflective class in [5, 3], a more general situation was studied in [4]. We follow the Börger's use of 'multi' instead of cone-reflective or cocone-coreflective used, e.g., in [1].

We shall deal with the *almost* concepts only; the *weak* concepts are really weak, and the corresponding *almost* class can be obtained from the *weak* one by adding the retracts of spaces from the class.

If one admits void sets $\mathcal{C}_K$ in Definition 2, the empty class is multicoreflective – that is a situation we want to exclude. The condition $C \neq \emptyset$ was added because of multicoreflectivity (otherwise every multicoreflective class would be coreflective).

DEFINITION 3. Let $\mathcal{N}$ be a class of morphisms of $\mathcal{K}$. The class $\mathcal{C}$ of all objects C having the property that, for every $m\colon X \to Y$, from $\mathcal{N}$, the mapping $\mathcal{K}(C,X) \to \mathcal{K}(C,Y)$ assigning mf to f, is surjective (or bijective) is called $\mathcal{N}$-*projective* ($\mathcal{N}$-*coorthogonal*, resp.) and is denoted by $\operatorname{Proj}\mathcal{N}$ ($\operatorname{Coorthog}\mathcal{N}$, resp.).

A class $\mathcal{C}$ of objects of $\mathcal{K}$ will be called *projective* (or *coorthogonal*) if $\mathcal{C} = \operatorname{Proj}\mathcal{N}$ (or $\operatorname{Coorthog}\mathcal{N}$, resp.) for some $\mathcal{N}$.

DEFINITION 4. Let $\mathbf{N}$ be a class of sinks in $\mathcal{K}$. By MultProj $\mathbf{N}$ we denote the class of those $\mathcal{K}$-objects X such that for every $\{f_i\colon S_i \to T\} \in \mathbf{N}$ and every $f\colon X \to T$ there exists i and $g\colon X \to S_i$ with $f_i g = f$. The class MultProj $\mathbf{N}$ is also called $\mathbf{N}$-*multiprojective*.

If the i and g from the definition are always unique whenever $X \neq \emptyset$, the class MultProj $\mathbf{N}$ is called $\mathbf{N}$-*multicoorthogonal* and denoted by MultCoorthog $\mathbf{N}$.

A class of $\mathcal{K}$-objects is called *multiprojective* (or multicoorthogonal) if it is of the form MultProj $\mathbf{N}$ (or MultCoorthog $\mathbf{N}$, resp.) for some $\mathbf{N}$.

The preceding concept was defined in [14].

If one omits the condition $X \neq \emptyset$ in the definition of multicoorthogonality, it may happen that MultCoorthog $\mathbf{N}$ is empty and that it need not contain $\emptyset$ even if it is nonempty; under such a definition, every multicoorthogonal class containing $\emptyset$ is coorthogonal. As in the case of multicoreflection, we want to exclude such a situation.

Classes having one of the properties defined above are called *generalized coreflective classes*. The dual notions: weak or almost reflectivity, multireflectivity, injectivity, orthogonality, multiinjectivity, multiorthogonality or generalized reflective classes (we shall assume that multiorthogonal classes always contain a nonempty object).

Orthogonal classes were investigated in [7] and from that time in many other papers; injectivity was defined in [13].

2. Basic Properties and Problems

By our definitions, for instance in `Top` and `Unif`, all generalized coreflectivity classes contain the empty space and need not contain singletons. For generalized reflective classes the situation is converse.

Interrelations among generalized coreflective (or reflective, resp.) properties shown in Diagrams 1 and 2 are almost clear (the arrows mean implications). We shall look at the situation in `Top` and `Unif` (our considerations are valid in categories over `Set` with enough constants).

First a summary of observations concerning properties of a generalized coreflective class $\mathcal{C}$ of topological or uniform spaces.

1. $\mathcal{C}$ is closed under retracts. If $\mathcal{C}$ is (multi)coreflective or (multi)coorthogonal (i.e., it has the unique factorizations), then it is closed under quotients. If $\mathcal{C}$ is not a "multi"-class then it is closed under sums.
2. An $\mathbf{M}$-multiprojective class contains singletons iff $\mathbf{M}$ consists of episinks (otherwise we get a coreflective class composed of the empty set).
3. If $\mathcal{C} =$ MultiCoorthog $\mathbf{M}$, then $\mathcal{C} \neq \{\emptyset\}$ iff for every $\{m_i\colon T_i \to S\}$ from $\mathbf{M}$ the m_i's are monomorphisms and $\{m_i(T_i)\}$ is a partition of S (then $\mathcal{C}$ contains all singletons). If $\mathbf{M}$ is a class of morphisms, we get a result for coorthogonal classes: Either $\mathbf{M}$ consists of bimorphisms (then $\mathcal{C}$ is coreflective) or $\mathcal{C} = \{\emptyset\}$ (and, thus, is also coreflective).

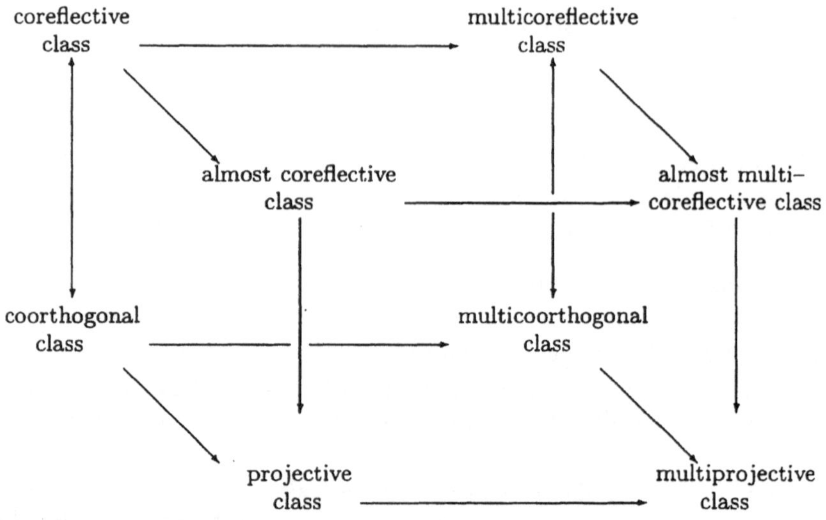

Diagram 1. Generalized coreflective classes.

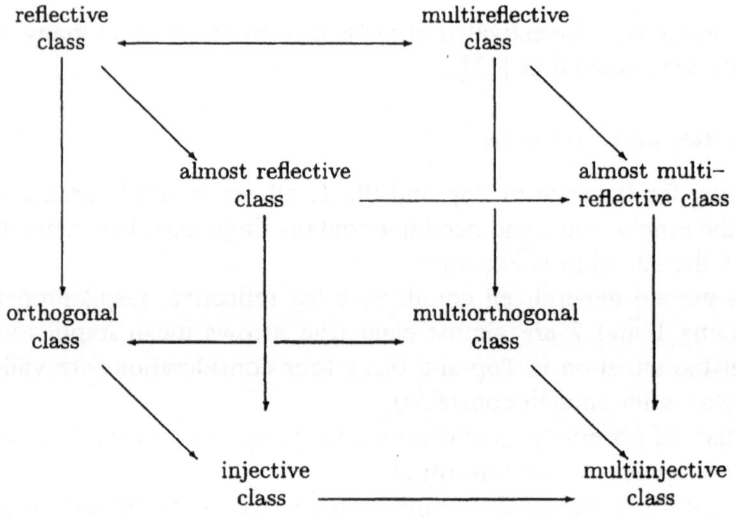

Diagram 2. Generalized reflective classes.

4. $\mathcal{C}$ is (multi)coreflective iff it is (multi)coorthogonal. (Indeed, for $\mathcal{C} \neq \emptyset$, the multicoreflection of X is composed of $\mathcal{C}$-components of X: take all monomorphisms $\{c_Y\colon Y \to X\}$, $Y \in \mathcal{C}$, a component of x is the union of those Y's containing x and endowed with the final structure.)

The situation for generalized reflective classes $\mathcal{C}$ of topological or uniform spaces is not completely dual:

1. $\mathcal{C}$ is closed under retracts. If $\mathcal{C}$ is not a "multi"-class then it is productive. Neither of the classes considered may be hereditary.
2. If $\mathcal{C}$ is multireflective (or multiorthogonal) then it is reflective (or orthogonal, resp.).
3. An $\mathbf{N}$-multiinjective (or multiorthogonal) class contains $\emptyset$ iff for every $\{m_i: \emptyset \to T_i\} \in \mathbf{N}$ there is some (a unique, resp.) i such that $T_i = \emptyset$.
4. The fact that every orthogonal class is injective is not so trivial as for coreflective properties (see, e.g., [1] or, implicitly, [16]).

We remind that the property of being any class in the lower halves of both preceding diagrams is closed under intersections. Also, both diagrams, regarded as ordered structures (of eight properties), are lower sublattices in all classes; of course, such a formulation is illegitimate – it simply means that, e.g., a class is coreflective iff it is multicoreflective and almost coreflective, etc. for other situations.

In the sequel, we shall assume for generalized reflective and coreflective classes to contain both the empty and a nonempty space (hence, a singleton). For MultProj(N) it means that N consists of episinks.

EXAMPLES. 1. Clearly, `Top` or `Unif` are the biggest generalized coreflective and reflective classes of any type, class `Singl` of at most one-point spaces is the smallest generalized coreflective class of multi-type, and generalized reflective class of any type.

2. `Dis` is the smallest generalized coreflective class of any nonmulti-type (by `Dis` we denote the class of discrete objects, i.e., of discrete spaces in `Top`, of uniformly discrete spaces in `Unif` – unlike discrete objects in `Unif`, discrete spaces in `Unif` are uniform spaces carrying the discrete topology). The class `Dis` is almost multireflective (neither almost reflective nor injective).

3. The class `Indis` of indiscrete spaces is multicoreflective and reflective (not almost coreflective).

4. Every class preserved by quotients is almost multicoreflective (e.g., the class of compact spaces in `Top`, the class `Prec` of precompact spaces in `Unif`, the class of spaces of cardinality less than a given cardinal, any class of connectedness). Realize that unlike, e.g., `Prec` in `Unif`, there is no nontrivial orthogonal class in `Top` closed under quotients (but pathwise connected spaces form an almost reflective class in `Top` closed under continuous images). Indeed, if $\mathcal{C}$ is productive, closed under quotients and different from `Indis`, then either the Sierpiński space $\tilde{2}$ or the two-point discrete space 2 belongs to $\mathcal{C}$, hence all the two-point spaces belong to $\mathcal{C}$; if $\mathcal{C} = \operatorname{Orthog} \mathcal{M}$ then $\mathcal{M}$ is a class of bimorphisms (since the two-point discrete space is in $\mathcal{C}$), thus $\mathcal{C} =$ `Top` since $\tilde{2} \in \mathcal{C}$.

5. Every hereditary class containing a nonempty space is almost multireflective. For instance, the class of subsequential spaces in `Top` or the class of proximally discrete spaces in `Unif` are coreflective and hereditary. The property to be hereditary may be weakened still to get almost multireflectivity: it suffices

to assume that there is a cardinal function ϕ such that every subspace A of a member X of the class is contained in some subspace C of X also contained in the class and having its cardinality at most $\phi(A)$.

6. The class of pathwise connected spaces is almost reflective in Top ([8]) – the same procedure works in Unif, too. This class is also multicoreflective (take the embeddings of pathwise components as the solution sets). Also this approach may be generalized: Take a space P and define $\mathcal{C}$ to be the class of spaces, where any two points can be joined by a (finite, intersecting) chain of (uniformly, resp.) continuous images of P. The class $\mathcal{C}$ is the left constant class of the right constant class of $\{P\}$. It is multicoreflective and almost reflective. Instead of a single space P we could take a set of spaces.

At the end of this section we want to recall the following important results implying that the classes in the lower part of the last diagram are intersections of the corresponding classes in the upper part.

THEOREM A. *In* Top *and* Unif, *every orthogonal (or injective, or multiinjective) class is an intersection of reflective (or almost reflective, or almost multireflective, resp.) classes.*

The result for orthogonal classes was proved in [7]; as was noticed in [1, 8] (see also [17]), the same proof works for injective classes, too. It is easy to modify the proof so that it works also for multiinjectivity. Up to now, it is not known whether the corresponding dual result is valid in Top and Unif. In the latter category at least a partial result is true, [2] (again, the proof given there for projective classes can be modified for multiprojective classes):

THEOREM B. *In* Unif, *every projective (or multiprojective) class generated by classes of mappings into separated spaces is an intersection of almost coreflective (or almost multicoreflective, resp.) classes.*

To get a final result in Unif, it remains to prove (or to disprove) that every projective class in Unif generated by a uniformly continuous mapping from a separated space into at least two-point indiscrete space, is almost coreflective (or, is an intersection of almost coreflective classes) in Unif.

H. Herrlich asked in his paper [8] whether there is a nontrivial almost reflective or injective subcategory of Top that is also almost coreflective or projective. Motivation for the question is the fact that in Top, Unif, ..., there are no nontrivial subcategories that are both coreflective and reflective. We can reformulate the question using the other generalized notions defined above.

PROBLEM 1. *Are there classes $\mathcal{C}$ of topological or uniform spaces that are both generalized coreflective of a given type and generalized reflective of a given type?*

For instance, are there nontrivial classes of topological spaces that are both coreflective and multiinjective?

3. Subcategories of Top

The main result from [12] asserts that *there is no class in* Top *that is both coreflective and reflective except* Top. Another known result from [6] says that it is consistent with ZFC that no nontrivial coreflective class in Top is productive (e.g., under GCH + there is no uncountable inaccessible cardinal) – thus (consistently) answering a problem of the first author of this paper, whether there is a nontrivial class of topological spaces that is closed under sums, products and quotients, see [10]. Consequently, *it is consistent with* ZFC *that no coreflective class in* Top *is injective*. That is practically all what is known about Problem 1 in Top. We shall add one more negative result. In [8] many reflective classes of Top were shown not to be almost coreflective. We can show that only trivial reflective classes in Top may be almost coreflective, in fact even more:

THEOREM 1. *There is no nontrivial multiprojective orthogonal class in* Top.

Of course, the classes **Singl**, **Indis**, Top are multicoreflective and orthogonal in Top. The proof will follow from several assertions that may be interesting to state separately. In the sequel we shall denote by $\tilde{2}$ the two-point connected T_0-space (the Sierpiňski space) with the underlying set $\{0, 1\}$, where the one-point set $\{0\}$ is open.

PROPOSITION 1. *If C is a multiprojective class in* Top *that does not contain the Sierpiňski space $\tilde{2}$, then there exists a cardinal κ such that $C \subset \{X$: every open set in X is an $F_{<\kappa}$-set$\}$.*

Proof. Let $C = \operatorname{MultProj} \mathbf{M}$, $\mathbf{M}$ a class of sinks. Since $\tilde{2} \notin C$, there exists $\{m_i\colon T_i \to S\} \in \mathbf{M}$ and a continuous map $\phi\colon \tilde{2} \to S$ that cannot be factorized continuously via some m_i. Thus ϕ is not constant and if $m_i(t) = \phi(0)$, then $m_i(\bar{t}) \subset \phi(0) \cup (S \setminus \phi(\tilde{2}))$. Let $\kappa > \sup_i |m_i^{-1}(\phi(0))|$, and $X \in C$; we shall show that every open set in X is a union of less than κ closed sets. Let G be open in X and $f\colon X \to \tilde{2}$ be such that $f^{-1}(0) = G$. There must be an index i and a continuous $g\colon X \to T_i$ such that the following diagram commutes

$$\begin{array}{ccc} X & \xrightarrow{\;f\;} & \tilde{2} \\ {\scriptstyle g}\downarrow & & \downarrow{\scriptstyle \phi} \\ T_i & \xrightarrow[\;m_i\;]{} & S \end{array} .$$

Then $G = \bigcup\{g^{-1}(\bar{t})\colon m_i(t) = \phi(0)\}$. □

COROLLARY. *If $\mathcal{C}$ is a multiprojective class in* Top *that does not contain the Sierpiński space $\tilde{2}$ and contains a nonindiscrete space, then $\mathcal{C}$ is not productive.*

Proof. Let X be a non-indiscrete member of $\mathcal{C}$. Its T_0-modification t_0X is a retract of X and, thus, belongs to $\mathcal{C}$. Moreover, t_0X is at least two-point space (since X is not indiscrete) and is T_1-space (otherwise, $\tilde{2} \in \mathcal{C}$ as retract of a T_0 non T_1-space). Take the cardinal κ from Proposition 1. Since the points of the power $(t_0X)^\kappa$ are closed and not $G_{<\kappa}$-sets, $\mathcal{C}$ cannot be productive by Proposition 1. □

PROPOSITION 2. *If $\mathcal{C}$ is a multiprojective class in* Top *that does not contain an indiscrete space, then there exists a cardinal κ such that $\mathcal{C} \subset \{X\colon 2^{|X|} \leqslant |\text{open}\, X|^\kappa\}$.*

Proof. Let $\mathcal{C} = \text{MultProj}\,\mathbf{M}$, $\mathbf{M}$ a class of sinks. Since a two-point indiscrete space does not belong to $\mathcal{C}$, there exists $\{m_i\colon T_i \to S\}_I \in \mathbf{M}$ such that S contains a two-point indiscrete subspace S_0 and no $P_i = m_i^{-1}(S_0)$ contains an indiscrete space mapped by m_i onto S_0. Therefore, every $m_i' = m_i \restriction P_i$ factorizes via the T_0-modification t_0P_i. Take a cardinal κ bigger than $|I|$ and all wP_i and X with $2^{|X|} > |\text{open}\, X|^\kappa$. There is $2^{|X|}$ of continuous maps from X to the indiscrete space S_0, and at most $|\text{open}\, X|^\kappa$ of continuous maps from X to all t_0P_i (indeed, we may assume that $t_0P_i \subset \tilde{2}^{wP_i}$ so that the number of continuous maps from X to t_0P_i is at most $|\text{open}\, X|^\kappa$). Consequently, not every map $X \to S_0$ can be factorized via some m_i. □

COROLLARY. *If $\mathcal{C}$ is a multiprojective class in* Top *that does not contain an indiscrete space, then for every at least two-point space X there exists a cardinal κ with $X^\kappa \notin \mathcal{C}$. In particular, every such a $\mathcal{C}$ different from* Singl *is not productive.*

We shall not need the following result to prove our Theorem 1 but it is connected to the previous two propositions and may be useful in some situations.

PROPOSITION 3. *If a multiprojective class in* Top *contains all T_1-spaces, then it contains all indiscrete spaces.*

Proof. Let $\mathcal{C} = \text{MultProj}\,\mathbf{M}$, $\mathbf{M}$ a class of sinks. If an indiscrete space X does not belong to $\mathcal{C}$, there exists $\{m_i\colon T_i \to S\} \in \mathbf{M}$ and a continuous $f\colon X \to S$ that cannot be factored via some m_i, i.e., no T_i contains an indiscrete space mapped by m_i onto $f(X)$. Take a cardinal $\kappa \geqslant |S|, \kappa \geqslant \sup_i\{|T_i|\}$ and suppose that a space Y has a cardinality at least κ^+. Take a map $\phi\colon Y \to f(X)$ such that every fiber $\phi^{-1}(t)$ has cardinality at least κ^+. Assume that ϕ factors via some m_i, $\phi = m_i\psi$. Then for each $t \in f(X)$ there is some $s_t \in m_i^{-1}(t)$ with $|\psi^{-1}(s_t)| \geqslant \kappa^+$. If Y has a topology such that $\bar{A} \cap B \neq \emptyset$ whenever $|A| = |B| = \kappa^+$, then ψ cannot be continuous because $\{s_t\colon t \in f(X)\}$ would be indiscrete and is mapped by m_i onto $f(X)$. Consequently, Y does not belong to $\mathcal{C}$. Clearly, the cofinite topology has the required property. □

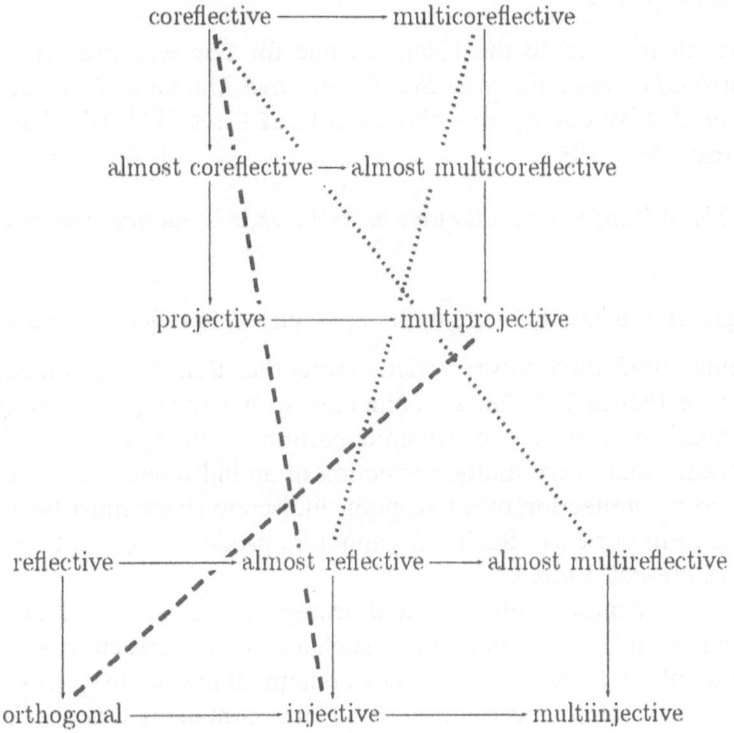

Diagram 3. The situation in Top.

Proof of Theorem 1. Suppose that C is multiprojective and productive class in Top that is different from Singl and Indis. By Corollary to Proposition 1, C contains the Sierpiński space $\tilde{2}$, and by Proposition 2 it contains all indiscrete spaces. If C is orthogonal, it is an intersection of reflective classes and every such a reflective class contains the two-point indiscrete space, hence it is bireflective. Every bireflective class in Top containing the Sierpiński space coincides with Top. □

The present situation of Problem 1 in Top is described in Diagram 3 (unlike in Diagrams 1, 2, we have omitted some equivalent properties now). The dashed lines mean that the connected properties can appear only by trivial classes of topological spaces (the same is then true for properties that are stronger than those connected, i.e., for "smaller" vertices). The dot-lines mean the opposite (and follow from Examples 5 and 6).

What remains to prove to make solutions of Problem 1 complete? Unfortunately, exactly the original Herrlich's question, i.e., whether there are nontrivial classes of topological spaces that are both almost coreflective or projective and almost reflective or injective. All the other cases are solved.

4. Subcategories of Unif

An analogous result in Unif to the Kannan's one for Top was proved in [9]: *There is no nontrivial coreflective class that is reflective.* Unlike in Top (see [6]), there are many productive coreflective classes in Unif – see [11]. We shall now generalize the result from [9]:

THEOREM 2. *There is no nontrivial class in* Unif *that is multicoreflective and orthogonal.*

Of course, Singl, Indis and Unif are multicoreflective and reflective in Unif.

Proof. If a multicoreflective class C strictly larger than Singl does not contain an indiscrete space (hence $C \cap$ Indis coincides with Singl), then every its member is proximally discrete, i.e., every finite partition of the space is a uniform cover. Indeed, realize that every multicoreflection of an indiscrete space must be a coreflection, and a coreflection of a two-point indiscrete space must be a two-point discrete space in our case. Such a C cannot be productive (infinite powers of discrete spaces are not discrete).

Consequently, every multicoreflective and orthogonal class in Unif different from Singl contains Indis. By Theorem A, such a C is an intersection of reflective classes, in fact of bireflective classes (they contain all indiscrete spaces), thus C is bireflective in Unif. Since every bireflective class containing a nonindiscrete space contains two-point discrete spaces, and multicoreflective class containing two-point discrete spaces are coreflective, C must coincide with Unif. □

We shall now show that there is a nontrivial almost coreflective, reflective class in Unif.

EXAMPLES. Denote by C the class of uniformly zerodimensional uniform spaces, i.e., spaces having a base composed of partitions. As is well-known, C is bireflective in Unif. We shall show that it is also almost coreflective. The class C is hereditary and closed under sums and so, it remains to show that every uniformly continuous mapping f from a uniformly zerodimensional space (Y, v) into a uniform space (X, u) factors via a uniformly zerodimensional space Z of a small cardinality depending on that of X (the uniform structures u, v are systems of uniform covers). We shall use the following two cardinal functions on a uniform space (X, u): the *covering character* $\operatorname{cov}(X, u)$ which is the smallest infinite cardinality κ such that every uniform cover has a uniform subcover of cardinality at most κ, and the *uniform weight* $w_u(X, u)$ which is the smallest infinite cardinality of a base of (X, u). If the space X is separated then $|X| \leqslant (\operatorname{cov} X)^{w_u X}$. We shall denote by sX the separated (Hausdorff) modification of X.

For every uniform cover $\mathcal{U} = \{U_\alpha\}_{\alpha<\tau} \in u$ there is a uniform partition $\mathcal{P}_U \in v$ refining the cover $\{f^{-1}(U)\colon U \in \mathcal{U}\}$. For every $U \in \mathcal{U}$ define $W_\alpha = \bigcup\{P \in \mathcal{P}_U\colon \alpha$ is the first index with $P \subset f^{-1}(U_\alpha)\}$. Then $\{W_\alpha\}_{\alpha<\tau}$ is a

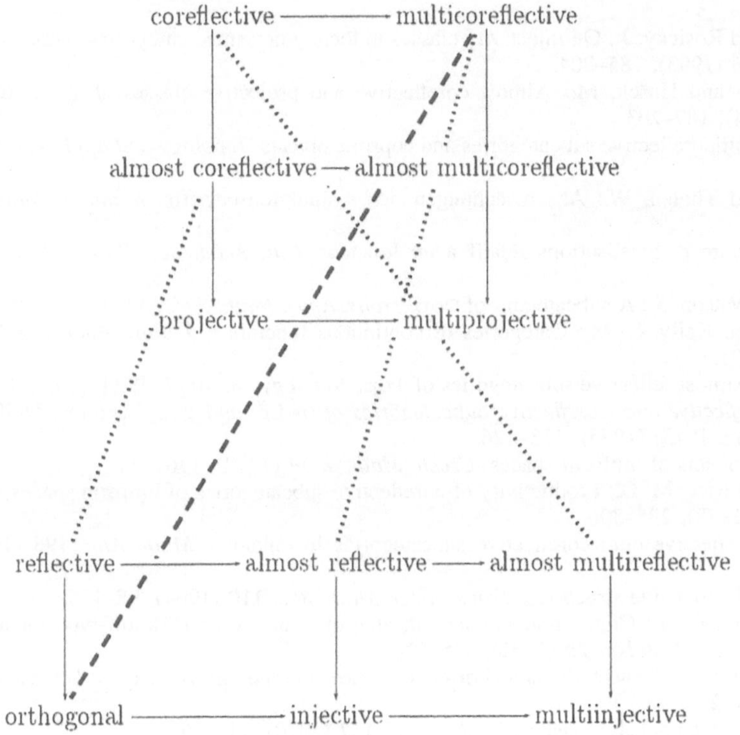

Diagram 4. The situation in Unif.

uniform partition of the space Y, and the space $\widetilde{Y}$ (the set Y endowed with the subbase $\{W_\alpha\}_{\alpha<\tau}$ of a uniformity) is uniformly zerodimensional and so is its separated modification $s\widetilde{Y}$. The cardinality of $s\widetilde{Y}$ is (by the preceding paragraph) at most $(\operatorname{cov} X)^{w_u X}$.

Clearly, the factorized map $g\colon \widetilde{Y} \to X$ is uniformly continuous and so is $sg\colon s\widetilde{Y} \to sX$. If X is separated we are ready and put $Z = s\widetilde{Y}$ since then $f = (sg)s_Y$, where s_Y is the canonical map from Y into $s\widetilde{Y}$. If X is not separated, we can find Z as a subspace of $s\widetilde{Y} \times \widetilde{X}$, where $\widetilde{X}$ is the set X endowed with the indiscrete uniformity (in fact, Z is a pullback of sg and the modification $X \to sX$). The required maps $Z \to X, Y \to Z$ are then liftings of sg, s_Y.

Because of Examples 4, 5 and 6, it remains to answer the following questions:

PROBLEM 2. *Are there nontrivial coreflective classes in* Unif *that are almost reflective* (*or injective*)?

The present situation in Unif is described in Diagram 4. The dashed line and dot-lines have the same meaning as in Diagram 3.

References

1. Adámek, J. and Rosický, J.: On injectivity classes in locally presented categories, *Trans. Amer. Math. Soc.* **336** (1993), 785–804.
2. Bentley, L. H. and Hušek, M.: Almost coreflective and projective classes, *J. Appl. Categ. Struct.* **2** (1994), 187–203.
3. Börger, R.: Multicoreflective subcategories and coprime objects, *Topology and Appl.* **33** (1989), 127–142.
4. Börger, R. and Tholen, W.: Abschwächungen des Adjunktionsbegriffs, *Manuscr. Math.* **19** (1976), 19–45.
5. Diers, Y.: Spectre et localisations relatif à un foncteur, *C.R. Acad. Sci. Paris* **287** (1978), A985–A988.
6. Dow, A. and Watson, S.: A subcategory of Top, *Trans. Amer. Math. Soc.* **337** (1993), 825–837.
7. Freyd, P. J. and Kelly, G. M.: Categories of continuous functors I, *J. Pure Appl. Algebra* **2** (1972), 169–191.
8. Herrlich, H.: Almost reflective subcategories of Top, *Topology and Appl.* **49** (1993), 251–264.
9. Hušek, M.: *Reflective and Coreflective Subcategories of* Unif (*and* Top), Seminar Unif. Sp. 1973/74 (Prague 1975) (1975), 113–126.
10. Hušek, M.: Products of uniform spaces, *Czech. Math. J.* **29** (1979), 130–141.
11. Hušek, M. and Rice, M. D.: Productivity of coreflective subcategories of uniform spaces, *Gen. Top. Appl.* **9** (1978), 295–306.
12. Kannan, V.: Reflective cum coreflective subcategories in topology, *Math. Ann.* **195** (1972), 168–174.
13. Maranda, J.-M.: Injective structures, *Trans. Amer. Math. Soc.* **110** (1964), 98–135.
14. Németi, I. and Sain, I.: Cone-implicational subcategories and some Birkhoff-type theorems, *Colloq. Math. Soc. J. Bolyai* **29** (1982), 535–578.
15. Petz, D.: Generalized connectedness and disconectedness in topology, *Ann. Univ. Sci. Budapest* **24** (1981), 247–252.
16. Ringel, C. M.: Diagonalisierungspaare I, *Math. Z.* **117** (1970), 248–266.
17. Tholen, W.: Factorizations, localizations, and the orthogonal subcategory problem, *Math. Nachr.* **114** (1983), 63–85.

Applied Categorical Structures **4**: 69–79, 1996.

On the Largest Coreflective Cartesian Closed Subconstruct of *Prtop*

E. LOWEN-COLEBUNDERS and G. SONCK*
Departement Wiskunde, Vrije Universiteit Brussel, Pleinlaan 2, 1050 Brussel, Belgium
**Aspirant NFWO*

(Received: 16 November 1994; accepted: 6 July 1995)

Abstract. We show that the subconstruct *Fing* of *Prtop*, consisting of all finitely generated pretopological spaces, is the largest Cartesian closed coreflective subconstruct of *Prtop*. This implies that in any coreflective subconstruct of *Prtop*, exponential objects are finitely generated. Moreover, in any finitely productive, coreflective subconstruct, exponential objects are precisely those objects of the subconstruct that are finitely generated. We give a counterexample showing that without finite productivity the previous result does not hold.

Mathematics Subject Classifications (1991). 54B30, 18D15, 54A05.

Key words: pretopological space, finitely generated space, exponential object, Cartesian closedness.

1. Introduction

It is well-known that the topological constructs *Top*, of topological spaces and continuous maps, and *Prtop*, of pretopological spaces and continuous maps, are not Cartesian closed. For *Top*, this fact was already observed in 1946 by Arens [2] and for *Prtop* the conclusion follows from essentially the same argument. Both these negative results were generalized by Schwarz [18] who proved that if an epireflective subconstruct of *Prtop* contains a non-indiscrete space, then it fails to be Cartesian closed.

In this paper we will study Cartesian closedness for coreflective subconstructs of *Prtop* (containing at least one nonempty space). Some coreflective subconstructs arise very naturally, in *Top* as well as in *Prtop*. We mention a few examples.

The functors F: $L^* \to Prtop$, mapping a sequential convergence space to the pretopological space on the same underlying set whose closure is derived from the convergent sequences, and G: $L^* \to Top$ mapping a sequential convergence space to the topological space on the same underlying set whose closed sets are derived from the convergent sequences, both functors F and G leaving morphisms unchanged as set functions, have as image the coreflective subconstruct *FrPrtop* of *Prtop* of all Fréchet pretopologies and the coreflective subconstruct *SeqTop* of *Top* of all sequential topologies, respectively ([7, 12]).

The functors H: *SGph* $\to$ *Prtop* mapping a spatial directed graph (i.e. a reflexive relation) to the pretopological space on the same underlying set whose closure is the up-closure as described in [6] and leaving morphisms unchanged as set functions, and its restriction K: *PrSet* $\to$ *Top* to the class of all pre-ordered sets, have as image the coreflective subconstruct of *Prtop* of all finitely generated pretopological spaces and of all finitely generated topological spaces, respectively.

For coreflective subconstructs of *Top* the situation with regard to Cartesian closedness is quite well understood. An overview of results can be found in recent survey papers by Herrlich and Hušek ([10, 11]). In particular, the coreflective subconstruct *SeqTop* of sequential topological spaces mentioned previously, is known to be Cartesian closed and the coreflective subconstruct of all (compact, Hausdorff)-generated topological spaces has been efficiently used by many authors as a nice framework for homotopy theory, topological algebra and duality theory because of its Cartesian closedness.

In [3] Činčura has shown that there exists no largest coreflective Cartesian closed subconstruct of *Top* and in [4] he proved that the subconstruct of all finitely generated topological spaces, mentioned earlier, is the largest among all subconstructs that are finitely productive, coreflective and Cartesian closed.

In this paper we show that the situation with regard to Cartesian closedness for coreflective subconstructs of *Prtop* is quite different. In our main result 2.3 we prove that the subconstruct of finitely generated pretopological spaces, mentioned earlier, is the largest coreflective Cartesian closed subconstruct of *Prtop*. Since the constructs of "(compact, Hausdorff)-generated pretopological spaces" and *FrPrtop* of Fréchet pretopologies mentioned previously are not contained in the construct of finitely generated pretopological spaces, it follows that in *Prtop* these constructs are not Cartesian closed.

This paper is a continuation of our work in [13] and our main Theorem 2.3 improves Theorem 4.8 and Corollary 4.10 in [13].

We also improve our previous results on exponential objects: we prove that exponential objects in a coreflective subconstruct of *Prtop* are finitely generated. Moreover, in a coreflective subconstruct that is finitely productive in *Prtop* the exponential objects are precisely those objects of the subconstruct that are finitely generated. We give a counterexample showing that without finite productivity the previous result does not hold.

The following notational conventions will be adopted.

When X is a set and $\mathcal{A} \subset \mathcal{P}(X)$ then

$$\operatorname{stack} \mathcal{A} = \{E \subset X; A \subset E \quad \text{for some} \quad A \in \mathcal{A}\}.$$

When x is an element of a set X, we shall denote $\dot{x}$ for the ultrafilter stack $\{\{x\}\}$. If f: $X \to Y$ is a map between sets, and $\mathcal{F}$ is a filter on X, then its image by f on Y is

$$f(\mathcal{F}) = \operatorname{stack}\{f(F); F \in \mathcal{F}\}.$$

Infinite ordinals are denoted by ω_α where α is an ordinal. When A is a set, $|A|$ denotes its cardinality.

A pretopological space is a structured set (X, q) where the structure q is a function assigning a neighborhood-filter $\mathcal{V}_q(x)$ to each point $x \in X$, and $\mathcal{V}_q(x)$ satisfies the condition $\mathcal{V}_q(x) \subset \dot{x}$. When no confusion can arise, we simply write X instead of (X, q) and $\mathcal{V}(x)$ instead of $\mathcal{V}_q(x)$. A function $f\colon (X,p) \to (Y,q)$ between pretopological spaces is continuous if $\mathcal{V}_q(f(x)) \subset f(\mathcal{V}_p(x))$ for each $x \in X$.

The construct of pretopological spaces and continuous maps is denoted by *Prtop*.

In any pretopological space X a Čech-closure operator *cl* is defined by

$$x \in clA \Leftrightarrow \forall V \in \mathcal{V}(x)\colon\ V \cap A \neq \emptyset$$

and the other way around, a Čech-closure operator *cl* on X determines a unique pretopology on X. *Prtop* is isomorphic to the construct of Čech closure spaces.

Prtop is a well-fibred topological construct. So in *Prtop* initial and final structures exist for arbitrary sources and sinks. We will make extensive use of final structures in *Prtop* and therefore recall their construction: if

$$(f_i\colon\ (X_i, q_i) \to Y)_{i \in I}$$

is a sink, then the final pretopological structure p on Y has as neighborhood filter in $y \in Y$

$$\mathcal{V}_p(y) = \dot{y} \quad \text{if} \quad y \notin \bigcup_{i \in I} f_i(X_i)$$

and

$$\mathcal{V}_p(y) = \bigcap_{\substack{i \in I, x \in X_i \\ f_i(x) = y}} f_i(\mathcal{V}_{q_i}(x)) \quad \text{if} \quad y \in \bigcup_{i \in I} f_i(X_i).$$

The product in *Prtop* will be denoted by $\times$.

Prtop is simple: *Prtop* is the epireflective hull of the class $\{\mathbf{3}\}$, where $\mathbf{3}$ is the pretopological space with underlying set $\{0, 1, 2\}$ and neighborhood-filters $\mathcal{V}(0) = \mathcal{V}(2) = \dot{0} \cap \dot{1} \cap \dot{2}$ and $\mathcal{V}(1) = \dot{1} \cap \dot{2}$.

All subconstructs are always assumed to be full and isomorphism closed and to contain at least one nonempty space.

A particular role will be played by the subconstruct *Fing* whose objects are finitely generated pretopological spaces, i.e. spaces in which every point has a smallest neighborhood. *Fing* is coreflective in *Prtop*.

Categorical terminology follows Adámek, Herrlich, Strecker [1]. Since we work only with well-fibred topological constructs $\mathcal{C}$, we use the following characterization of an exponential object: X is exponential in $\mathcal{C}$ if and only if the functor $X \times -$ preserves final epi-sinks.

$\mathcal{C}$ is Cartesian closed if and only if every $\mathcal{C}$-object is exponential.

2. Cartesian Closedness and Coreflectivity

Fing, the subconstruct of *Prtop* consisting of all finitely generated pretopological spaces, is coreflective and Cartesian closed. In order to investigate whether *Fing* is the largest subconstruct with both properties, we prove a preliminary result on coreflective subconstructs of *Prtop*, not included in *Fing*.

DEFINITION 2.1. If α is an ordinal number, X_α is the set $[0, \omega_\alpha]$, i.e.

$$\{\beta; \beta \text{ ordinal number, } 0 \leqslant \beta \leqslant \omega_\alpha\}.$$

$\mathcal{X}_\alpha$ is the filter on X_α generated by the sets $\{0\} \cup [\beta; \omega_\alpha]$, with $\beta < \omega_\alpha$; $F(\alpha)$ is the pretopological space with X_α as underlying set, with $\{X_\alpha\}$ as neighborhood filter of each $\beta < \omega_\alpha$, and $\mathcal{X}_\alpha$ as neighborhood filter of ω_α.

THEOREM 2.2. *A coreflective subconstruct of Prtop that is not included in Fing, always contains a space $F(\alpha)$, with ω_α a regular ordinal.*

Proof. Let $\mathcal{C}$ be a coreflective subconstruct of *Prtop*, not included in *Fing*. Then $\mathcal{C}$ contains an object (X, λ) in which there is a point x with $\bigcap \mathcal{V}_\lambda(x) \notin \mathcal{V}_\lambda(x)$. Let

$$\alpha = \min\{\gamma; \exists \mathcal{V} \subset \mathcal{V}_\lambda(x), |\mathcal{V}| = |\omega_\gamma|, \bigcap \mathcal{V} \notin \mathcal{V}_\lambda(x)\}.$$

It is easily seen that ω_α is a regular ordinal and satisfies

$$\mathcal{V} \subset \mathcal{V}_\lambda(x), |\mathcal{V}| < \omega_\alpha \Rightarrow \bigcap \mathcal{V} \in \mathcal{V}_\lambda(x), \tag{1}$$

$$\exists \{V_\beta; \beta < \omega_\alpha\} \subset \mathcal{V}_\lambda(x): \bigcap_{\beta < \omega_\alpha} V_\beta \notin \mathcal{V}_\lambda(x). \tag{2}$$

Now (2) implies that there exists $x_0 \in X \setminus \{x\}$ with $x \notin cl_\lambda\{x_0\}$, and hence

$$\{x_0\} \cup \bigcap_{\beta < \omega_\alpha} V_\beta \notin \mathcal{V}_\lambda(x).$$

Modifying the pretopology λ on X by taking the final pretopology for the sink

$$(f_i\colon (X, \lambda) \to X)_{i \in \{1,2\}}$$

with $f_1 = \mathbf{1}_X$ and

$$f_2\colon X \to X\colon z \to \begin{cases} x & \text{if } z = x, \\ x_0 & \text{if } z \neq x \end{cases}$$

the resulting space still satisfies (1) and (2). So we may assume that (X, λ), besides (1) and (2) also satisfies

$$\exists x_0 \in X \setminus \{x\}\colon \mathcal{V}_\lambda(x) \subset \dot{x}_0 \tag{3}$$

In view of (1), we may also assume that the family $(V_\beta)_{\beta<\omega_\alpha}$ given in (2) is strictly decreasing and satisfies $V_0 = X$. The proof now consists of the following parts:

a. construction of a quotient $f\colon (X, \lambda) \to (X_\alpha, q)$ with $\mathcal{X}_\alpha \subset \mathcal{V}_q(\omega_\alpha)$,
b. construction of a final sink $((X_\alpha, q) \to (X_\alpha, p))$ with $\mathcal{X}_\alpha = \mathcal{V}_p(\omega_\alpha)$,
c. construction of a final sink $((X_\alpha, p) \to F(\alpha))$.

Combination of these parts will lead to the conclusion that $F(\alpha) \in \mathcal{C}$.

a. The function $f\colon X \to X_\alpha$ defined by

$$f(x) = \omega_\alpha, \quad f|_{\bigcap_{\beta<\omega_\alpha} V_\beta} = 0$$

and

$$f(z) = \min\{\gamma; z \notin V_{\gamma+1}\}$$

for $z \in X \setminus \bigcap_{\beta<\omega_\alpha} V_\beta$ is onto. Because $f(V_\beta) = \{0\} \cup [\beta, \omega_\alpha]$ for all $\beta < \omega_\alpha$, the final pretopology q on X_α for $(f\colon (X, \lambda) \to X_\alpha)$ satisfies $\mathcal{X}_\alpha \subset \mathcal{V}_q(\omega_\alpha)$.

b. Let I be the set of all strictly increasing functions $X_\alpha \to X_\alpha$ leaving 0 and ω_α fixed and let p be the final pretopology on X_α for the sink

$$(g\colon (X_\alpha, q) \to X_\alpha)_{g\in I}.$$

Then $\mathcal{X}_\alpha = \mathcal{V}_p(\omega_\alpha)$.

c. We introduce the following classes of functions for $\beta, \gamma \in X_\alpha$:

$$f_{\beta,\gamma}\colon X_\alpha \to X_\alpha\colon z \to \begin{cases} \beta & \text{if } z = \omega_\alpha, \\ \gamma & \text{if } z \neq \omega_\alpha, \end{cases}$$

$$g_{\beta,\gamma}\colon X_\alpha \to X_\alpha\colon z \to \begin{cases} \beta & \text{if } z = \gamma, \\ z & \text{if } z \neq \gamma, \end{cases}$$

$$h_\beta\colon X_\alpha \to X_\alpha\colon z \to \begin{cases} 0 & \text{if } z = 0, \\ \omega_\alpha & \text{if } z \notin \{0, \omega_\alpha\}, \\ \beta & \text{if } z = \omega_\alpha. \end{cases}$$

Let $J' = \{\mathbf{1}_{X_\alpha}\} \cup \{f_{\beta,\gamma}; 0 \leqslant \beta, \gamma < \omega_\alpha\}$. We consider three cases:

i. if $cl_p\{\omega_\alpha\} = \omega_\alpha$, let $J = J' \cup \{f_{\beta,\omega_\alpha}; 0 \leqslant \beta < \omega_\alpha\}$,
ii. if $\exists \delta \in]0, \omega_\alpha[\colon \delta \in cl_p\{\omega_\alpha\}$, let $J = J' \cup \{g_{\beta,\delta}; 0 \leqslant \beta < \omega_\alpha\}$,
iii. if $0 \in cl_p\{\omega_\alpha\}$ and $\forall \delta \in]0, \omega_\alpha[\colon \delta \notin cl_p\{\omega_\alpha\}$, let $J = J' \cup \{h_\beta; 0 \leqslant \beta < \omega_\alpha\}$.

Then, in all cases,

$$(h\colon\ (X_\alpha, p) \to F(\alpha))_{h\in J}$$

is a final sink in *Prtop*. □

As usual, subconstructs of *Prtop* are partially ordered by inclusion on their object-classes. The objects $F(\alpha)$ generate the minimal coreflective subconstructs of *Prtop*, not included in the subconstruct of finitely generated spaces. The corresponding minimal coreflective subconstruct of *Top*, not included in the construct of finitely generated topological spaces, are generated by the spaces $B(\alpha)$, introduced in [8]. The tails $[\beta, \omega_\alpha]$ are basic neighborhoods of ω_α in $B(\alpha)$. Contrary to the situation in *Top*, taking the tails $[\beta, \omega_\alpha]$ as neighborhoods of ω_α instead of the sets $\{0\} \cup [\beta, \omega_\alpha]$, gives a pretopological space for which the generated coreflective hull is strictly finer than the hull of $F(\alpha)$.

THEOREM 2.3. *A coreflective subconstruct of Prtop that is not included in Fing is not Cartesian closed.*

Proof. Let $\mathcal{C}$ be a coreflective subconstruct of *Prtop*, not included in *Fing*. Then $\mathcal{C}$ contains a space $F(\alpha)$ for some regular ordinal ω_α. Let p be the final pretopology on X_α for the sink

$$(f_{\beta,\gamma}\colon\ F(\alpha) \to X_\alpha)_{\beta\in X_\alpha,\, \gamma\in X_\alpha\setminus\{\omega_\alpha\}},$$

where the functions $f_{\beta,\gamma}$ are taken as in the proof of Theorem 2.2. Then p is the indiscrete pretopology on X_α and the space $Y = (X_\alpha, p)$ belongs to $\mathcal{C}$. If □ denotes the product in the construct $\mathcal{C}$, we prove that

$$(\mathbf{1}_{X_\alpha} \Box f_{\beta,\gamma}\colon\ F(\alpha) \Box F(\alpha) \to F(\alpha) \Box Y)_{\beta\in X_\alpha,\, \gamma\in X_\alpha\setminus\{\omega_\alpha\}}$$

is not final in $\mathcal{C}$. Let q be the final pretopology on $X_\alpha \times X_\alpha$ for the sink

$$(\mathbf{1}_{X_\alpha} \Box f_{\beta,\gamma}\colon\ F(\alpha) \Box F(\alpha) \to X_\alpha \times X_\alpha)_{\beta\in X_\alpha,\, \gamma\in X_\alpha\setminus\{\omega_\alpha\}}.$$

Then

$$\mathcal{V}_q(\omega_\alpha, \omega_\alpha) = \bigcap_{\beta<\omega_\alpha} \mathbf{1}_{X_\alpha} \Box f_{\beta,\gamma}\big(\mathcal{V}_{F(\alpha)\Box F(\alpha)}(\omega_\alpha, \omega_\alpha)\big).$$

Since $F(\alpha) \times F(\alpha) \leqslant F(\alpha) \Box F(\alpha)$, clearly

$$\begin{aligned} W &= \bigcup_{\beta<\omega_\alpha} \mathbf{1}_{X_\alpha} \Box f_{\beta,\gamma}\big(\{0\} \cup [\beta, \omega_\alpha] \times X_\alpha\big) \\ &= \bigcup_{\beta<\omega_\alpha} \{0\} \cup [\beta, \omega_\alpha] \times \{\beta, \omega_\alpha\} \end{aligned}$$

is a q-neighborhood of $(\omega_\alpha, \omega_\alpha)$. Since the functions

$$h\colon\ F(\alpha) \to Y\colon \begin{cases} \omega_\alpha \to \omega_\alpha & \\ \beta \to \beta + 1 & \text{for all } \beta < \omega_\alpha \end{cases}$$

and

$$i\colon F(\alpha) \to F(\alpha)\Box F(\alpha)\colon x \to (x, x)$$

clearly are continuous, also $g\colon F(\alpha) \to F(\alpha)\Box Y$ defined by $g = (1_{X_\alpha}\Box h) \circ i$ is continuous. Now suppose that $F(\alpha)\Box Y$ coincides with $(X_\alpha \times X_\alpha, q)$ then we would have

$$g^{-1}(W) \in \mathcal{X}_\alpha$$

and then surely

$$\exists \eta \in]0, \omega_\alpha[\colon\ \eta \in g^{-1}(W).$$

This is impossible. So we can conclude that $F(\alpha)\Box Y$ does not coincide with $(X_\alpha \times X_\alpha, q)$. □

COROLLARY 2.4. *Fing is the largest Cartesian closed coreflective subconstruct of Prtop.*

Remark the difference between the previous result and the situation in *Top* where Čincura proved that there exists no largest Cartesian closed coreflective subconstruct ([3]).

The following proposition shows that Theorem 2.3 also gives information on the Cartesian closedness of initially structured subconstructs of *Prtop* that contain **3**.

The term initially structured is used in the sense of [15] and [17]. The proof of the next result is based on some ideas of [14] and of Proposition 3 in [9].

PROPOSITION 2.5. *For a subconstruct $\mathcal{A}$ of Prtop that contains* **3** *the following are equivalent:*

1. *$\mathcal{A}$ is initially structured,*
2. *$\mathcal{A}$ is coreflective in* Prtop,
3. *$\mathcal{A}$ is topological.*

Proof. It suffices to show that an initially structured subconstruct $\mathcal{A}$ of *Prtop* that contains **3** is coreflective in *Prtop*. Let $\mathcal{C}$ be the coreflective hull of $\mathcal{A}$ in *Prtop*, and take a $\mathcal{C}$-object (X, λ). Since *Prtop* is the epireflective hull of the one-object class $\{\mathbf{3}\}$, we can choose a point-separating initial source $(f_i\colon (X, \lambda) \to \mathbf{3})_{i\in I}$ in *Prtop* (and so in $\mathcal{C}$). Now $\mathcal{A}$ is initially structured and contains **3**, so there exists an initial $\mathcal{A}$-structure μ on X for the source $(f_i\colon X \to \mathbf{3})_{i\in I}$. The fact that $\mathcal{A}$ is finally dense in $\mathcal{C}$ implies that $(f_i\colon (X, \mu) \to \mathbf{3})_{i\in I}$ also is an initial source in $\mathcal{C}$, and so $\lambda = \mu$. This shows that (X, λ) belongs to $\mathcal{A}$ and hence we can conclude that $\mathcal{A} = \mathcal{C}$. □

EXAMPLES 2.6.

1. As an immediate consequence of Theorem 2.3 we find the well-known fact that *Prtop* is not Cartesian closed.
2. Whereas the minimal coreflective subconstructs of *Top* not included in the construct of finitely generated topological spaces are Cartesian closed ([4]), the coreflective hull of a pretopological space $F(\alpha)$ in *Prtop* never is Cartesian closed.
3. Let α be some ordinal and let $C(\alpha)$ be the (pre)topological space on the set X_α as introduced in [8]: the tails form the neighborhood filter in ω_α and all points $\beta < \omega_\alpha$ are isolated. Then the coreflective hull of $C(\alpha)$ in *Prtop* consists of all ω_α-sequentially-determined pretopological spaces. In particular the coreflective hull of $C(0)$ is the construct of all Fréchet pretopologies. By Theorem 3.2, none of these subconstructs of *Prtop* are Cartesian closed. This result is again quite the opposite of what is known for *Top*. The coreflective hull of $C(\alpha)$ in *Top* consists of all ω_α-sequentially-determined topological spaces and $C(0)$ generates the construct of sequential topological spaces. All these subconstructs of *Top* are Cartesian closed ([3]).
4. In view of Proposition 2.5, Theorem 2.3 can be reformulated in terms of initially structured subconstructs containing **3**. The condition "containing **3**" can not be dropped in this formulation: the construct of strongly Fréchet pretopologies described in [13] is Cartesian closed topological and hence is a counterexample.

3. Exponential Objects

THEOREM 3.1. *The exponential objects of a coreflective subconstruct $\mathcal{C}$ of Prtop are finitely generated. If, in addition, $\mathcal{C}$ is finitely productive in Prtop, then the exponential objects of $\mathcal{C}$ are precisely those $\mathcal{C}$-objects that are finitely generated.*

Proof. Let $\mathcal{E}$ be the full subconstruct of $\mathcal{C}$ consisting of all exponential objects of $\mathcal{C}$. Then $\mathcal{E}$ is an exponential subcategory of $\mathcal{C}$ (in the sense of L. D. Nel) and by Theorem 5 in [16] the coreflective hull $\mathcal{H}$ of $\mathcal{E}$ in $\mathcal{C}$ is Cartesian closed. Since $\mathcal{H}$ also is coreflective in *Prtop*, Theorem 2.3 implies that $\mathcal{H}$ (and hence also $\mathcal{E}$) is contained in *Fing*.

Now suppose moreover that $\mathcal{C}$ is finitely productive in *Prtop* and take a $\mathcal{C}$-object X that is finitely generated. Then by Theorem 3.4 in [13] X is an exponential object in *Prtop* and so the functor $X \times -$: *Prtop* $\rightarrow$ *Prtop* preserves final sinks. Since $\mathcal{C}$ is coreflective in *Prtop*, and the product $\square$ in $\mathcal{C}$ coincides with the usual product in *Prtop*, $X \times -\colon \mathcal{C} \to \mathcal{C}$ also preserves final sinks. So X is an exponential object of $\mathcal{C}$. □

EXAMPLES AND COUNTEREXAMPLES 3.2.

1. For an ordinal number α, the coreflective subconstruct $Prtop_\alpha$ of *Prtop* consists of all pretopological spaces in which an intersection of strictly less than ω_α neighborhoods of a point again is a neighborhood of that point. $Prtop_\alpha$ is coreflective and finitely productive in *Prtop*. By Theorem 3.1 the exponential objects of $Prtop_\alpha$ are precisely the finitely generated pretopological spaces. In particular for $Prtop_0 = Prtop$ we obtain Theorem 3.4 in [13].
2. In view of Proposition 2.5, Theorem 3.1 can be reformulated in terms of initially structured subconstructs containing **3**. The condition "containing **3**" can not be droped: in the initially structured subconstruct *Top* the class of exponential objects coincides with the class of corecompact topological spaces ([5]).
3. We produce an example of a coreflective subconstruct $\mathcal{C}$ of *Prtop*, not finitely productive in *Prtop* and for which $\mathcal{C} \cap Fing$ contains an object, not exponential in $\mathcal{C}$. In order to construct the example the following result concerning the spaces $F(\alpha)$ will be used:

   If α is an ordinal number, $f\colon F(\alpha) \to F(\alpha)$ is a continuous function and there exists $\delta \in [0, \omega_\alpha[$ such that $f(\delta) = \omega_\alpha$, then $f(X_\alpha) \subset \{0, \omega_\alpha\}$.

   Let α be an ordinal number and let $\mathcal{C}$ be the coreflective hull of $F(\alpha)$ in *Prtop*. Then $\mathcal{C}$ is coreflective in *Prtop*. Moreover, $\mathcal{C}$ contains **3** for $q\colon F(\alpha) \to \mathbf{3}$ defined by $q(1) = 0$, $q(\omega_\alpha) = 1$ and $q(z) = 2$ for $x \in X_\alpha \setminus \{1, \omega_\alpha\}$ is a quotient. If □ denotes the product in $\mathcal{C}$, then

   $$V = \bigcup_{\substack{f\colon F(\alpha) \to F(\alpha) \times \mathbf{3} \\ (\omega_\alpha, 0) \in f(X_\alpha) \\ f \text{ continuous}}} f(X_\alpha)$$

   is a neighborhood of $(\omega_\alpha, 0)$ in $F(\alpha)\square\mathbf{3}$. The above mentioned result on $F(\alpha)$ shows that V contains no point $(\beta, 1)$ with $0 < \beta < \omega_\alpha$. This means that $F(\alpha) \times \mathbf{3}$ doesn't belong to $\mathcal{C}$ and consequently that $\mathcal{C}$ is not finitely productive in *Prtop*.

   If, for $\beta, \gamma \in X_\alpha$, we define the functions $f^*_{\beta,\gamma}\colon X_\alpha \to X_\alpha$ by modifying the functions $f_{\beta,\gamma}$ used in the proof of Theorem 2.2 as follows:

   $$f^*_{\beta,\gamma} = \left(f_{\beta,\gamma} \setminus \{(0,\gamma)\}\right) \cup \{(0,0)\}$$

   then

   $$\left(f^*_{\beta,\gamma}\colon F(\alpha) \to Y\right)_{\beta \in X_\alpha, \gamma \in X_\alpha \setminus \{\omega_\alpha\}}$$

   is a final sink in $\mathcal{C}$, where Y is the indiscrete pretopology on X_α.

We now show that each neighborhood of $(\omega_\alpha, 0)$ in $Y\Box\mathbf{3}$ contains the point $(1,1)$.
Therefore take the quotient q: $F(\alpha) \to \mathbf{3}$ defined earlier and the continuous mapping

$$g'\colon\ F(\alpha) \to Y\colon\ z \to \begin{cases} 1 & \text{if } z \neq 1, \\ \omega_\alpha & \text{if } z = 1. \end{cases}$$

Then the function $Q = (q', q)$: $F(\alpha) \to Y \times \mathbf{3}$ is continuous. This implies

$$\mathcal{V}_{Y\Box\mathbf{3}}(\omega_\alpha, 0) \subset Q(\mathcal{V}_{F(\alpha)}(1)) = Q(\{X_\alpha\}) = \text{stack}\,\{Q(X_\alpha)\}$$

and so each neighborhood of $(\omega_\alpha, 0)$ in $Y\Box\mathbf{3}$ contains the point $(1,1) = Q(\omega_\alpha)$.
Now let φ be the final pretopology on $X_\alpha \times (0,1,2)$ for the sink

$$(f^*_{\beta,\gamma}\Box\mathbf{1}\colon\ F(\alpha)\Box\mathbf{3} \to X_\alpha \times \{0,1,2\})_{\beta\in X_\alpha,\gamma\in X_\alpha\setminus\{\omega_\alpha\}}.$$

Now

$$f^*_{\beta,\gamma}\Box\mathbf{1}(z) = (\omega_\alpha, 0) \Leftrightarrow \beta = \omega_\alpha \text{ and } z = (\omega_\alpha, 0)$$

and so

$$V' = \bigcup_{\gamma\in X_\alpha\setminus\{\omega_\alpha\}} f^*_{\omega_\alpha,\gamma}\Box\mathbf{1}(V)$$

is a φ-neighborhood of $(\omega_\alpha, 0)$ (V is the neighborhood of $(\omega_\alpha, 0)$ in $F(\alpha)\Box\mathbf{3}$ constructed earlier in this counterexample). Suppose $(1,1) \in V'$, then

$$\exists\gamma \in X_\alpha \setminus \{\omega_\alpha\}\colon\ (1,1) \in f_{\omega_\alpha,\gamma}\Box\mathbf{1}(V).$$

Now

$$f^*_{\omega_\alpha,\gamma}\Box\mathbf{1}(z) = (1,1) \Leftrightarrow \gamma = 1 \text{ and } \exists\beta \in]0, \omega_\alpha[\colon\ z = (\beta, 1)$$

and so $(1,1) \in V'$ would imply

$$\exists\beta \in]0, \omega_\alpha[\colon\ (\beta, 1) \in V$$

contradicting the choice of V. So we have proved $Y\Box\mathbf{3} \neq (X_\alpha \times \{0,1,2\}, \varphi)$ and

$$(f^*_{\beta,\gamma}\Box\mathbf{1}\colon\ F(\alpha)\Box\mathbf{3} \to Y\Box\mathbf{3})_{\beta\in X_\alpha,\gamma\in X_\alpha\setminus\{\omega_\alpha\}}$$

is not a final sink in $\mathcal{C}$. This shows that $\mathbf{3}$ is not an exponential object in $\mathcal{C}$. So not all objects of $\mathcal{C}\cap$ *Fing* are exponential in $\mathcal{C}$.

References

1. Adámek, J., Herrlich, H., and Strecker, G. E.: *Abstract and Concrete Categories*, Wiley, New York, 1990.
2. Arens, R. F.: A topology for spaces of transformations, *Ann. Math.* **47** (1946), 480–495.

3. Činčura, J.: Cartesian closed coreflective subcategories of the category of topological spaces, *Topology Appl.* **41** (1991), 205–212.
4. Činčura, J.: Products in Cartesian closed coreflective subcategories of the category of topological spaces, Preprint, 1992.
5. Day, B. J. and Kelly, G. M.: On topological quotient maps preserved by pullbacks or products, *Proc. Camb. Phil. Soc.* **67** (1970), 553–558.
6. Dikranjan, D. and Tholen, W.: Categorical structure of closure operators, Preprint, 1993.
7. Frič, R. and Koutník, V.: Sequential convergence: iteration, extension, completion, enlargement, in *Recent Progress in General Topology*, Elsevier Science Publishers, 1992, pp. 201–213.
8. Herrlich, H.: *Topologische Reflexionen und Coreflexionen*, Springer Lect. Notes Math. **78**, 1968.
9. Herrlich, H.: Are there convenient subcategories of TOP?, *Topology Appl.* **15** (1983), 263–271.
10. Herrlich, H. and Hušek, M.: Categorical topology, in *Recent Progress in General Topology*, Elsevier Science Publishers, 1992, pp 369–403.
11. Herrlich, H. and Hušek, M.: Some open categorical problems in TOP, *Appl. Cat. Struct.* **1** (1993), 1–19.
12. Kent, D. C.: Decisive convergence spaces, Fréchet spaces and sequential spaces, *Rocky Mountain J. Math.* **1**(2) (1971), 367–374.
13. Lowen-Colebunders, E. and Sonck, G.: Exponential objects and cartesian closedness in the construct Prtop, *Appl. Cat. Struct.* **1** (1993), 345–360.
14. Müller, H.: Uber die Vertauschbarkeit von Reflexionen und Coreflexionen, Unpublished Manuscript, Bielefeld, 1974.
15. Nel, L. D.: Initially structured categories and Cartesian closedness, *Canadian J. Math.* **27** (1975), 1361–1377.
16. Nel, L. D.: Cartesian closed coreflective hulls, *Quaestiones Math.* **2** (1977), 269–333.
17. Preuss, G.: *Theory of Topological Structures*, Reidel, Dordrecht, 1988.
18. Schwarz, F.: Cartesian closedness, exponentiality and final hulls in pseudotopological spaces, *Quaestiones Math.* **5** (1982), 289–304.

Applied Categorical Structures **4:** 81–85, 1996.

Topological Spaces and Quasi-Varieties*

MICHAEL BARR and M. CRISTINA PEDICCHIO
McGill University, Department of Mathematics, Burnside Hall, 805 Sherbrooke St. West, P.Q. H3A 2K6, Montreal, Canada; and Dipartimento di Mathematica, Università 34100 Trieste, Italy

(Received: 7 December 1994; accepted: 27 September 1995)

Abstract. We describe Top^{op} and Sob^{op} as quasi-varieties by means of suitable "schizophrenic" objects.

Mathematics Subject Classifications (1991). 18A40, 18C15, 54B10.

Key words: category, topological space, variety, quasi-variety.

Introduction

The fact that the dual category of sober spaces is a quasi-variety, where quasi-varieties are intended in an infinitary sense, can be easily proved by using the duality between the category Sob of sober spaces and that one of spatial frames [7]. In such a duality theorem, the role played by the Sierpinski space 2 is essential, in fact the adjointness between spaces and frames is determined by representable functors, represented by 2, where 2 is thought of as a topological space or as a frame depending on the context. This double nature of 2 as a topological frame (schizophrenic object [11, 6]) allows to define a topological or algebraic enrichment on the induced hom sets.

Furthermore, the simple remark that 2 is an injective cogenerator [9] in Sob, will suffice to reconstruct the whole category of frames from 2 itself.

It appears quite surprising that arguments, similar to those developped for Sob, still apply to arbitrary topological spaces, simply by replacing 2 by 3, where 3 is a three points space $\{0,1,2\}$ with $\{1,2\}$ as the only non-trivial open subset.

The fact that 3 is an injective cogenerator in the category Top of topological spaces [1] will suffice to define a variety, containing Top^{op} as a (regular epi)-reflective subcategory (= quasi-variety). Again, all the information are given by representable functors represented by 3 and 3 will have a double nature as a space and as an algebra, more precisely as a topological grid [4].

If we replace 3 by any other injective cogenerator of Top, we get a new triple on Set; we show that the corresponding varieties of algebras are categorically equivalent.

---

* Research of the first author supported by grants from the NSERC of Canada and the FCAR du Québec. Research of the second author supported by the Topology grant 40% and by the NATO grant CRG 941330.

We are grateful to H. Herrlich for interesting suggestions on the subject.

Duality Theorems

We say that a category $\mathcal{E}$ is a *variety* iff it is monadic over Set, i.e. it is equivalent to a category Set^T of T-algebras for a triple T over Set (see [8]); similarly, $\mathcal{E}$ is a *quasi-variety* or quasi-monadic over Set iff, up to equivalences, it is a (regular epi)-reflective subcategory of a variety.

Algebraicity and quasi-algebraicity for a category can be easily characterized in terms of properties of the "free object on one generator" as in Theorems 2 and 3.

We will need the following definition, where $P \in \mathcal{E}$ admits all copowers $\coprod_J P$, $i_j\colon P \to \coprod_J P$ denotes the canonical inclusion and $\mathcal{E}(P, X)$ is the *hom* set in $\mathcal{E}$.

DEFINITION 1. An object $P \in \mathcal{E}$ is a regular generator (or regular separator) iff for any $X \in \mathcal{E}$ the morphism $\epsilon_X\colon \coprod_{\mathcal{E}(P,X)} P \twoheadrightarrow X$, defined by $i_f \epsilon_X = f$, for any $f \in \mathcal{E}(P, X)$, is a regular epi.

THEOREM 2 ([12]). *$\mathcal{E}$ is a variety iff it satisfies the following properties*:

(1) *$\mathcal{E}$ is exact in the sense of Barr* [3];
(2) *there exists a regular projective, $P \in \mathcal{E}$, with copowers*;
(3) *P is a regular generator.*

THEOREM 3 ([10]). *$\mathcal{E}$ is a quasi-variety iff it satisfies the following properties*:

(1) *$\mathcal{E}$ has finite limits and coequalizers of equivalence relations*;
(2) *there exists a regular projective, $P \in \mathcal{E}$, with copowers*;
(3) *P is a regular generator.*

In particular, we are interested to the case where $\mathcal{E}$ is a dual category, $\mathcal{E} = \mathcal{C}^{op}$; then conditions (2) and (3) dualize to

(2′) there exists a regular injective $I \in \mathcal{C}$, with powers, and I is a regular cogenerator.

Now, let $\mathcal{C}$ be Top, the category of topological spaces and continuous maps; our aim is to show that Theorem 3 applies to Top^{op} (see also [4]), and moreover that the proof reduces to a simple computation of contravariant adjoint functors determined by I considered as a "schizophrenic object".

Recall that a topological space is a regular cogenerator iff it contains an indiscrete subspace with two elements and a Sierpinski subspace, and that a topological space is a regular injective iff it is a retract of a power of 3, where $3 = \{0, 1, 2\}$ with opens $\Omega(3) = \{\phi, \{1, 2\}, \{0, 1, 2\}\}$ (see [1]).

Choose $I = 3$, and define $G\colon Top^{op} \to Set$ by $G(X) = Top(X, 3)$ for any $X \in Top$, and $F\colon Set \to Top^{op}$ by $F(Y) = \prod_Y 3$, for any $Y \in Set$, where the product is clearly the topological one.

By construction $F \dashv G$, with ϵ_X as counit, for any $X \in Top$.

Observe that F has an underlying representable functor given by $F = Set(-, 3)$ where 3 now denotes the three points set. In the following we will often use the same notation for an object and its corresponding underlying set; the hom set notation will denote in which category the object must be considered.

Now, define the triple $T = GF$ on Set induced by the adjunction and consider the corresponding category Set^T of T-algebras, with forgetful and free functors given by $G^T\colon Set^T \to Set$ and $F^T\colon Set \to Set^T$ respectively.

Let $\Phi\colon Top^{op} \to Set^T$ be the canonical comparison functor defined by $\Phi(X) = (G(X), G(\epsilon_X))$.

If we apply Φ to the topological space $\{*\}$, we get a T-algebra with underlying set 3 and structural map $h\colon Top(\prod_3 3, 3) \to 3$ defined by $h(u) = u(0, 1, 2)$, for any continuous map $u\colon 3 \times 3 \times 3 \to 3$. For any algebra $A \in Set^T$ we can then define a functor $\Psi\colon Set^T \to Top^{op}$ by putting $\Psi(A)$ equal to the set of T-homomorphisms $Set^T(A, 3)$, topological as a subspace of the topological product $\prod_A 3$.

The double nature of 3 as a space and as an algebra (more precisely as a topological algebra) implies the following:

LEMMA 4. *The functors Φ and Ψ are adjoint functors: $\Psi \dashv \Phi$.*

Proof. It suffices to show that

$$Top(X, \Psi(A)) \simeq Set^T(A, \Phi(X)).$$

By applying the definition of Φ and Ψ, straightforward computations show that both sides correspond to maps $v\colon X \times A \to 3$ such that, for all $x \in X$, $v(x, -)\colon A \to 3$ is a T-algebra morphism, and for all $a \in A$, $v(-, a)\colon X \to 3$ is a continuous function. □

Up to now we did not apply the hypotheses on 3.

The fact that 3 is a regular cogenerator in Top is equivalent to Φ being full and faithful (see [2, Theorem 9, p. 111]), so we can consider – up to equivalences – Top^{op} as a full reflective subcategory of Set^T, with the evaluation map $\eta_A\colon A \to \Phi\Psi(A) = Top(Set^T(A, 3), 3)$ as unit.

The second assumption that 3 is regular injective is equivalent, by definition, to G preserves regular epimorphisms, and this implies that Φ preserves regular epis, since G^T reflects regular epis. Since any T-algebra A is a quotient $q\colon F^T(Y) \twoheadrightarrow A$ of a free one, and free T-algebras are in Top^{op}, we get the following commutative diagram:

$$\begin{array}{ccc} F^T(Y) & \xrightarrow{\quad q \quad} & A \\ {\scriptstyle \simeq}\big\downarrow & & \big\downarrow{\scriptstyle \eta_A} \\ \Phi\Psi(F^T(Y)) & \xrightarrow{\ \Phi\Psi(q)\ } & \Phi\Psi(A) \end{array}.$$

Both Φ and Ψ preserves regular epis, so $\Phi\Psi(q)$ is a regular epi, hence η_A is a regular epi for all A. Finally, we have proved the following:

THEOREM 5. *Top^{op} is a quasi-variety.*

We remark that Top^{op} corresponds to the (regular epi)-reflective hull in Set^T of the full subcategory given by all spaces $\{\prod_Y 3, Y \in Set\}$. In fact, being Top^{op} (regular epi)-reflective, it suffices to check it is the smallest possible one containing all $\{\prod_Y, 3, Y \in Set\}$; this easily follows, by the characterization of reflective hulls (see [1, Proposition 16.22]) since, for any $X \in Top$, there exists a surjective family of constant morphisms h_x: $\{*\} = \prod_\phi 3 \to X$ with $x \in X$.

If we replace 3 by any other regular injective, regular cogenerator I for Top (for example I could be the three points space $\{0,1,2\}$ with topology $\Omega(\{0,1,2\}) = \{\phi, \{0\}, \{0,1,2\}\})$, we can repeat the same construction we did, by choosing I as "schizophrenic object". In this way we define a new triple T_I: $Set \to Set$ on Set, by $T_I(Y) = Top(\prod_Y, I, I)$.

PROPOSITION 6. *The categories Set^T and Set^{T_I} are equivalent.*

Proof. If I is a regular injective, regular cogenerator in Top, then the full subcategory $\mathcal{I}$ of Top, given by all retracts of powers of I, is the subcategory of all regular injective topological spaces.

By applying Φ_I: $Top^{op} \to Set^{T_I}$ to objects in $\mathcal{I}$, we get regular projective T_I-algebras since, by definition of the comparison functor Φ_I, $\Phi_I(\mathcal{I})$ corresponds to free T_I-algebras and retracts of free T_I-algebras in Set^{T_I}.

Hence $\Phi_I(\mathcal{I})$ is a projective cover for Set^{T_I}, i.e. not only all $A \in \Phi_I(\mathcal{I})$ are regular projective, but also every algebra in Set^{T_I} is a regular epimorphic image of a certain object $B \in \Phi_I(\mathcal{I})$. Being $\mathcal{I}$ equivalent to $\Phi_I(\mathcal{I})$ then, for any regular injective cogenerator I, we get $\Phi_I(\mathcal{I}) \simeq \Phi(\mathcal{I})$, and this equivalence between projective covers can be easily extended to an equivalence between the corresponding categories of T-algebras Set^T and Set^{T_I} [12]. □

In [4] the category of T-algebras corresponding to the choice $I = 3$, has been explicitely constructed as the category of *Grids*. A grid is defined as a frame $(G, \bigvee, \wedge)$ with an additional unary operator $'$ satisfying, for $u, v \in G$, the following axioms (in addition to the frames ones):

(1) $u'' = u$;
(2) the unary operation $u \vee u'$ and $u \wedge u'$ are $\bigvee$-homomorphisms;
(3) the unary operation $u \vee u'$ is a $\wedge$-homomorphism;
(4) $(u \wedge v) \wedge (u \wedge v)' = u \wedge v \wedge v'$;
(5) the interval $[u \wedge u', u \vee u']$ is a complete atomic boolean algebra with the operation $\bigvee$ and $'$.

Moreover in [4] it is shown that Top^{op} coincides with the full subcategory of *Grids* defined by the Horn clause:

$$(u \vee u') \vee (t \wedge t') = (v \vee v') \vee (t \wedge t') \Rightarrow u \vee u' = v \vee v'$$

where t denotes the terminal object in $Grids$.

In this approach, a topological space X, corresponds to the grid $Top(X,3)$ of all pairs (U,W) with U open in X and $W \subseteq U$, with pointwise frame structure and $'$ given by the boolean complement on the second component.

REMARK 7. The construction given in this note applies to the category $C = Sob$ of sober spaces and continuous maps, by choosing as regular injective, regular cogenerator the Sierpinski space 2. The triple T: $Set \to Set$, obtained in this way is the "free frame" triple defined, for any $Y \in Set$, by $T(Y) = Top(\prod_Y 2, 2) \simeq \Omega(\prod_Y 2)$, so Set^T is equivalent to the category Frm of frames. The duality between sober spaces and spatial frames is realized by the representable functors $Top(-,2) \simeq \Omega(-)$: $Sob^{op} \to Frm$ and $Frm(-,2) \simeq pt(-)$: $Frm \to Sob^{op}$ [7]. Observe that 2 does not maintain the regular cogenerator property if we enlarge Sob to T_0-spaces [9].

References

1. Adamek, J., Herrlich, H., and Strecker, G. E.: *Abstract and Concrete Categories*, Wiley, 1990.
2. Barr, M. and Wells, C.: *Toposes, Triples and Theories*, Grundlehren der math. Wissenschaften 278, Springer-Verlag, Berlin, 1985.
3. Barr, M., Grillet, P. A., and van Osdal, D. H.: *Exact Categories and Categories of Sheaves*, LNM 236, Springer-Verlag, Berlin, 1971.
4. Barr, M. and Pedicchio, M. C.: Top^{op} is a quasi-variety, *Cahiers de Topologie et Géométrie Différentielle Catégoriques* **36** (1995), 3–10.
5. Borceux, F.: *A Handbook in Categorical Algebra*, Cambridge Univ. Press, 3 volumes, 1994.
6. Isbell, J. R.: General functorial semantics I, *Amer. J. Math.* **94** (1972), 535–596.
7. Johnstone, P. T. J.: *Stone Spaces*, Cambridge Press, 1982.
8. Mac Lane, S.: *Categories for the Working Mathematicians*, Graduate Texts in Mathematics 5, Springer-Verlag, Berlin, 1972.
9. Nel, L. D. and Wilson, R. G.: Epireflections in the category of T_0 spaces, *Fund. Math.* **75** (1972), 69–74.
10. Pedicchio, M. C.: On k-permutability for categories of T-algebras, *J. Proceedings Mag. Conf.* (to appear).
11. Simmons, H.: A couple of triples, *Topology Appl.* **XIII** (1982), 201–223.
12. Vitale, E.: On the characterization of monadic categories over set, *J. Cahiers de Topologie et Géométrie Différentielle Catégoriques* **36** (1995).

Applied Categorical Structures **4:** 87–95, 1996.

α-Sober Spaces via the Orthogonal Closure Operator

LURDES SOUSA*
Escola Superior de Tecnologia, Instituto Politécnico de Viseu, Campus Politécnico, Repeses, 3500 Viseu, Portugal

(Received: 3 November 1994; accepted: 6 July 1995)

Abstract. Each ordinal α equipped with the upper topology is a T_0-space. It is well known that for $\alpha = 2$ the reflective hull of α in $\mathcal{T}op_0$ is the subcategory of sober spaces. Here, we define α-sober space for each $\alpha \geqslant 2$ in such a way that the reflective hull of α in $\mathcal{T}op_0$ is the subcategory of α-sober spaces. Moreover, we obtain an order-preserving bijective correspondence between a proper class of ordinals and the corresponding (epi)reflective hulls. Our main tool is the concept of orthogonal closure operator, first introduced in [12].

Mathematics Subject Classifications (1991). 18A40, 18B30, 18B35, 54B30, 54F65, 04A10.

Key words: orthogonal closure operator, strongly closed object, (epi)reflective hull, α-sober space.

1. Introduction

It is well-known that the conglomerate of all $\mathcal{E}$-reflective subcategories of an $(\mathcal{E}, \mathbb{M})$-category is a complete "lattice" with respect to the inclusion order. Several authors have contributed to the existing knowledge of the "lattice" of epireflective subcategories of everyday categories (see, e.g., [5] and references there). In particular, in what concerns the category of T_0-spaces, we refer to, e.g., [6, 7, 10] and [8].

Under convenient conditions, the orthogonal closure operator gives a characterization of the reflective hull of a subcategory in terms of closedness ([12]). In this paper, we consider each ordinal α endowed with the upper set topology and characterize the orthogonal closure operator induced by $\alpha \geqslant 1$ in $\mathcal{T}op_0$ (which for $\alpha = 2$ coincides with the b-closure). This gives us the means to obtain a topological characterization of the spaces which belong to the reflective hull of α (for $\alpha \geqslant 2$), the α-sober spaces. Furthermore, we determine for which ordinals the corresponding hulls coincide and we prove that the subcategories of α-sober spaces form a well-ordered proper class which is contained in the "lattice" of epireflective subcategories of $\mathcal{T}op_0$.

In Section 2 we collect all needed notions and facts concerning the orthogonal closure operator.

* The author acknowledges financial support from Instituto Politécnico de Viseu and from Centro de Matemática da Universidade de Coimbra.

2. The Orthogonal Closure Operator Versus Reflectivity

We consider an $(\mathcal{E}, \mathbb{M})$-category $\mathcal{X}$, with $\mathbb{M}$ a conglomerate of monosources, and a full and replete subcategory $\mathcal{A}$ of $\mathcal{X}$. We denote by $\mathcal{A}^{\perp}$ the class of all morphisms $f: X \to Y$ of $\mathcal{X}$ such that, for every $A \in \mathcal{A}$, we have that $f \perp A$, i.e., that every map $\mathcal{X}(f, A): \mathcal{X}(Y, A) \to \mathcal{X}(X, A)$ is bijective. The subcategory of $\mathcal{X}$ which consists of all objects Z such that $f \perp Z$ for each $f \in \mathcal{A}^{\perp}$, is the orthogonal hull of $\mathcal{A}$ in $\mathcal{X}$. It is the smallest orthogonal subcategory of $\mathcal{X}$ containing $\mathcal{A}$ and will be denoted by $\mathcal{O}(\mathcal{A})$. It is well-known that $\mathcal{O}(\mathcal{A})$ is contained in every reflective subcategory of $\mathcal{X}$ which contains $\mathcal{A}$. Consequently, if the orthogonal hull of $\mathcal{A}$ is reflective in $\mathcal{X}$, then it is the reflective hull.

Let $\mathbb{M}(\mathcal{A})$ denote the $\mathcal{E}$-reflective hull of $\mathcal{A}$ in $\mathcal{X}$ (see [1]). Then $\mathcal{A}^{\perp}$ consists of all morphisms of $\mathcal{X}$ whose reflection in $\mathbb{M}(\mathcal{A})$ belongs to $\mathcal{A}^{\perp}$ (i.e., whose reflection is orthogonal to $\mathcal{A}$ in $\mathbb{M}(\mathcal{A})$). Furthermore, the orthogonal hull of $\mathcal{A}$ in $\mathcal{X}$ coincides with the orthogonal hull of $\mathcal{A}$ in $\mathbb{M}(\mathcal{A})$ (see [11]). Consequently, in order to characterize the orthogonal hull of $\mathcal{A}$ and to investigate conditions under which it is the reflective hull, we can assume, without loss of generality, that the $\mathcal{E}$-reflective hull of $\mathcal{A}$ in $\mathcal{X}$ is $\mathcal{X}$ itself. In the two theorems below we shall consider this assumption. Let $\mathcal{M}$ be the class of all morphisms (i.e., singleton sources) in $\mathbb{M}$. From now on we assume further that $\mathcal{X}$ has pushouts and is $\mathcal{M}$-complete (see [1]). We are going to define a closure operator which enables us to characterize $\mathcal{A}^{\perp}$ and $\mathcal{O}(\mathcal{A})$ in terms of denseness and closedness, respectively.

For each morphism $m: X \to Y$ in $\mathcal{M}$, we consider a morphism $c_{\mathcal{A}}(m): c_{\mathcal{A}}(X) \to Y$ in $\mathcal{M}$ obtained as follows: For each morphism g with domain X and codomain in the subcategory $\mathcal{A}$, we form the pushout $(\overline{m}, g')$ of (m, g) in $\mathcal{X}$. Let $m' \cdot e$ be the $(\mathcal{E}, \mathcal{M})$-factorization of $\overline{m}$ and let m_g be the pullback of m' along g'.

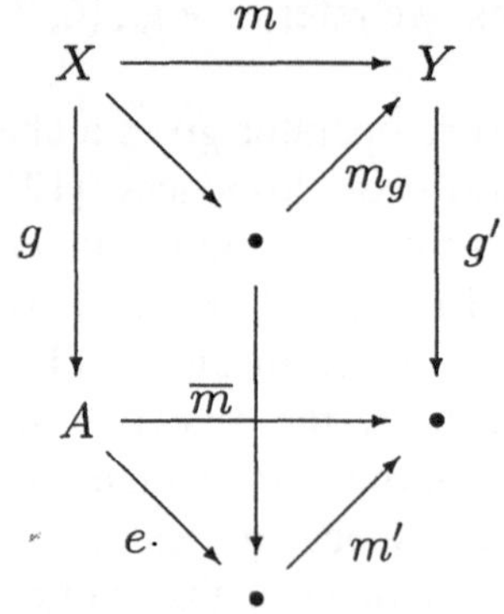

The morphism $c_{\mathcal{A}}(m): c_{\mathcal{A}}(X) \to Y$ is the intersection of all morphisms m_g obtained this way. It is easy to check that there is a unique morphism $d_{\mathcal{A}}(m)$ such that $c_{\mathcal{A}}(m) \cdot d_{\mathcal{A}}(m)$ is a factorization of m. The function $c_{\mathcal{A}}$ which assigns, to each morphism $m \in \mathcal{M}$, the morphism $c_{\mathcal{A}}(m)$ determines a closure operator on $\mathcal{X}$ with respect to $\mathcal{M}$ in the sense of Dikranjan and Giuli [3]. It will be called the orthogonal closure operator induced by $\mathcal{A}$.

The following propositon and two theorems were proved in [12].

Let $PS(\mathcal{M})$ denote the subclass of $\mathcal{M}$ consisting of all morphisms for which the pushout along any morphism belongs to $\mathcal{M}$. It is just the greatest pushout-stable subclass of $\mathcal{M}$.

PROPOSITION 2.1. *Under the assumption that* $\mathbb{M}(\mathcal{A}) = \mathcal{X}$, *the class* $\mathcal{A}^{\perp}$ *consists of all morphisms in* $PS(\mathcal{M})$ *which are dense with respect to the orthogonal closure operator induced by* $\mathcal{A}$.

Having in mind to characterize the orthogonal hull of $\mathcal{A}$, let us consider the following definition:

DEFINITION 2.2. An object X is said to be $\mathcal{A}$-*strongly closed* provided that every morphism in $PS(\mathcal{M})$ with domain X is closed with respect to the orthogonal closure operator induced by $\mathcal{A}$.

THEOREM 2.3. *Under the assumption that* $\mathbb{M}(\mathcal{A}) = \mathcal{X}$, $c_{\mathcal{A}}$ *preserves* $PS(\mathcal{M})$-*morphisms if and only if, for each* $m \in PS(\mathcal{M})$, $d_{\mathcal{A}}(m)$ *is* $c_{\mathcal{A}}$-*dense. In this case,* $c_{\mathcal{A}}$ *considered as a closure operator with respect to* $PS(\mathcal{M})$ *is idempotent and weakly hereditary and* $\mathcal{O}(\mathcal{A})$ *is the subcategory of all* $\mathcal{A}$-*strongly closed objects.*

THEOREM 2.4. *Under the assumption that* $\mathbb{M}(\mathcal{A}) = \mathcal{X}$ *and that* $c_{\mathcal{A}}$ *preserves* $PS(\mathcal{M})$-*morphisms, the orthogonal hull of* $\mathcal{A}$ *is reflective in* $\mathcal{X}$ *if and only if, for each object* X, *there exists some morphism in* $PS(\mathcal{M})$ *with domain* X *and codomain in the orthogonal hull. In this case, the orthogonal hull is the (epi)reflective hull of* $\mathcal{A}$ *in* $\mathcal{X}$.

3. The Orthogonal Closure Operators c_α in $\mathcal{T}op_0$

Let $\mathcal{T}op_0$ be the category of T_0-spaces and continuous maps and let $\mathcal{A}$ be a full and replete subcategory of $\mathcal{T}op_0$.

Let $\mathcal{M}$ be the class of all embeddings in $\mathcal{T}op_0$. Thus, since in $\mathcal{T}op_0$ embeddings are pushout stable, we have that the orthogonal closure operator in $\mathcal{T}op_0$ with respect to $\mathcal{M}$ and induced by $\mathcal{A}$ assigns, to each subspace X of a space Y, another subspace $c^Y_{\mathcal{A}}(X)$ which is the intersection of all subspaces of Y which are pullbacks of m along some $g \in \mathcal{T}op_0(X, \mathcal{A})$, where m is the inclusion of X in Y.

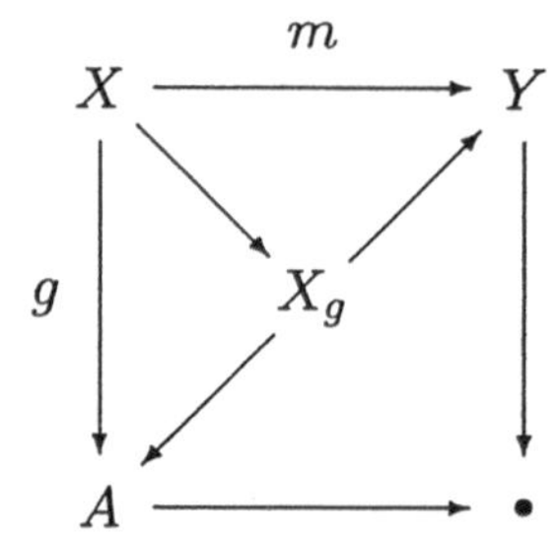

Now, using the theorems of the previous section, it is easy to deduce the following

PROPOSITION 3.1. *If Top_0 is the epireflective hull of A in Top, then the closure operator c_A is idempotent and weakly hereditary, and the (epi)reflective hull of A in Top_0 consists of all A-strongly closed spaces.*

Let α be a nonempty element of the class Ord of all ordinals. We consider α as a topological space endowed with the upper set topology, i.e., the non trivial open sets of α are all $\beta \uparrow= \{\delta \in \alpha : \delta \geqslant \beta\}$, $\beta \in \alpha$. It is clear that $\alpha \in Top_0$ and that, for $\alpha \geqslant 2$, the epireflective hull of α in Top is Top_0. If A is the full and replete subcategory of Top_0 generated by α, we denote by c_α the respective orthogonal closure operator and we speak about α-*strongly closed spaces*.

The set of all open sets of a space X will be denoted by $\Omega(X)$.

PROPOSITION 3.2. *If X is a subspace of Y in Top_0 and $y \in Y$, then $y \in c_\alpha(X)$ if and only if*

(A) *for each continuous map $g \colon X \to \alpha$ there exists $\beta_0 \in \alpha$ such that for every $H \in \Omega(Y)$ and every $\beta \in \alpha + 1$ with $H \cap X = g^{-1}(\beta \uparrow)$, one has $y \in H$ iff $\beta \leqslant \beta_0$.*

Proof. Since for $y \in X$ the result is trivial, we assume that $y \in Y \backslash X$.

Let y satisfy condition (A). Firstly, we show that $y \in \overline{X}$ (where $\overline{X}$ denotes the usual closure of X in Y). In fact, let $H \in \Omega(Y)$ be such that $H \cap X = \emptyset$; define $g \colon X \to \alpha$ by $g(x) = 0$, $x \in X$. Since $g^{-1}(1 \uparrow) = \emptyset = \emptyset \cap X$, the ordinal β_0 required by (A) must be smaller than 1, thus $\beta_0 = 0$ and, as $H \cap X = g^{-1}(1 \uparrow)$, $y \notin H$.

Now, let $g \colon X \to \alpha$ be an arbitrary continuous map and let us consider the following pushout in Top, where $m \colon X \to Y$ is the embedding of X in Y.

$$
\begin{array}{ccc}
X & \xrightarrow{\;\;m\;\;} & Y \\
{\scriptstyle g}\downarrow & & \downarrow{\scriptstyle g'} \\
\alpha & \xrightarrow[\;\;m'\;\;]{} & W
\end{array}
$$

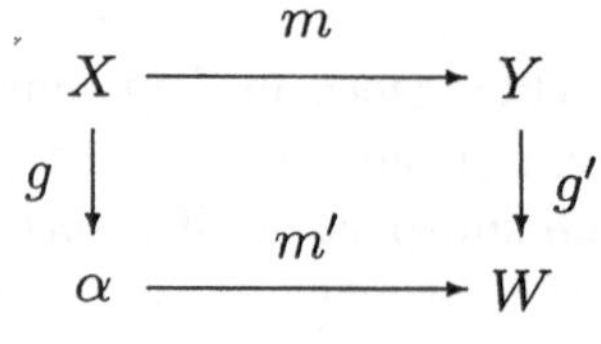

* For $g \colon X \to \alpha$, $g^{-1}(\alpha \uparrow)$ denotes the empty set.

We assume that m': $\alpha \to W$ is the inclusion of α into $\alpha \dot{\cup} (Y \backslash X)$. Let β_0 be the ordinal whose existence is guaranteed by (A). We are going to show that, for every $U \in \Omega(W)$, $y \in U$ iff $\beta_0 \in U$, so that $r_W(y) = r_W(\beta_0)$ (where r_W is the reflection of W in $\mathcal{T}op_0$) and, consequently, $y \in X_g$. Let $U \in \Omega(W)$, i.e., $(g')^{-1}(U) \in \Omega(Y)$ and $(m')^{-1}(U) \in \Omega(\alpha)$. If $y \in U$, $y \in (g')^{-1}(U)$ and, then, since $y \in \overline{X}$, $(g')^{-1}(U) \cap X \neq \emptyset$. Thus $(g')^{-1}(U) \cap X = g^{-1}((m')^{-1}(U))$ with $(m')^{-1}(U) = \beta \uparrow$ for some $\beta \in \alpha$. So we have that

$$\begin{aligned} y \in U \quad & \text{iff } y \in (g')^{-1}(U), \\ & \text{iff } (g')^{-1}(U) \cap X = g^{-1}((m')^{-1}(U)) \\ & \qquad \text{with } (m')^{-1}(U) = \beta \uparrow \text{ for some } \beta \leqslant \beta_0, \\ & \text{iff } \beta_0 \in (m')^{-1}(U) = \beta \uparrow \text{ for some } \beta, \\ & \text{iff } \beta_0 \in U. \end{aligned}$$

Therefore, since $y \in X_g$ for each $g \in \hom(X, \alpha)$, we have that $y \in c_\alpha(X)$.

Conversely, let $y \in c_\alpha(X)$ and let $g \in \hom(X, \alpha)$. Then $y \in X_g$, which is equivalent to saying that there exists $\beta_0 \in \alpha$ such that $r_W(y) = r_W(\beta_0)$ (and this β_0 is unique from the pushout-stability of embeddings in $\mathcal{T}op_0$). Let $H \in \Omega(Y)$ and $\beta \in \alpha$ be such that $g^{-1}(\beta \uparrow) = H \cap X$. Put $U = (H \backslash X) \dot{\cup} \beta \uparrow$; then $U \in \Omega(W)$ and, since $r_W(y) = r_W(\beta_0)$, $y \in (g')^{-1}(U)$ iff $\beta_0 \in \beta \uparrow$, i.e., $y \in H$ iff $\beta \leqslant \beta_0$. □

COROLLARY 3.3. *If α and β are ordinals such that $\alpha \leqslant \beta$ then $c_\beta \leqslant c_\alpha$.*

Proof. Let X be a subspace of Y in $\mathcal{T}op_0$ and $y \in c_\beta(X)$. Let g: $X \to \alpha$ be a continuous map and let e: $\alpha \to \beta$ be the inclusion of α in β. Then there is $\beta_0 \in \beta$ such that for every $H \in \Omega(Y)$ and every $\delta \in \beta$ which fulfil the equality $H \cap X = (e \cdot g)^{-1}(\delta \uparrow)$, $y \in H$ iff $\delta \leqslant \beta_0$. Since $(e \cdot g)^{-1}(\delta \uparrow) = \emptyset$ for $\delta \geqslant \alpha$, it must be $\beta_0 \in \alpha$. Consequently, $y \in c_\alpha(X)$. □

REMARK 3.4. It is easy to check that, for $n \in \omega_0 \backslash 2$, the closure operator c_n coincides with the b-closure (cf. [2]). The same does not hold for $n = 1$. In fact, c_2 is strictly smaller than c_1. In order to show this, consider the embedding m: $2 \to 3$ defined by $m(0) = 0$ and $m(1) = 2$; then $c_2(m) = m$ and $c_1(m) = 1_3$. This is easily verified by taking into account that we can rewrite Proposition 3.2 for the case $n = 1$ as follows:

> $y \in c_1(X)$ *if and only if* $y \in \overline{X}$ *and, for every* $H \in \Omega(Y)$, *the inclusion* $X \subseteq H$ *implies that* $y \in H$.

4. α-Sober Spaces

DEFINITIONS 4.1.

1. Let $X \in \mathcal{T}op_0$ and let $\alpha \geqslant 1$. A closed subset F of X is said to be α-*irreducible* if it satisfies the following conditions

   (i_0) F is irreducible, i.e., if F is the union of two closed sets then F is equal to one of them;

   (i_α) for every continuous map $g\colon X \to \alpha$, the set $g(F)$ has a maximum.

2. A T_0-space X is said to be α-*sober* if every its α-irreducible closed set is the closure of a unique point.

It is clear that, for every finite ordinal $n \neq 0$, a nonempty closed set is n-irreducible iff it satisfies the condition (i_0) (since (i_n) trivially holds). Consequently, in this case, to be an n-sober space means to be a sober space. However, it is not interesting for our purpose to consider the definition of α-sober space for $\alpha = 1$, since the characterizations obtained in 4.2 and 4.3 below for $\alpha \in Ord\backslash 2$ do not hold for $\alpha = 1$. We recall that $\mathcal{T}op_0$ is the (epi)reflective hull of α in $\mathcal{T}op$ only if $\alpha > 1$.

PROPOSITION 4.2. *For $\alpha \in Ord\backslash 2$, a T_0-space X is α-sober if and only if it is α-strongly closed.*

Proof. Let us assume that X is not an α-sober space. This means that X has an α-irreducible closed set F which is not the closure of a singleton. Let us define a space Y as follows: $Y = X \dot{\cup} \{a\}$ and $\Omega(Y)$ consists of $\emptyset$, all sets H such that $H \in \Omega(X)$ and $H \cap F = \emptyset$, and all sets $H \dot{\cup} \{a\}$ such that $H \in \Omega(X)$ and $H \cap F \neq \emptyset$. The fact that F is irreducible guarantees that $\Omega(Y)$ is a topology. Let us show that $Y \in \mathcal{T}op_0$. It is clear that every two distinct points of Y are "separated" by some open set; the only interesting case is when we consider the point a and some $x \in F$. But, if $x \in F$, we have that $\overline{\{x\}} \neq F$, by hypothesis on F. Then there is some $x' \in F$ and $G \in \Omega(X)$ such that $x' \in G$ but $x \notin G$. Hence $G \dot{\cup} \{a\} \in \Omega(Y)$ "separates" a from x.

It is obvious that X is a subspace of Y. Now, let us see that $a \in c_\alpha(X)$, so that X is not α-strongly closed. Let $g\colon X \to \alpha$ be a continuous map. By hypothesis on F, there exists an ordinal $\beta_0 \in \alpha$ such that $\beta_0 = \max g(F)$. In order to show that β_0 fulfils the requirement of (A), let $H \in \Omega(Y)$ and $\beta \in \alpha$ be such that $H \cap X = g^{-1}(\beta\uparrow)$. Hence, on the one hand, if $a \in H$, then $H = g^{-1}(\beta\uparrow) \dot{\cup} \{a\}$ with $g^{-1}(\beta\uparrow) \cap F \neq \emptyset$ and, by definition of β_0, we have that $\beta \leqslant \beta_0$. On the other hand, if $a \notin H$, then $H = g^{-1}(\beta\uparrow)$ and $g^{-1}(\beta\uparrow) \cap F = \emptyset$; thus, $\beta_0 \notin \beta\uparrow$, i.e., $\beta_0 < \beta$.

Conversely, let us assume that X is α-sober. Let X be a subspace of an arbitrary T_0-space Y and $y \in Y$ such that $y \in c_\alpha(X)$. We want to show that y must be a point of X.

Let $\overline{\{y\}}$ be the closure of $\{y\}$ in Y. Firstly, let us notice that, from the above corollary, $y \in c_2(X)$ and, then, since c_2 is the b-closure operator (i.e., $y \in c_2(X)$ iff for each $H \in \Omega(Y)$ such that $y \in H$, $\overline{\{y\}} \cap X \cap H \neq \emptyset$, cf. [2]), it follows that $\overline{\{y\}} \cap X$ is a closed set of X which satisfies condition (i_0). On the other hand, it satisfies condition (i_α); in fact, since $y \in c_\alpha(X)$, for each continuous map $g\colon X \to \alpha$, let $\beta_0 \in \alpha$ be the ordinal whose existence is guaranteed in condition (A). We are going to show that $\beta_0 = \max g(\overline{\{y\}} \cap X)$. Let $x \in \overline{\{y\}} \cap X$; then, for some $H \in \Omega(Y)$, $g^{-1}(g(x)\uparrow) = H \cap X$, and, since $x \in H \cap \overline{\{y\}}$, y must belong to H, hence $g(x) \leqslant \beta_0$. Now, let $H \in \Omega(Y)$ be such that $H \cap X = g^{-1}(\beta_0\uparrow)$; then $y \in H$ and, since $y \in c_2(X)$, $\overline{\{y\}} \cap X \cap H \neq \emptyset$, that is, there is some $x \in \overline{\{y\}} \cap X$ such that $g(x) \in \beta_0\uparrow$. But, as we have seen, $g(x) \leqslant \beta_0$; then $g(x) = \beta_0$ and β_0 is the wanted maximum.

Therefore, since X is an α-sober space and $\overline{\{y\}} \cap X$ is α-irreducible, $\overline{\{y\}} \cap X = \overline{\{x\}} \cap X$ for some $x \in X$, from which $\overline{\{y\}} = \overline{\{x\}}$ follows and, thus, $y = x$. □

COROLLARY 4.3. *For each ordinal $\alpha \in Ord\backslash 2$, the (epi)reflective hull of α in $\mathcal{T}op_0$ is the full subcategory of all α-sober spaces.*

Proof. It is an immediate consequence of the above proposition and Proposition 3.1. □

From now on, for each $\alpha \in Ord\backslash 2$, we denote the full subcategory of α-sober spaces by $Sob(\alpha)$.

COROLLARY 4.4. *For α, $\beta \in Ord\backslash 2$ such that $\alpha \leqslant \beta$, $Sob(\alpha) \subseteq Sob(\beta)$.*

Proof. It is a consequence of the above proposition and of Corollary 3.3. □

REMARK 4.5. Let α_* be the T_0-space obtained by equipping the ordinal α with the "open half lines" topology, i.e., the non trivial open sets are of the form $\{\delta \in \alpha\colon \delta < \beta\}$, $\beta \in \alpha$. S. Mantovani [8] characterized the epireflective hulls of these spaces in $\mathcal{T}op_0$ and proved that the epireflective hulls of α_* and β_* coincide iff $cf(\alpha) = cf(\beta)$. She also showed that these epireflective hulls are not comparable in the "lattice" of epireflective subcategories of $\mathcal{T}op_0$. It is clear that, for each ordinal α, the "open half line" topology and the upper set topology coincide iff $\alpha \leqslant \omega_0$. It is also easy to see that, for $\alpha > \omega_0$, the epireflective hulls obtained in this paper are different from Mantovani's ones. Furthermore, the characterization of its objects given here provides a more natural generalization of the concept of sober space. Namely (and in contrast with Mantovani's epireflective hulls) we have that:

1. The function $Ord\backslash 2 \to \mathcal{L}(\mathcal{T}op_0)$ (where $\mathcal{L}(\mathcal{T}op_0)$ denotes the "lattice" of epireflective subcategories of $\mathcal{T}op_0$) which assigns, to each ordinal α, the subcategory $Sob(\alpha)$ is order-preserving (from the latter corollary).

In the next section we shall see that, by convenient restriction and corestriction of this function, we obtain an order-preserving bijection between a proper class of ordinals and subcategories of α-sober spaces.

2. Let $\mathcal{O}rd$ denote the category whose objects are all ordinals and whose morphisms are all order-preserving maps. Then, assigning to each ordinal α the T_0-space consisting of the corresponding underlying set equipped with the upper set topology, we get a concrete full embedding of $\mathcal{O}rd$ into $\mathcal{T}op_0$.

5. The Chain of Subcategories $Sob(\alpha)$

As we have seen, for $n \in \omega_0 \backslash 2$, $Sob(n) = Sob(2)$. Next we deal with the question of knowing for which ordinals $\alpha < \beta$ we have that $Sob(\alpha)$ is strictly contained in $Sob(\beta)$. The following theorem enables us to conclude that there exists a well-ordered proper class of subcategories $Sob(\alpha)$ with $\alpha \in Ord$.

THEOREM 5.1. *For $\alpha, \beta \in Ord\backslash 2$ such that $\alpha < \beta$, we have that $Sob(\alpha)$ is strictly contained in $Sob(\beta)$ if and only if there is some infinite regular cardinal λ such that $\alpha \leqslant \lambda \leqslant \beta$.*

Proof. Let α, β, $\lambda \in Ord\backslash 2$ be such that $\alpha < \beta$ and $\alpha \leqslant \lambda \leqslant \beta$ with λ an infinite regular cardinal. The closed set λ of β trivially satisfies (i_0); we shall show that it also satisfies (i_α), so that λ is α-irreducible. Let $g\colon \beta \to \alpha$ be a continuous map. We may have two cases: $\lambda \in \beta$ or $\lambda = \beta$. In the first case, let $\delta \in \alpha$ be such that $g(\lambda) = \delta$; hence, since the continuity of g is equivalent to the preservation of order (see 4.5.2), it follows that $\lambda \subseteq g^{-1}(\downarrow \delta)$. In the second case, since $\alpha = \bigcup_{\delta \in \alpha} \downarrow \delta$, we have that $\lambda \subseteq \bigcup_{\delta \in \alpha} g^{-1}(\downarrow \delta)$ and, then, as $\alpha < \lambda$ and λ is regular, $\lambda \subseteq g^{-1}(\downarrow \delta)$ for some $\delta \in \alpha$. Thus, in the two cases, there exists $\delta_0 = \min\{\delta \in \alpha \,|\, \lambda \subseteq g^{-1}(\downarrow \delta)\}$. Moreover, $\lambda \cap g^{-1}(\{\delta_0\}) \neq \emptyset$, so that $\delta_0 = \max g(\lambda)$. Indeed, if $\lambda \cap g^{-1}(\{\delta_0\}) = \emptyset$, then $\lambda \subseteq \bigcup_{\delta \in \delta_0} g^{-1}(\downarrow \delta)$. But, since λ is regular, it follows that $\lambda \subseteq g^{-1}(\downarrow \delta)$ for some $\delta \in \delta_0$, which contradicts the definition of δ_0. Hence, one must have $\delta_0 \in g(\lambda)$. Therefore, we have shown that λ is an α-irreducible closed set of β. But it is not the closure of a single point. Then β is not an α-sober space, that is, $\beta \notin Sob(\alpha)$, so the inclusion $Sob(\alpha) \subseteq Sob(\beta)$ is strict.

Now, let us assume that there exists no infinite regular cardinal between α and β. The only closed subsets of β which are not the closure of a single point are the limit ordinals. We shall show that they are not α-irreducible, so that $\beta \in Sob(\alpha)$ and $Sob(\alpha) = Sob(\beta)$. Let γ be an infinite ordinal in β, let λ be its cofinality character (which is an infinite regular cardinal, see, e.g., [9]) and let f be an existing strictly increasing function with domain λ such that $\gamma = \bigcup_{\delta \in \lambda} f(\delta)$. By hypothesis, λ must be smaller than α, and, according to the assumptions over f, we have that, for every $\phi \in \gamma$, the set $\{\delta \in \lambda \,|\, \phi \leqslant f(\delta)\}$ is not empty. Thus, let $g\colon \beta \to \alpha$ be defined as follows: if $\phi \in \gamma$, $g(\phi) = \min\{\delta \in \lambda \,|\, \phi \leqslant f(\delta)\}$;

otherwise, $g(\phi) = \lambda$. It is obvious that g is continuous, since it is nondecreasing, and that γ fails (i_α) with respect to g. □

COROLLARY 5.2. *The family $(Sob(\alpha))$ of reflective hulls of ordinals α, such that α is an infinite cardinal or α is successor of an infinite regular cardinal, form a well-ordered proper class which is contained in the "lattice" of epireflective subcategories of $\mathcal{T}op_0$.*

Proof. If α is an infinite regular cardinal and $\beta = \alpha + 1$ then it is clear, by the above theorem, that $Sob(\alpha) \neq Sob(\beta)$. On the other hand, if α and β are infinite cardinals and $\alpha < \beta$ then there is some infinite regular cardinal between them, since we have that, for every infinite cardinal α, α^+ is regular (where α^+ is the least cardinal which is greater than α, see, for instance, [9]). Thus, the inequality $Sob(\alpha) \neq Sob(\beta)$ follows. Now, using 4.4, we get the claimed result. □

It is clear that the ordinals for which the corresponding reflective hull "jumps" are just those mentioned in the above corollary. In the remaining cases the equality holds. Thus, we have that

$$\begin{aligned} &Sob(2) \subset Sob(\omega_0) \subset Sob(\omega_0 + 1) \\ &\quad = Sob(\omega_0 + \omega_0) = \cdots = Sob(\omega_0 \cdot \omega_0) = \cdots \\ &\quad \subset Sob(\omega_1) \subset Sob(\omega_1 + 1) = \cdots \subset Sob(\omega_\omega) = Sob(\omega_\omega + 1) = \cdots . \end{aligned}$$

References

1. Adámek, J., Herrlich. H., and Strecker, G. S.: *Abstract and Concrete Categories*, Wiley, Inc., New York, 1990.
2. Baron, S.: Note on epi in T_0, *Canad. Math. Bull.* **11** (1968), 503–504.
3. Dikranjan, D. and Giuli, E.: Closure operators I, *Topology Appl.* **27** (1987), 129–143.
4. Dikranjan, D., Giuli, E., and Tholen, W.: Closure operators II, in *Categorical Topology and its Relations to Analysis, Algebra and Combinatorics*, Proceedings Int. Conf. Categorical Topology Prague 1988, World Scientific, Singapore, 1989, pp. 297–335.
5. Herrlich, H.: Categorical topology 1971–1981, in *General Topology and its Relations to Modern Analysis and Algebra V*, Proceedings Fifth Prague Topological Symposium 1981, Heldermann-Verlag, Berlin, 1982, pp. 279–383.
6. Hoffmann, R.-E.: Charakterisierung nüchterner Räume, *Manuscripta Math.* **15** (1975), 185–191.
7. Hong, S. S.: Extensive subcategories of the category of T_0-spaces, *Canad. J. Math.* **27** (1975), 311–318.
8. Mantovani, S.: Epireflective hulls of spaces of ordinals in T_0, *Quaestiones Math.* **7** (1984), 203–211.
9. Monk, J. D.: *Introduction to Set Theory*, International Series in Pure and Applied Mathematics, McGraw-Hill, Inc., New York, 1969.
10. Nel, L. D. and Wilson, R. G.: Epireflections in the category of T_0-spaces, *Fund. Math.* **LXXV** (1972), 69–74.
11. Sousa, L.: Solid hulls of concrete categories, *Appl. Categ. Structures* **3** (1995), 105–118.
12. Sousa, L.: Orthogonality and closure operators, *Cahiers Topologie Géom. Différentielle Catégoriques*, to appear.

Applied Categorical Structures **4:** 97–106, 1996.

Some Aspects of Topological Descent

MANUELA SOBRAL*
Departamento de Matemática, Universidade de Coimbra, 3000 Coimbra, Portugal
(e-mail: sobral@mat.uc.pt)

(Received: 3 November 1994; accepted: 6 July 1995)

Abstract. The paper deals with (effective) descent morphisms for subfibrations $\mathbb{E}(X)$ of the basic fibration $\mathcal{T}op/X$, for topological spaces X and classes $\mathbb{E}$ of continuous functions stable under pullback. For a category with pullbacks, we prove the stability under pullback of effective $\mathbb{E}$-descent morphisms for a class $\mathbb{E}$ satisfying some suitable conditions. This plays a rôle in relating effective $\mathbb{E}$-descent to effective global descent and enables us to obtain a criterion for effective étale-descent. We also show that the inclusion of the class of effective global-descent maps in the class surjective effective étale-descent is strict.

Mathematics Subject Classifications (1991). 54B30, 18A30, 18C20, 18A40, 54C10.

Key words: descent data, (effective) descent map, monad, monadic functor, universal regular epimorphism, effective equivalence relation, étale-descent.

1. Introduction

The paper gives solutions to three problems in descent theory. First, a general criterion for pullback stability of effective descent morphisms, with respect to a subfibration of the basic fibration of a category, is established; it generalizes a corresponding result in [10] for the basic fibration. Then the paper gives a characterization in $\mathcal{T}op$ of the maps which are effective descent with respect to the subfibration given by the local homeomorphisms. Finally, it is shown that such maps need not be effective descent with respect to the basic fibration, even when they are surjective. This solves a problem left open in the recent article [6].

Let $\mathbb{E}$ be a class of continuous functions stable under pullback. For a space X, $\mathbb{E}(X)$ is the full subcategory of the category $\mathcal{T}op/X$ with objects all $\mathbb{E}$-bundles over X.

For a continuous map $p\colon E \to B$ and an $\mathbb{E}$-bundle (C,γ) over E, let $(E \times_B C; \pi_1, \pi_2)$ be the pullback of $(p, p\gamma)$, i.e. $E \times_B C = \{(e,c) \mid p(e) = p\gamma(c)\}$, and, for $p(e) = p(e')$, let $i_{e,e'}\colon \gamma^{-1}(e') \to E \times_B C$, with $i_{e,e'}(c) = (e,c)$, be the embedding.

* Partial financial support by Centro de Matemática da Universidade de Coimbra is gratefully acknowledged.

Descent data for (C,γ) relative to p are given by maps $\xi_{e,e'}\colon \gamma(e) \to \gamma(e')$, for $p(e) = p(e')$, satisfying the following conditions:

1. $\xi_{e,e} = \mathrm{id}$ and $\xi_{e',e''}.\xi_{e,e'} = \xi_{e,e''}$, for e, e' and e'' in the same p-fiber;
2. the unique function $\bar{\xi}\colon E \times_B C \to E \times_B C$, which makes all diagrams

$$\begin{array}{ccc} \gamma^{-1}(e) & \xrightarrow{\xi_{e,e'}} & \gamma^{-1}(e') \\ {\scriptstyle i_{e',e}}\downarrow & & \downarrow{\scriptstyle i_{e,e'}} \\ E \times_B C & \xrightarrow{\bar{\xi}} & E \times_B C \end{array}$$

commute, is continuous.

Thus, "les donnés de descente" $\xi_{e,e'}$ are homeomorphisms and so $\bar{\xi}$ is an homeomorphism too.

The category of $\mathbb{E}$-bundles over a space E equipped with descent data, $(C,\gamma;\bar{\xi})$, are the objects of a category $Des_{\mathbb{E}}(p)$ with morphisms

$$f\colon (C,\gamma;\bar{\xi}) \to (C',\gamma';\bar{\xi}')$$

all maps $f\colon (C,\gamma) \to (C',\gamma')$ in $\mathbb{E}(E)$ compatible with descent data, i.e. such that

$$f(\xi_{e,e'}(c)) = \xi'_{e,e'}(f(c))$$

for $p(e) = p(e')$ and $c \in \gamma^{-1}(e)$.

There is a comparison functor

$$\Phi^p_{\mathbb{E}}\colon \mathbb{E}(B) \to Des_{\mathbb{E}}(p)$$

defined by $\Phi^p_{\mathbb{E}}(A,\alpha) = (E \times_B A, \pi_1; \bar{\varphi})$, with $\bar{\varphi}\colon E \times_B (E \times_B A) \to E \times_B (E \times_B A)$ such that $\bar{\phi}(e,(e',a)) = (e',(e,a))$, and $\Phi^p_{\mathbb{E}}(f) = 1 \times_B f$. It gives rise to a commutative diagram of functors

$$\begin{array}{ccc} \mathbb{E}(B) & \xrightarrow{\Phi^p_{\mathbb{E}}} & Des_{\mathbb{E}}(p) \\ & {\scriptstyle p^*}\searrow \quad \swarrow{\scriptstyle U^p} & \\ & \mathbb{E}(E) & \end{array}$$

where p^* is defined by pulling back along p and U^p is the forgetful functor.

A map p is $\mathbb{E}$-*descent* if $\Phi^p_{\mathbb{E}}$ is full and faithful and p is *effective* $\mathbb{E}$-*descent* if $\Phi^p_{\mathbb{E}}$ is an equivalence.

In case $\mathbb{E}$ is the class of all continuous maps, the prefix $\mathbb{E}$ is dropped and, using the terminology of [5], we sometimes speak of (effective) global-descent.

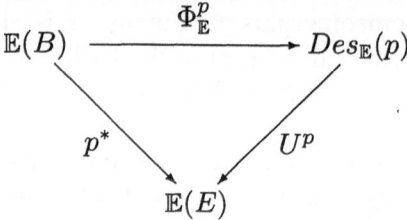

The functor $p^*\colon \mathcal{T}op/B \to \mathcal{T}op/E$ has a left adjoint $p!$ defined by composition on the left with p. Let $\mathbb{T}$ be the monad induced in $\mathcal{T}op/E$ by this adjunction. There is a bijective correspondence H between descent data and $\mathbb{T}$-algebra structures on $(C,\gamma) \in \mathcal{T}op/E$ defined by

$$H(\bar{\xi}) = \pi_2 \cdot \bar{\xi}$$

which we will denote by ξ, and $\bar{\xi} = H^{-1}(\xi)$: $E \times_B C \to E \times_B C$ defined by

$$\bar{\xi}(e,c) = (\gamma(c), \xi(e,c)).$$

This is the object function of an isomorphism $Des(p) \cong (\mathcal{T}op/E)^{\mathbb{T}}$, with $H(f) = f$ since f is a morphism of $Des(p)$ if and only if it is a morphism between the corresponding $\mathbb{T}$-algebras.

Then we have the following commutative diagram

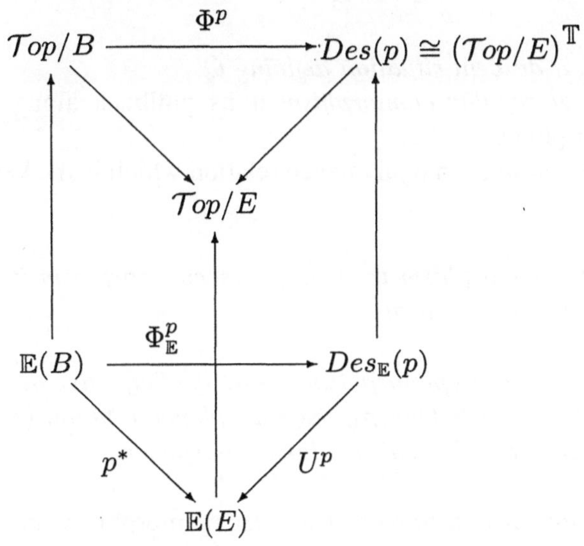

where the vertical arrows are full embeddings.

This is an instance of a more general result due to Bénabou and Roubaud [1]: for every bifibration, over a category with pullbacks, which satisfies the Beck–Chevalley condition (like the "codomain" functor $\mathcal{T}op^{\mathbf{2}} \to \mathcal{T}op$ we are considering here), descent data can be interpreted as structure maps of algebras over a monad and $Des(p)$ is isomorphic to the category of Eilenberg–Moore algebras for this monad.

2. Descent in a Category with Pullbacks and Coequalizers

Let us recall first some facts about global-descent in a category $\mathcal{X}$ with pullbacks and coequalizers. Indeed, we just need to assume the existence of pullbacks and coequalizers of equivalence relations as explained in [6] and already implicit in [10].

For a morphism $p\colon E \to B$ in $\mathcal{X}$ and using the isomorphism $Des(p) \cong (\mathcal{X}/E)^{\mathbb{T}}$ we will refer to objects $(C, \gamma; \xi) \in Des(p)$, where ξ is a $\mathbb{T}$-structure on (C, γ).

We recall that $(\pi_2, \xi = \pi_2 \cdot \bar{\xi})$ is an equivalence relation on C and $\bar{\xi}$ is its symmetry (see, e.g., [10, 2.2]).

Let $q = coeq(\pi_2, \xi)$ and δ be the unique morphism such that $\delta \cdot q = p \cdot \gamma$. Then $(Q, \delta) = \Psi^p(C, \gamma; \xi)$, for the left adjoint Ψ^p to $\Phi^p\colon \mathcal{X}/B \to Des(p)$.

The diagram

$$\begin{array}{ccccc} E \times_B C & \overset{\pi_2}{\underset{\xi}{\rightrightarrows}} & C & \xrightarrow{\;q\;} & Q \\ & & \downarrow{\scriptstyle \gamma} & & \downarrow{\scriptstyle \delta} \\ & & E & \xrightarrow[\;p\;]{} & B \end{array}$$

will be called, following [3], *a descent situation defining* Q.

A morphism is a *universal regular epimorphism* if its pullback along any morphism is a regular epimorphism.

An *effective equivalence relation* is an equivalence relation which is the kernel pair of some morphism.

PROPOSITION 2.1 ([4, 2.2]). *A morphism in $\mathcal{X}$ is a descent morphism if and only if it is a universal regular epimorphism.*

THEOREM 2.2 ([10, 2.5]). *A regular epimorphism p in $\mathcal{X}$ is of effective descent type if and only if, for each $(C, \gamma; \xi) \in Des(p)$, the equivalence relation (π_2, ξ) is effective and its coequalizer is a universal regular epimorphism.*

The above conditons imply that p is a universal regular epimorphism. Indeed, the terminal object $(E, 1_E; p_1)$ of $Des(p)$, (p_1, p_2) being the kernel pair of p, is a descent situation defining B.

In $\mathcal{T}op$ all such equivalence relations are effective, as it follows from 2.4 in [10]. Hence a continuous function p is an effective descent morphism if and only if it is a universal regular epimorphism (characterized by Day and Kelly in [2]) and, for each descent situation defining Q, the square is a pullback (see proof of 2.5 in [10]).

Let us denote by $\mathbb{M}$ the class of effective descent morphisms in $\mathcal{X}$, i.e. the class of all morphisms p such that the functor p^* is monadic. The following inclusions hold,

$$SplitEpi(\mathcal{X}) \subseteq \mathbb{M} \subseteq UnivRegEpi(\mathcal{X})$$

and, like the other two classes, $\mathbb{M}$ is closed under composition ([8, Thm. 4]) and stable under pullback ([10, 3.1]).

These inclusions are strict in $\mathcal{T}op$: open surjections belong to $\mathbb{M}$ ([7], see also [10]) and an example of a non-effective descent map is given in [9] which also gives a complete characterization of effective global descent maps.

For a class $\mathbb{E} \subseteq Mor(\mathcal{X})$ stable under pullback, the compositivity of effective $\mathbb{E}$-descent morphisms was proved in Section 5 of [8] (see also [6]).

Let us now investigate its stability under pullback. For that we consider the classes

$$\mathbb{E}_1 = \{f \mid gf \in \mathbb{E} \text{ whenever } g \in \mathbb{E}\},$$

$$\mathbb{E}_2 = \{f \mid h^*(f) \in \mathbb{E}_1 \text{ for any } h\}$$

and $\mathbb{E}^*(p)$ of all morphisms which are pullbacks of p along some morphism in $\mathbb{E}$. Then $\mathbb{E}_2$, being the class of those morphisms for which every pullback is in $\mathbb{E}_1$, is the greatest subclass of $\mathbb{E}_1$ which is stable under pullback.

PROPOSITION 2.3. *For every $\mathbb{E}$-descent morphism p, all morphisms $p' \in \mathbb{E}^*(p)$ are $\mathbb{E}_2$-descent morphisms.*

Proof. Let $(M, \mu) \in \mathbb{E}(B)$, (μ', p') be the pullback of (p, μ) and $P = E \times_B M$. With no loss of generality, we identify $E \times_B D$ with $P \times_M D$, for each $(D, \delta) \in \mathbb{E}_2(M)$.

If $\Phi^p_{\mathbf{E}}\colon \mathbb{E}(B) \to Des_{\mathbf{E}}(p)$ is full and faithful then the same holds for $\Phi^{p'}_{\mathbf{E}_2}\colon \mathbb{E}_2(M) \to Des_{\mathbf{E}_2}(p')$. Indeed, if

$$\Phi^{p'}_{\mathbf{E}_2}(f) = \Phi^{p'}_{\mathbf{E}_2}(g)$$

for $f, g\colon (D, d) \to (D', d')$ in $\mathbb{E}_2(M)$ then, since $f, g\colon (D, \mu \cdot d) \to (D', \mu \cdot d')$ belong to $\mathbb{E}(B)$ and $\Phi^p_{\mathbf{E}}(f) = \Phi^p_{\mathbf{E}}(g)$, we conclude that $f = g$. Also,

$$\text{if } t\colon \Phi^{p'}_{\mathbf{E}_2}(D, d) = \Phi^{p'}_{\mathbf{E}_2}(D', d') \quad \text{then} \quad t\colon \Phi^p_{\mathbf{E}}(D, \mu \cdot d) \to (D', \mu \cdot d')$$

and so there is a unique $f\colon (D, \mu \cdot d) \to (D', \mu \cdot d')$ such that $\Phi^p_{\mathbf{E}}(f) = t$. It remains to be proved that $d' f = d$, i.e. that $f \in \mathbb{E}_2(M)$. Since

$$d' f, d\colon (D, \mu \cdot d) \to (M, \mu)$$

in $\mathbb{E}(B)$ and $\Phi^p_{\mathbf{E}}(d' f) = \Phi^p_{\mathbf{E}}(d)$ then $d' f = d$ and so $t = \Phi^{p'}_{\mathbf{E}_2}(f)$. Therefore, p' is $\mathbb{E}_2$-descent whenever p is $\mathbb{E}$-descent. □

Let us now assume that p is effective $\mathbb{E}$-descent. Given $(C, \gamma; \xi) \in Des_{\mathbf{E}_2}(p')$ then $(C, \mu'\gamma; \xi) \in Des_{\mathbf{E}}(p)$. If $\Phi^p_{\mathbf{E}}$ is an equivalence then there is an object (D, δ) in $\mathbb{E}(B)$ and an isomorphism $\Phi^p_{\mathbf{E}}(D, \delta) \cong (C, \mu'\gamma; \xi)$. Considering also strict equality here, then

$$\gamma\colon \Phi^p_{\mathbf{E}}(D, \delta) \to \Phi^p_{\mathbf{E}}(M, \mu)$$

and so there is a unique $d\colon (D, \delta) \to (M, \mu)$ such that $\Phi^p_{\mathbf{E}}(d) = \gamma$. Thus, $(C, \gamma; \xi) \cong \Phi^{p'}_{\mathbf{E}_2}(D, d)$, if $d \in \mathbb{E}_2$. □

If $\mathbb{E}$ contains the isomorphisms of $\mathcal{X}$ then $\mathbb{E}_1$ is contained in $\mathbb{E}$. If, in addition, $\mathbb{E}$ is closed under composition then $\mathbb{E} = \mathbb{E}_1 = \mathbb{E}_2$. Assuming further that $\mathbb{E}$ is weakly left cancellable, that is $f \in \mathbb{E}$ whenever $gf,\ g \in \mathbb{E}$, we have proved the following relative version of Theorem 3.1 in [10]:

THEOREM 2.4. *If $\mathbb{E}$ contains Iso($\mathcal{X}$), is closed under composition and weakly left cancellable then the class of effective $\mathbb{E}$-descent morphisms is stable under pullback.*

3. Effective Étale-Descent

In the last section of [5], Janelidze and Tholen investigate $\mathbb{E}$-descent for some classes $\mathbb{E}$ of continuous functions stable under pullback, namely for open-subspace embeddings (open-descent) and for local homeomorphisms (étale-descent). They characterize there open and effective open-descent (4.2) and étale-descent (4.4).

From [5], we have the following:

PROPOSITION 3.1. *If $p\colon E \to B$ is étale-descent then p is surjective or $B \setminus p(E)$ is a non-open subspace of B. Furthermore, there is a one-to-one correspondence between the open subsets of B and the ones of its subspace $p(E)$.*

Proof. Since open embeddings are local homeomorphisms, if p is étale-descent then it is also open-descent ([5, 2.6]) and open-descent maps are exactly those $p\colon E \to B$ for which the function $p^*\colon \mathcal{O}(B) \to \mathcal{O}(E)$, between the corresponding lattices of open sets, defined by $p^*(U) = p^{-1}(U)$, is injective ([5, 4.2]). □

From now on $\mathbb{E}$ will denote the class of local homeomorphisms and $p\colon E \to B$ a continuous surjection.

PROPOSITION 3.2. *A surjective map p is a quotient whenever it is effective étale-descent.*

Proof. If $p^{-1}(U) \in \mathcal{O}(E)$ then $(p^{-1}(U), i; \xi) \in Des_{\mathbb{E}}(p)$, where i is the embedding, hence a local homeomorphism, and

$$\xi\colon E \times_B p^{-1}(U) \to p^{-1}(U)$$

is defined by $\xi(e, x) = \pi_1(e, x) = e$. Since p is effective étale-descent, there exists an $\mathbb{E}$-bundle (M, μ) over B such that $\Phi^p_{\mathbb{E}}(M, \mu) \cong (p^{-1}(U), i; \xi)$. Since p is surjective, μ is a monomorphism and $\mu(M) = U$. Consequently, U is an open subset of B, because μ is a local homeomorphism and so an open map. □

A morphism is said to be an $\mathbb{E}$-*universal regular epimorphism* if its pullback along any morphism in $\mathbb{E}$ is a regular epimorphism.

PROPOSITION 3.3. *If p is a surjective effective étale-descent morphism then it is an étale-universal regular epimorphism.*

Proof. It follows from 2.4 and 3.2. □

This enables us to conclude that if p is effective $\mathbb{E}$-descent then the pseudo-inverse to the comparison functor $\Phi^p_{\mathbb{E}}\colon \mathbb{E}(B) \to Des_{\mathbb{E}}(p)$ is defined by restricting the left adjoint Ψ^p of Φ^p to $\mathbb{E}(B)$.

PROPOSITION 3.4. *For the class $\mathbb{E}$ of étale morphisms, the surjective map p is effective $\mathbb{E}$-descent if and only if the adjunction $\Psi^p \dashv \Phi^p\colon \mathcal{T}op/B \to Des(p)$ restricts to an equivalence between $\mathbb{E}(B)$ and $Des_{\mathbb{E}}(p)$.*

Proof. We have just to prove that Ψ^p restricted to $\mathbb{E}(B)$ has its image in $Des_{\mathbb{E}}(p)$ if p is effective étale-descent. If $(C,\gamma;\xi) \in Des_{\mathbb{E}}(p)$ and $(M,\mu) \in \mathbb{E}(B)$ is such that $\Phi^p_{\mathbb{E}}(M,\mu) = (E \times_B M, \widetilde{\pi_1}; \varphi) \cong (C,\gamma;\xi)$

$$\begin{array}{ccc} E \times_B M & \xrightarrow{\widetilde{\pi_2}} & M \\ {\scriptstyle \widetilde{\pi_1}}\downarrow & & \downarrow{\scriptstyle \mu} \\ E & \xrightarrow[p]{} & B \end{array}$$

then $\widetilde{\pi_2}$ is a regular epimorphism because p is an $\mathbb{E}$-universal regular epimorphism. Thus, it is the coequalizer of its kernel pair (π_2, ξ) and so $\Psi^p(C,\gamma;\xi) = (M,\mu) \in \mathbb{E}(B)$. □

We note that 3.4 still holds for classes $\mathbb{E}$ and morphisms p such that $\mathbb{E}^*(p)$ is contained in the class of regular epimorphisms of an arbitrary category with pullbacks and coequalizers as considered in the previous section.

THEOREM 3.5. *A surjective morphism p is effective étale-descent if and only if p is an étale-universal regular epimorphism and for each descent situation defining Q*

$$\begin{array}{ccccc} E \times_B C & \overset{\pi_2}{\underset{\xi}{\rightrightarrows}} & C & \xrightarrow{q} & Q \\ & & {\scriptstyle \gamma}\downarrow & & \downarrow{\scriptstyle \delta} \\ & & E & \xrightarrow[p]{} & B \end{array}$$

δ is a local homeomorphism if γ is a local homeomorphism.

Proof. If p is an $\mathbb{E}$-universal regular epimorphism, for $\mathbb{E}$ the class of étale maps, then, by 1.6 in [5], we conclude that p is $\mathbb{E}$-descent.

Since $\alpha = \langle \gamma, q \rangle\colon C \to E \times_B Q$, the unit of the adjunction $\Psi^p \dashv \Phi^p$, is a bijective map for each $(C,\gamma;\xi)$, if the above transferability condition holds then

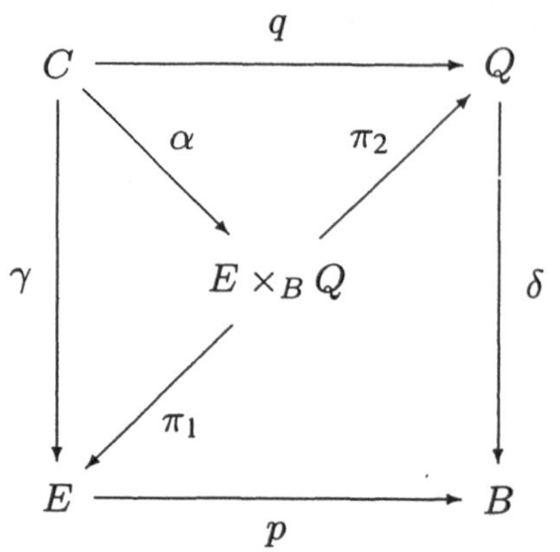

$\pi_1 \in \mathbb{E}$, by the stability under pullback of $\mathbb{E}$, and $\alpha \in \mathbb{E}$ because $\pi_1 \cdot \alpha = \gamma$ and γ belongs to $\mathbb{E}$, by hypothesis. Therefore, α is an homeomorphism, since it is a continuous open bijective map, and so we complete the proof that the two conditions are sufficient. Their necessity follows from 3.3 and 3.4. □

Effective global-descent morphisms are effective étale-descent ([5, 4.7]). The converse is not true: if p is a surjective map which is of effective étale-descent type then its compositions from the right with embeddings i: $B \to B'$, such that i^*: $\mathcal{O}(B') \to \mathcal{O}(B)$ are bijective, are effective étale-descent morphisms. Indeed, by Proposition 4.3 in [5], the effective étale-descent morphisms are exactly the composition on the right of effective étale quotients with such embeddings.

We present a counterexample to prove that the converse also fails for surjective maps.

PROPOSITION 3.6. *The class of effective global-descent maps is strictly contained in the class of surjective effective étale-descent maps.*

Proof. For the topological spaces

$$E = (\{e, e', e_1, e_2\}, \{\emptyset, E, \{e, e_1\}, \{e', e_2\}\}),$$

$$B = (\{b, b_1, b_2\}, \{\emptyset, B\}).$$

We consider the continuous morphism p: $E \to B$ defined by $p(e) = p(e') = b$ and $p(e_i) = b_i$ for $i = 1, 2$. Then p is an étale-universal regular epimorphism since it is a universal one. It remains to be proved that the transferability condition holds.

First we observe that, for $(C, \gamma; \xi) \in Des_{\mathbb{E}}(p)$, not only $\gamma^{-1}(e)$ and $\gamma^{-1}(e')$ are homeomorphic subspaces of $E \times_B C$, but they are also homeomorphic to the fibres $\gamma^{-1}(e_i)$ for $i = 1, 2$, because γ is a local homeomorphism. Indeed, it is enough to notice that, for each $x \in \gamma^{-1}(e)$, there is an open set U_x in C and an homeomorphism

$$\gamma_x\colon\ U_x \cong \gamma(U_x) = \{e, e_1\}$$

and that, if $U_x = \{x, d\}$ and $U_y = \{y, d'\}$, with $x, y \in \gamma^{-1}(e)$ and $d, d' \in \gamma^{-1}(e_1)$, then $d \neq d'$ whenever $x \neq y$. For $q = coeq(\pi_2, \xi)$ and $c_i \in \gamma^{-1}(e)$, let

$$q(c_i) = q(\xi(e', c_i)) = q(c_i')$$

be denoted by $[c_i]$. Analogously, for $d_i \in \gamma^{-1}(e_1)$ and $f_i \in \gamma^{-1}(e_2)$ with $\{c_i, d_i\} \cong \gamma(\{c_i, d_i\}) = \{e, e_1\}$ and $\{c', f_i\} \cong \gamma(\{c_i', f_i\}) = \{e', e_2\}$, we take $q(d_i) = [d_i]$ and $q(f_i) = [f_i]$.

Since the q-saturated open subsets of C are

$$W_i = \{c_i, d_i\} \cup \{c_i', f_i\} = q^{-1}\{[c_i], [d_i], [f_i]\},$$

then Q is the disjoint sum $\coprod\{B_x; x \in \gamma(e)\}$ of copies of the space B and δ is the map induced by the identities $B_x \to B$, for each $x \in \gamma^{-1}(e)$. Thus, δ is a local homeomorphism and so p is an effective étale-descent map.

To prove that p is not an effective global-descent map it is enough to find a descent situation defining Q

$$\begin{array}{ccccc} E \times_B C & \overset{\pi_2}{\underset{\xi}{\rightrightarrows}} & C & \xrightarrow{\;q\;} & Q \\ & & \downarrow{\scriptstyle \gamma} & & \downarrow{\scriptstyle \delta} \\ & & E & \xrightarrow[\;p\;]{} & B \end{array}$$

for which the square is not a pullback (see remarks following 2.2). Let C be the topological space with the same underlying set as E, whose topology has the base $\{\mathcal{O}(E) \cup \{e\}\}$. Then, for γ the identity function and $\xi(e, x) = \pi_1(e, x) = e$, $(C, \gamma; \xi) \in Des(p)$. Indeed, ξ is continuous because

$$\xi^{-1}(e) = \{(e, e), (e, e')\} = \{e, e_1\} \times_B \{e, e', e_2\}$$

is an open subset of $E \times_B C$, and the equalities

$$\gamma \cdot \xi = \pi_1, \qquad \xi \cdot 1 \times_B \xi = \xi \cdot 1 \times_B \pi_2$$

hold. This descent situation defines $Q = B$, with δ the identity and $q = p$ at the level of sets. Thus the square is not a pullback in $\mathcal{T}op$. □

References

1. Bénabou, J. and Roubaud, J.: Monades et descente, *C.R. Acad. Sc.* **270** (1970), 96–98.
2. Day, B. J. and Kelly, G. M.: On topological quotient maps preserved by pullbacks, *Proc. Cambridge Phil. Soc.* **67** (1970), 553–558.
3. Kock, A.: Fibre bundles in general categories, *J. Pure Appl. Algebra* **56** (1989), 233–245.
4. Janelidze, G. and Tholen, W.: How algebraic is the change-of-base functor?, in *Lect. Notes in Math.* **1448**, Springer, Berlin, 1991, pp. 157–173.
5. Janedidze, G. and Tholen, W.: Facets of descent I, *Applied Cat. Struct.* **2** (1994), 245–281.

6. Janelidze, G. and Tholen, W.: Facets of descent II, *Applied Cat. Struct.* (to appear).
7. Moerdijk, I.: Descent theory for toposes, *Bull. Soc. Math. Belgique* **41** (1989), 373–391.
8. Reiterman, J., Sobral, M., and Tholen, W.: Composites of effective descent maps, *Cahiers Topologie Géom. Differentiele Catégoriques* **XXXIV**(3) (1993), 193–207.
9. Reiterman, J. and Tholen, W.: Effective descent maps of topological spaces, *Topology Appl.* **57** (1994), 53–69.
10. Sobral, M. and Tholen, W.: Effective descent morphisms and effective equivalence relations, in *CMS Conference Proceedings*, Vol. 13, AMS, Providence, RI, 1992, pp. 421–433.

Applied Categorical Structures **4**: 107–120, 1996.

The Pullback Closure Operator and Generalisations of Perfectness

DAVID HOLGATE*
Department of Mathematics and Applied Mathematics, University of Cape Town, Rondebosch 7700, South Africa (e-mail:holgate@maths.uct.ac.za)

(Received: 18 November 1994; accepted: 27 September 1995)

Abstract. A categorical closure operator induced via pullback by a pointed endofunctor is introduced. Various notions of a perfect morphism relative to a pointed endofunctor and the induced closure are then considered. The main result explores how these notions are interrelated, linking also with earlier notions of perfectness.

Mathematics Subject Classifications (1991). Primary: 18A20; secondary: 18A32, 18B30, 54B30, 54C10.

Key words: pointed endofunctor, closure operator, perfect morphism.

1. Introduction

Since their formal introduction in [8], categorical closure operators have been enjoying ever increasing usage, particularly in generalising topological concepts to abstract categories. In this paper we introduce a closure operator and apply it to the problem of generalising to an abstract category the topological notion of a perfect continuous map. Thus far categorical investigations into perfectness have been twofold, using factorisation structures and more recently closures. This new closure illuminates some links between these approaches. Also, the results are not restricted to constructs as most closure related investigations of perfectness have been until now.

As is highlighted later, some of the ideas in this paper are thanks to my PhD supervisor Professor Guillaume Brümmer. I am also grateful to Walter Tholen, Manuel Clementino, Eraldo Giuli and Dikran Dikranjan for valuable discussion on these topics, as well as the referee whose comments helped improve the exposition.

* The author acknowledges financial support from the University of Cape Town, from the Foundation for Research Development through the Categorical Topology Research Group at the University of Cape Town, and from the University of L'Aquila.

2. Preliminaries

Categorical notation and terminology is as in [1]. We work in a category $\mathbf{X}$. The class $\mathcal{M} \subseteq Mor\,\mathbf{X}$ is the second component of a factorisation structure $(\mathbf{E}, \mathcal{M})$ for sinks in $\mathbf{X}$. It is fixed throughout and represents *subobjects* in $\mathbf{X}$. $(\mathcal{E}, \mathcal{M})$ will denote the restriction of this factorisation structure to morphisms in $\mathbf{X}$. The strength of the sink factorisation is such that (amongst other things) $\mathcal{M} \subseteq Mono\,\mathbf{X}$ and pullbacks of $\mathcal{M}$-morphisms along arbitrary $\mathbf{X}$-morphisms exist and are again in $\mathcal{M}$. We make regular use of these properties.

The pair (R, r) will always denote a pointed endofunctor on $\mathbf{X}$, that is, an endofunctor R: $\mathbf{X} \to \mathbf{X}$ and a natural transformation r: $1_{\mathbf{X}} \to R$.

We talk of a *closure operator* C on $\mathbf{X}$ with respect to $\mathcal{M}$ in the sense introduced in [8]. In particular, any $m \in \mathcal{M}$ has a canonical factorisation:

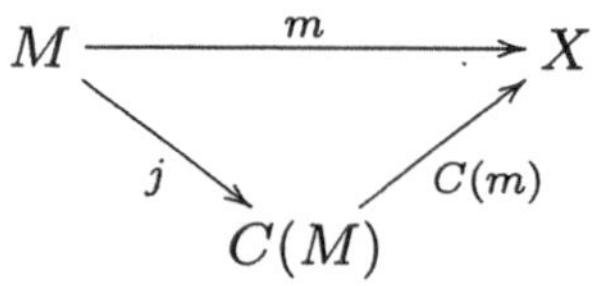

A subobject

$$M \xrightarrow{m} X \in \mathcal{M}$$

is *C-closed* in X if j is an isomorphism. It is *C-dense* if $C(m)$ is an isomorphism.

A closure operator, C, is said to be *weakly hereditary* if any $m \in \mathcal{M}$ is dense in its closure (i.e. in the above diagram j is C-dense), while C is called *idempotent* if for any $m \in \mathcal{M}$,

$$C(C(m)) \cong C(m)$$

(i.e. $C(m)$ is C-closed). Weakly hereditary, idempotent closures are in one to one correspondence with morphism factorisation structures $(\mathcal{F}, \mathcal{N})$ on $\mathbf{X}$ for which $\mathcal{N} \subseteq \mathcal{M}$. (More information about closure operators can be found in [8, 10] and [16].)

We also make use of the following fairly common notations. For a class $\mathcal{A} \subseteq Ob\,\mathbf{X}$, $\mathcal{A}_\perp$ denotes all $\mathbf{X}$-morphisms that are orthogonal to every $\mathcal{A}$-object. For any class $\mathcal{A} \subseteq Mor\,\mathbf{X}$, $\mathcal{A}^\downarrow$ is the class of all morphisms g for which if we have a commutative square $vf = gu$ with $f \in \mathcal{A}$ then there exists a unique d such that

$$df = u \quad \text{and} \quad gd = v.$$

Dually we will speak of the class $\mathcal{A}^\uparrow$.

3. Pullback Closure

DEFINITION 1. Let (R, r) be a pointed endofunctor on $\mathbf{X}$. For $M \xrightarrow{m} X \in \mathcal{M}$, construct the diagram below, where $ne = Rm$ is the $(\mathcal{E}, \mathcal{M})$-factorisation of Rm and $\overline{m}$ is the pullback of n along r_X.

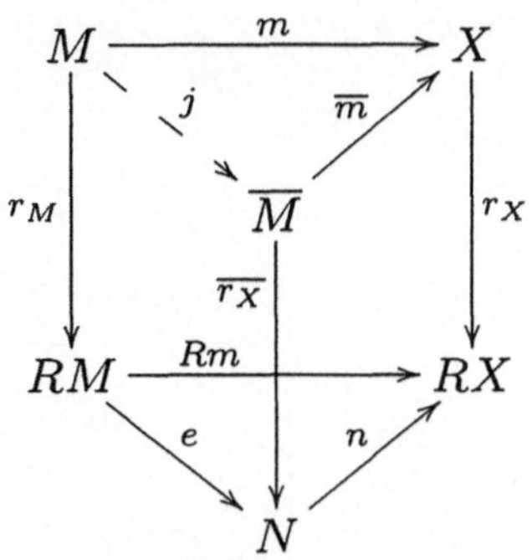

Put $\Phi_{(R,r)}(m) := \overline{m}$.

PROPOSITION 2. *If (R, r) is any pointed endofunctor on $\mathbf{X}$ then $\Phi_{(R,r)}$ is a closure operator on $\mathbf{X}$ with respect to $\mathcal{M}$.*

The proof is routine and is left to the reader.

DEFINITION 3. We use the following notions associated with a pointed endofunctor (R, r) on $\mathbf{X}$.

(1) $\mathrm{Fix}(R, r) = \{X \in Ob\,\mathbf{X} \mid r_X\colon X \to RX$ is an isomorphism$\}$.

(2) $\Sigma_R = \{f \in Mor\,\mathbf{X} \mid Rf$ is an isomorphism$\}$, in the notation of [3].

(3) We say that (R, r) is *idempotent* if $RX \in \mathrm{Fix}(R, r)$ for every $X \in Ob\,\mathbf{X}$.

(4) [Brümmer and Giuli] (R, r) will be called *direct* if for any $X \xrightarrow{f} Y \in Mor\,\mathbf{X}$ the pullback $(P, (p, q))$ below of Rf along r_Y exists and the uniquely induced morphism $u\colon X \to P$ is in Σ_R.

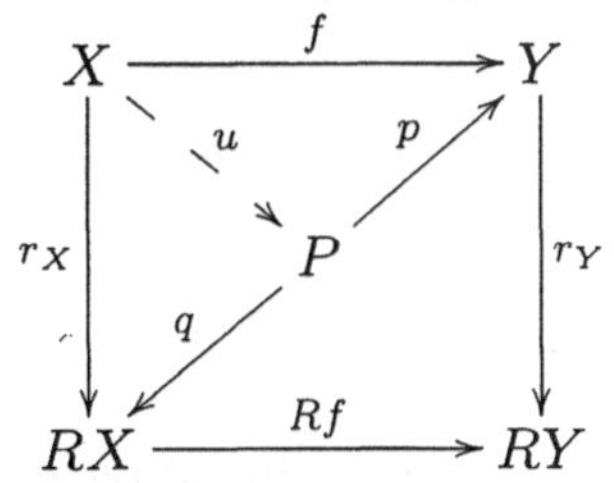

$(*)$

LEMMA 4. *If the pullback in diagram $(*)$ above exists for any $f\colon X \to Y$ in $\mathbf{X}$ and the induced u is such that $Ru \in \mathcal{E}$, then if $\mathcal{E}$ is closed under pullback, $\Phi_{(R,r)}$ is weakly hereditary.*

Proof. Take $M \xrightarrow{m} X \in \mathcal{M}$ and form its $\Phi_{(R,r)}$-closure. By assumption we can take the pullback $(P, (p, q))$ of Rm along r_X. As always, $ne = Rm$ is the $(\mathcal{E}, \mathcal{M})$ factorisation needed to construct $\Phi_{(R,r)}(m)$.

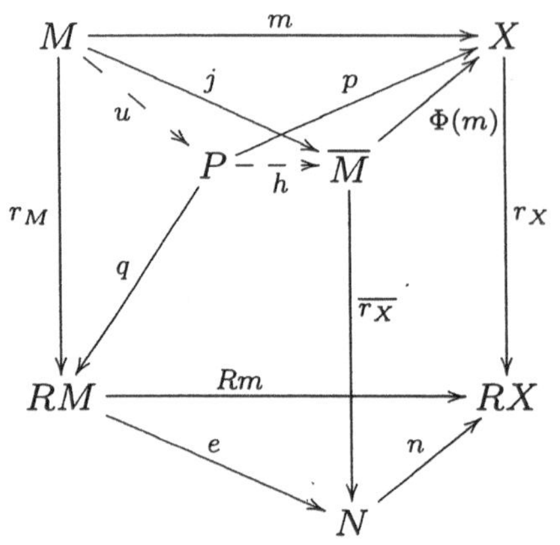

Since $(\overline{M}, (\Phi_{(R,r)}(m), \overline{r_X}))$ is a pullback, there is a unique h: $P \to \overline{M}$ such that

$$\Phi_{(R,r)}(m)h = p \quad \text{and} \quad \overline{r_X}h = eq.$$

Considering the two pullbacks $(\overline{M}, (\Phi_{(R,r)}(m), \overline{r_X}))$ and $(P, (p, q))$ we see that h is the pullback of e along $\overline{r_X}$ and so since $\mathcal{E}$ is closed under pullback, $h \in \mathcal{E}$.

Now, $\Phi_{(R,r)}(m)j = m = pu = \Phi_{(R,r)}(m)hu$ so $hu = j$. But since $Ru \in \mathcal{E}$ it follows that u is $\Phi_{(R,r)}$-dense, and then since $h \in \mathcal{E}$, $j = hu$ is $\Phi_{(R,r)}$-dense too. From this we conclude that $\Phi_{(R,r)}$ is weakly hereditary. □

COROLLARY 5. *If $\mathcal{E}$ is closed under pullback and (R, r) is direct, then $\Phi_{(R,r)}$ is weakly hereditary.*

LEMMA 6. *If (R, r) is idempotent, and for any $m \in \mathcal{M}$ the $(\mathcal{E}, \mathcal{M})$-factorisation below of Rm gives $N \in$ Fix(R, r), then $\Phi_{(R,r)}$ is idempotent.*

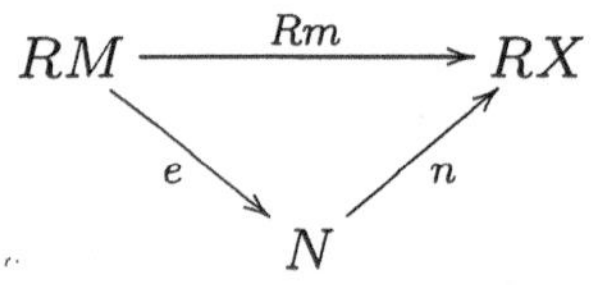

Proof. Since (R, r) is idempotent, RM, RX and N in the above diagram are all in Fix(R, r). Thus it is easy to see that n is $\Phi_{(R,r)}$-closed. But for any closure operator C, C-closed $\mathcal{M}$-morphisms are closed under pullback, so $\Phi_{(R,r)}(m)$ being the pullback of n along r_X is $\Phi_{(R,r)}$-closed and hence $\Phi_{(R,r)}$ is idempotent. □

COROLLARY 7. *If (R, r) is a reflection to a full subcategory of* **X** *which is closed under $\mathcal{E}$-images or $\mathcal{M}$-subobjects, then $\Phi_{(R,r)}$ is idempotent.*

EXAMPLE 8. The following are some $\Phi_{(R,r)}$ for well known (R, r).

(1) If (R, r) is the pointed endofunctor on $\underline{\text{TYCH}}$ induced by the Stone–Čech compactification β: $\underline{\text{TYCH}} \to \underline{\text{HCOMP}}$, then the induced $\Phi_{(R,r)}$ is the usual topological closure.

(2) If $(-)_0$: $\underline{\text{TOP}} \to \underline{\text{TOP}_0}$ is the $\underline{\text{TOP}_0}$ reflection, then for the induced pointed endofunctor (R, r), $\Phi_{(R,r)}$ is the operator that acts as follows: For $M \subseteq X \in \underline{\text{TOP}}$, $\Phi_{(R,r)}(M) = \{x \in X \mid \exists\, m \in M$ for which $cl_X\{m\} = cl_X\{x\}\}$. This is idempotent, hereditary and fully additive but is not the b-closure as we might have expected.
   This example shows that the regular closure induced by Fix(R, r) does not coincide with $\Phi_{(R,r)}$ in general, and certainly $\Phi_{(R,r)}$ is not always regular.

(3) The composition of the $\underline{\text{TOP}_0}$ reflection and the sobrification of a $\underline{\text{TOP}_0}$ space gives a reflection, σ: $\underline{\text{TOP}} \to \underline{\text{SOB}}$. If (R, r) is the induced pointed endofunctor on $\underline{\text{TOP}}$, then $\Phi_{(R,r)}$ is the *b*-closure.

(4) The torsion free reflection of an abelian group, T: $\underline{\text{ABGRP}} \to \underline{\text{TFAB}}$ induces the pullback closure Φ on $\underline{\text{ABGRP}}$ and for a subgroup $M \subseteq G \in \underline{\text{ABGRP}}$, $g \in \Phi(M) \Leftrightarrow \exists h \in M$ and $1 \leqslant n < \infty$ such that $g^n = h^n$. This is an idempotent, hereditary and fully additive closure.

(5) If R: $\underline{\text{GRP}} \to \underline{\text{ABGRP}}$ is the abelian reflection and Φ is the induced pullback closure, then $g \in \Phi(M)$ for a subgroup $M \subseteq G \in \underline{\text{GRP}} \Leftrightarrow \exists h \in M$ such that $gC = hC$ where C is the commutator subgroup of G. This closure is idempotent and fully additive, but not weakly hereditary.

REMARK 9. This pullback closure has not been systematically studied in the theory of closure operators. In [7], Definition 3.5, the modification of a closure operator along a reflection is introduced. If (R, r) is an $\mathcal{E}$-reflection and the discrete closure is being modified, then one has the pullback closure. We understand this concept will be taken further in the forthcoming book [11]. Recently in [2] the *splitting closure* was introduced. Again, this coincides with the pullback closure if (R, r) is an $\mathcal{E}$-reflection (hence in Example 8(2), (4) and (5)), but not in general.

If Fix(R, r) is closed under $\mathcal{M}$-subobjects, then $\Phi_{(R,r)}$ restricted to Fix(R, r) is discrete. In particular this is true if (R, r) is an $\mathcal{E}$-reflection to a subcategory **A** of **X**. In such a case, the regular closure induced by **A** coincides with $\Phi_{(R,r)}$ iff the regular closure is discrete when restricted to **A**.

As annotated, the concept of directness is due to Brümmer and Giuli. It is being investigated further in their joint work with ourselves. In [3] the term "simple" is used for a reflection in a finitely complete category that has the directness property.

4. Perfect Morphisms

In the 1970's a number of investigations were made into generalisations of the topological notion of a perfect map (cf. [13, 14, 17, 18] and [19]). The theory was tidily settled with the notion of a *perfect class of morphisms* $\mathcal{P}$ being one that is of the form $\mathcal{P} = (Epi\,\mathbf{X} \cap \mathcal{A}_{\perp})^{\downarrow}$ for some $\mathcal{A} \subseteq Ob\,\mathbf{X}$.

More recently, there have been efforts to generalise perfect maps by looking at their closure properties. In topology we have that a map $f\colon X \to Y$ is perfect iff f is closed and preimages of points are compact iff for any space Z, $f \times 1_Z\colon X \times Z \to Y \times Z$ is closed. Also, a space X is compact iff for any space Z, the natural projection $\pi_2\colon X \times Z \to Z$ is closed.

In [15] these results were explored in constructs with the use of morphism factorisation structures. Using closure operators they can easily be stated in categorical language, and a fair theory is developing around generalisations of compactness and perfectness (cf. [4, 5, 6, 9] and [12]). It seems that Φ is an appropriate closure to use in linking these different approaches to categorical perfectness.

One useful result in factorisation theory is Theorem 3.3 from [3] which can be summarised as follows.

PROPOSITION 10. *If* $(\eta, \epsilon) : S \dashv T\colon \mathbf{A} \to \mathbf{X}$ *is an adjoint situation and* $(R, r) = (TS, \eta)$ *then:*

(i) $\Sigma_R = \Sigma_S = (T(Mor\,\mathbf{A}))^{\uparrow}$.

(ii) *If* $\mathbf{X}$ *is finitely well complete then* $(\Sigma_R, \Sigma_R^{\downarrow})$ *is a factorisation structure for morphisms in* $\mathbf{X}$.

In the situations where these factorisations exist, the class $\Sigma_R^{\downarrow}$ provides one possible generalisation of perfect maps. If we look at Φ-closed $\mathcal{M}$-morphisms in this context, we begin to see the ties between $\Phi_{(R,r)}$ and perfectness.

PROPOSITION 11. *Let* $\Phi_{(R,r)}$ *be induced by* (R, r) *on* $\mathbf{X}$ *and let* $m \in \mathcal{M}$.

(i) m *is* $\Phi_{(R,r)}$*-closed* $\Rightarrow m \in \Sigma_R^{\downarrow}$.

(ii) *If* (R, r) *is direct and* $\mathcal{E}$ *is closed under pullback, then:* $m \in \Sigma_R^{\downarrow} \Rightarrow m$ *is* $\Phi_{(R,r)}$*-closed.*

Proof. (i) As was noted in the proof of Lemma 4, if for an $\mathbf{X}$-morphism f, $Rf \in \mathcal{E}$ then f is $\Phi_{(R,r)}$-dense. In particular $\Sigma_R \subseteq \{\Phi_{(R,r)}\textit{-dense morphisms}\}$. It is well known that $\{\Phi_{(R,r)}\textit{-closed morphisms}\} \subseteq \{\Phi_{(R,r)}\textit{-dense}\}^{\downarrow}$ (cf. [8] or [16]), and hence since $(-)^{\downarrow}$ reverses order we conclude that $\{\Phi_{(R,r)}\textit{-closed morphisms}\} \subseteq \{\Phi_{(R,r)}\textit{-dense morphisms}\}^{\downarrow} \subseteq \Sigma_R^{\downarrow}$.

(ii) Take $m \in \Sigma_R^{\downarrow} \cap \mathcal{M}$. Form both the pullback $(P, (p, q))$ of Rm along r_X and $\overline{m} = \Phi_{(R,r)}(m)$. u is the unique morphism such that $pu = m$ and $qu = r_M$ and h is the unique morphism such that $\overline{r_X}h = eq$ and $\overline{m}h = p$.

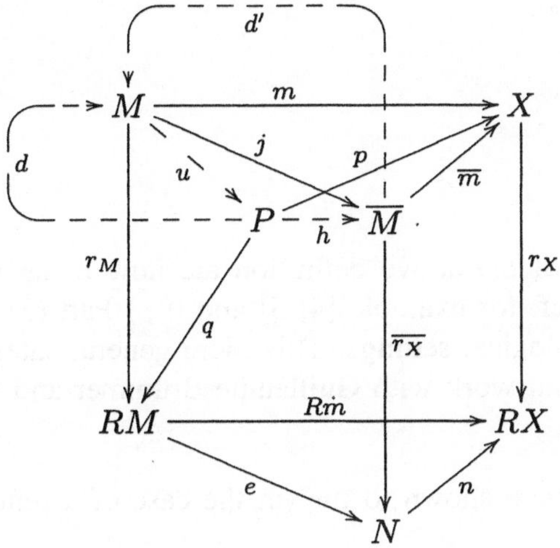

Since (R, r) is direct, $u \in \Sigma_R$, so because $pu = m \in \Sigma_R^{\downarrow}$ there is a diagonal d: $P \to M$ such that $md = p$ and $du = 1_M$. But $md = p = \overline{m}h$ and since $\mathcal{E}$ is closed under pullback, $h \in \mathcal{E}$. Thus by the $(\mathcal{E}, \mathcal{M})$ diagonalisation property there is a diagonal d': $\overline{M} \to M$ such that $md' = \overline{m}$ and $d'h = d$. Now $d'hu = du = 1_M$ and $md' \in \mathcal{M} \Rightarrow d' \in \mathcal{M}$ so d' is a retraction and a monomorphism, hence an isomorphism and m is $\Phi_{(R,r)}$-closed. □

REMARK 12. So $\Sigma_R^{\downarrow} \cap \mathcal{M} = \{\Phi_{(R,r)}\text{-closed}\} \cap \mathcal{M}$ if (R, r) is direct and $\mathcal{E}$ is closed under pullback. In this case the $\Phi_{(R,r)}$-closed subobjects are just those subobjects that we like to think of as "perfect". (In TOP the closed embeddings are just the perfect embeddings.)

$((\Sigma_R^{\downarrow} \cap \mathcal{M})^{\uparrow}, \Sigma_R^{\downarrow} \cap \mathcal{M})$ is always a factorisation structure for morphisms in **X**. In this case it induces the idempotent hull of $\Phi_{(R,r)}$ in **X**. So if (R, r) is induced by an adjunction (i.e. as in Proposition 10) and **X** is finitely well complete, then the factorisation structure $(\Sigma_R, \Sigma_R^{\downarrow})$ induces, along with $\mathcal{M}$, the idempotent hull of $\Phi_{(R,r)}$.

DEFINITION 13. (1) If C is a closure operator on **X** with respect to $\mathcal{M}$, then a morphism f: $X \to Y$ is called *C-preserving* if for any $M \xrightarrow{m} X \in \mathcal{M}$,

$$f(C(m)) \cong C(f(m)).$$

(Where $f(m)$ is the $\mathcal{M}$-component of the $(\mathcal{E}, \mathcal{M})$-factorisation of fm.)
(2) If **X** has finite products, an object X in **X** is called *C-compact* if for any

$Z \in Ob\,\mathbf{X}$ the projection morphism π_2: $X \times Z \to Z$ is C-preserving.
(3) A morphism f: $X \to Y$ will be called (R, r)-*perfect* if the square below is a pullback.

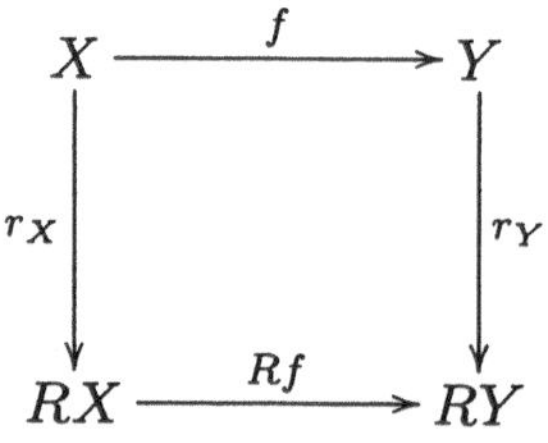

REMARK 14. Parts (1) and (2) of the above definition are now being widely used in closure operator theory (cf. for example [4, 5] and [9].) Part (3) has in the past been investigated in topological settings. This more general categorical formulation is the basis of our joint work with Guillaume Brümmer and Eraldo Giuli.

The following two propositions were shown to me (in the case of a reflection) by Guillaume Brümmer.

PROPOSITION 15. *If* $f \in Mor\mathbf{X}$ *is* (R, r)*-perfect then* $f \in \Sigma_R^{\downarrow}$.
Proof. Let f: $X \to Y$ be (R, r)-perfect and $A \xrightarrow{h} B \in \Sigma_R$ with morphisms u and v such that $vh = fu$.

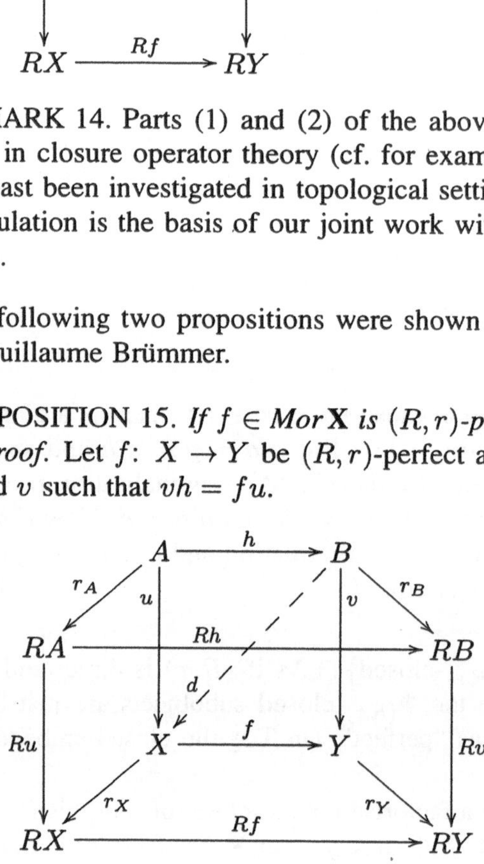

$RfRu(Rh)^{-1}r_B = Rvr_B = r_Yv$, so since we have a pullback there is a unique d: $B \to X$ such that $fd = v$ and $r_Xd = Ru(Rh)^{-1}r_B$. This gives $fdh = vh = fu$ and $r_Xdh = Ru(Rh)^{-1}r_Bh = Rur_A = r_Xu$, so since (f, r_X) is a monosource, $dh = u$ and d is a diagonal for the square $fu = vh$. It is a unique diagonal, since any other diagonal d' would give $R(d'h) = Ru \Rightarrow Rd' = Ru(Rh)^{-1} \Rightarrow Rd'r_B = Ru(Rh)^{-1}r_B \Rightarrow r_Xd' = Ru(Rh)^{-1}r_B$ and thus by the uniqueness condition on d, $d' = d$. Hence $f \in \Sigma_R^{\downarrow}$. □

PROPOSITION 16. *Let* (R, r) *in* $\mathbf{X}$ *be idempotent, then* (R, r) *is direct* $\Leftrightarrow$ $\mathbf{X}$ *has* $(\Sigma_R, (R, r)$*-perfect) factorisations.*

Proof.
⇒: For $f\colon X \to Y$ a morphism in $\mathbf{X}$, in the diagram $(*)$ of Definition 3, $u \in \Sigma_R$ so

$$\begin{aligned} Ruqu = r_P u &\Rightarrow R(Ruqu) = R(r_P u) \\ &\Rightarrow R(Ruq) = Rr_P \\ &\Rightarrow R(Ruq)r_P = (Rr_P)r_P \\ &\Rightarrow r_{RP}Ruq = r_{RP}r_P. \end{aligned}$$

But since (R, r) is idempotent r_{RP} is an isomorphism and we conclude that $Ruq = r_P$. Thus the square $Rpr_P = r_Y p$ is a pullback, and p is (R, r)-perfect. So $f = pu$ is a (Σ_R, (R, r)-perfect) factorisation.
⇐: In the diagram below, $ts = f$ is the (Σ_R, (R, r)-perfect) factorisation of f.

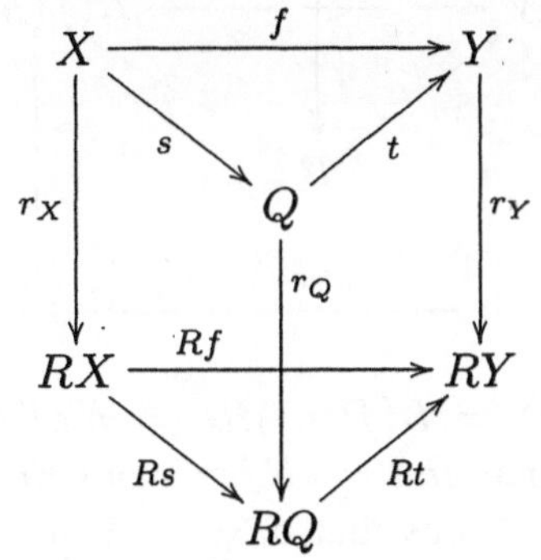

Rs is an isomorphism, so

$$r_Y t = Rf(Rs)^{-1} r_Q$$

is a pullback square (t is (R, r)-perfect). Thus t is the pullback of Rf along r_Y. Also s is the unique morphism such that $(Rs)^{-1} r_Q s = r_X$ and $ts = f$. Since $s \in \Sigma_R$, (R, r) is direct. □

COROLLARY 17. *If (R, r) is idempotent, then (R, r) is direct ⇔ (Σ_R, (R, r)-perfect) is a factorisation structure for morphisms in* $\mathbf{X}$.

Proof. If (R, r) is direct then by Proposition 16 $\mathbf{X}$ has (Σ_R, (R, r)-perfect) factorisations. Then by Proposition 15 above we have unique (Σ_R, (R, r)-perfect) diagonals. The reverse implication is clear from Proposition 16. □

PROPOSITION 18. *If $\mathcal{E}$ is closed under pullback and $Re \in \mathcal{E}$ for all $e \in \mathcal{E}$, then if $f \in Mor\,\mathbf{X}$ is (R, r)-perfect, f is $\Phi_{(R,r)}$-preserving.*

Proof. Let $f\colon X \to Y$ be (R, r)-perfect, and $M \xrightarrow{m} X \in \mathcal{M}$. We must show that

$$f(\Phi_{(R,r)}(m)) \cong \Phi_{(R,r)}(f(m)).$$

The constructions are shown in the diagram below. ($f(\overline{m}) = f(\Phi_{(R,r)}(m))$ and $\overline{f(m)} = \Phi_{(R,r)}(f(m))$.) The morphism h is induced by the pullback that forms $\overline{f(m)}$, so $\overline{f(m)}h = f(\overline{m})$. We must show that it is an isomorphism.

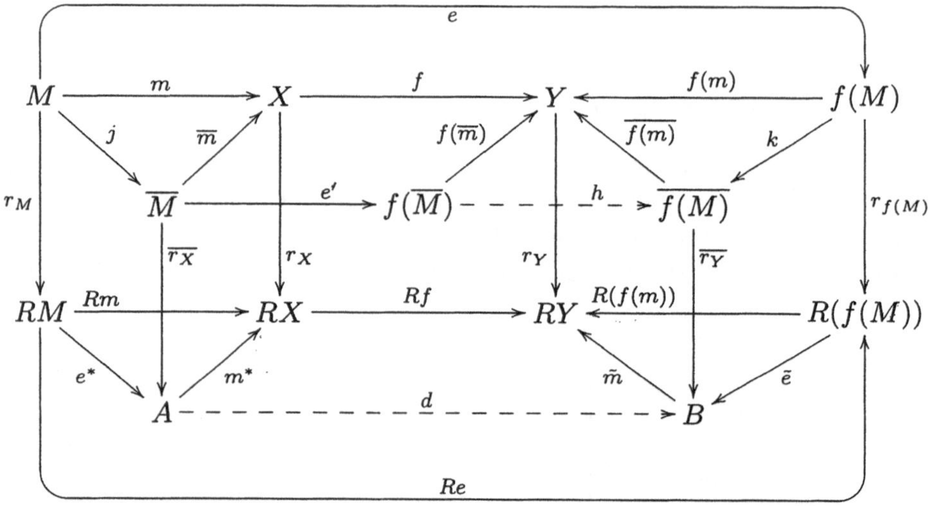

$Rfm^*e^* = RfRm = R(fm) = R(f(m)e) = R(f(m))Re = \tilde{m}\tilde{e}Re$ so there is an $(\mathcal{E},\mathcal{M})$-diagonal d: $A \to B$ such that $\tilde{m}d = Rfm^*$ and $de^* = \tilde{e}Re$. It is easy to see that $\tilde{m}d\overline{r_X} = \tilde{m}\overline{r_Y}he'$, and thus that $d\overline{r_X} = \overline{r_Y}he'$. But this is a pullback square. (Since f is (R,r)-perfect and both $(\overline{M},(\overline{m},\overline{r_X}))$ and $(\overline{f(M)},(\overline{f(m)},\overline{r_Y}))$ are pullbacks.) Also $de^* = \tilde{e}Re \in \mathcal{E} \Rightarrow d \in \mathcal{E}$, so since $\mathcal{E}$ is closed under pullback $he' \in \mathcal{E}$. This gives $h \in \mathcal{E}$, but $h \in \mathcal{M}$ so h is an isomorphism and the result follows. □

PROPOSITION 19. *Let* **X** *have finite products. If* (R,r) *is direct and idempotent and* $\{(R,r)\text{-perfect}\} \subseteq \{\Phi_{(R,r)}\text{-preserving}\}$ *then* $\mathrm{Fix}(R,r) \subseteq \{\Phi_{(R,r)}\text{-compact}\}$.

Proof. Let $X \in \mathrm{Fix}(R,r)$, then since r_X is an isomorphism, it is easy to see that if T is a terminal object in **X** then for the unique morphism t_X: $X \to T$, the square below is a pullback. Hence t_X: $X \to T$ is (R,r)-perfect.

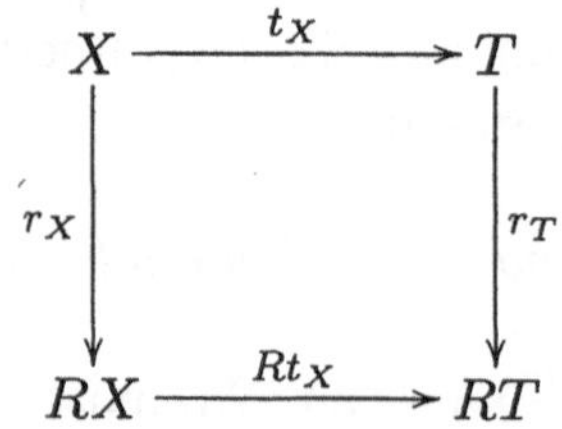

Since (R,r) is direct and idempotent, (R,r)-perfect morphisms are closed under products and so $(\pi_2\colon X \times Z \to Z) \cong (t_X \times 1_Z\colon X \times Z \to T \times Z)$ is

(R,r)-perfect. But then by assumption π_2: $X \times Z \to Z$ is $\Phi_{(R,r)}$-preserving and so X is $\Phi_{(R,r)}$-compact. □

LEMMA 20. *Let* **X** *have finite products. For* $f \in Mor\mathbf{X}$, $f \times 1_Z$ *is* (R,r)*-perfect for all* $Z \in Ob\mathbf{X} \Rightarrow f \times f$ *is* (R,r)*-perfect.*

Proof. The proof is routine, using the observation that for any f: $X \to Y$ in **X**, $f \times f \cong (f \times 1_Y)h(f \times 1_X)$, where h: $Y \times X \to X \times Y$ is the natural isomorphism between the two products. □

LEMMA 21. *Let* **X** *have finite products. For* $f \in Mor\mathbf{X}$, $f \times f$ *is* (R,r)*-perfect* $\Rightarrow f$ *is* (R,r)*-perfect.*

Proof. Straightforward. □

The final Lemma is well known (in a slightly different setting cf. [9], Proposition 3.2).

LEMMA 22. *For* $g, f \in Mor\mathbf{X}$ *and* C *any closure operator on* **X** *with respect to* $\mathcal{M}$, *if* gf *is* C*-preserving and* $g \in \mathcal{M}$ *then* f *is* C*-preserving.*

THEOREM 23. *For* f: $X \to Y$ *in* **X** *the following implications hold.*

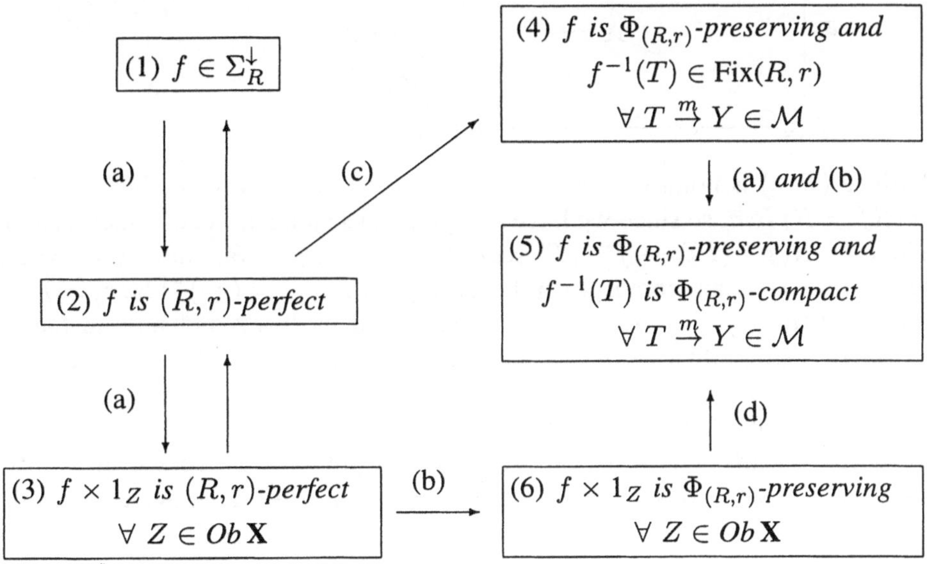

For (3) *and* (6) *we assume that* **X** *has finite products and for* (4) *and* (5) *that* **X** *has a terminal object* T. $f^{-1}(T)$ *is the pullback object formed when taking the pullback of* $T \overset{m}{\longrightarrow} Y$ *along* f. *The letters indicate the following conditions which are sufficient to ensure the validity of the associated implication.*

(a) (R,r) *is direct and idempotent.*

(b) $\mathcal{E}$ *is closed under pullback and* $Re \in \mathcal{E}$ *for all* $e \in \mathcal{E}$.

(c) (b) *and* (R, r) *is pointwise epimorphic.*
(d) $\mathcal{E}$ *is closed under pullback,* (R, r) *is direct and idempotent and every* $T \xrightarrow{m} Y \in \mathcal{M}$ *is* $\Phi_{(R,r)}$*-closed.*

Proof.
(1) ⇒ (2): Under (a), Corollary 17 gives that $\{(R, r)\text{-perfect}\} = \Sigma_R^{\downarrow}$.
(2) ⇒ (1): Proposition 15.
(2) ⇒ (3): Since under (a) $\{(R, r)\text{-perfect}\} = \Sigma_R^{\downarrow}$, $\{(R, r)\text{-perfect}\}$ is closed under products. Thus since both f and (for any Z) 1_Z are (R, r)-perfect, so is $f \times 1_Z$.
(3) ⇒ (2): Combine Lemmas 20 and 21.
(2) ⇒ (4): Under (c), Proposition 18 holds, so f is $\Phi_{(R,r)}$-preserving. Let $T \xrightarrow{m} TY \in \mathcal{M}$ and consider the following diagram. $((f^{-1}(T), (\bar{f}, \overline{m}))$ is the pullback of m along f.) We want to show that $r_{f^{-1}(T)}$ is an isomorphism.

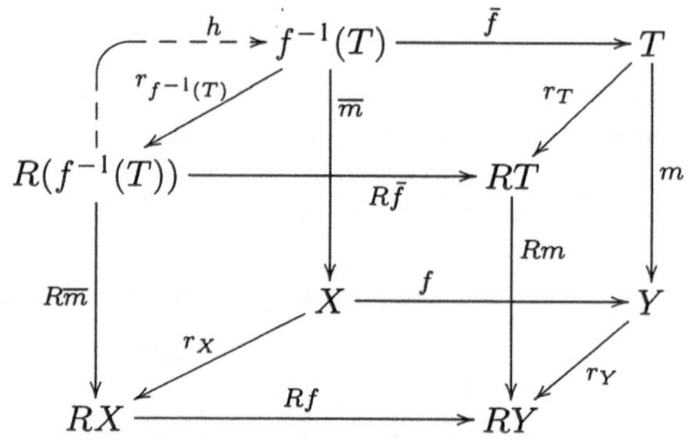

Since (R, r) is pointwise epimorphic, $T \in \text{Fix}(R, r)$. Now, $r_Y m (r_T)^{-1} R\bar{f} = RmR\bar{f} = Rf R\overline{m}$, so since we have two pullbacks alongside each other, there is a unique h: $R(f^{-1}(T)) \to f^{-1}(T)$ such that $\bar{f}h = (r_T)^{-1}R\bar{f}$ and $r_X\overline{m}h = R\overline{m}$. Hence $hr_{f^{-1}(T)}$is an isomorphism because $\bar{f}(hr_{f^{-1}(T)} = (r_T)^{-1}R\bar{f}r_{f^{-1}(T)} = \bar{f}$ and $r_X\overline{m}(hr_{f^{-1}(T)}) = R\overline{m}r_{f^{-1}(T)} = r_X\overline{m}$ and the back and bottom faces of the diagram are pullback squares. Thus $r_{f^{-1}(T)}$ is an epimorphism and a section, hence an isomorphism.
(3) ⇒ (6): Apply Proposition 18 under condition (b).
(6) ⇒ (5): Clearly f is $\Phi_{(R,r)}$-preserving, just put $Z = T$. To see that $f^{-1}(T)$ is $\Phi_{(R,r)}$-compact, consider the diagram below for $T \xrightarrow{m} Y \in \mathcal{M}$ and any $Z \in Ob\,\mathbf{X}$.

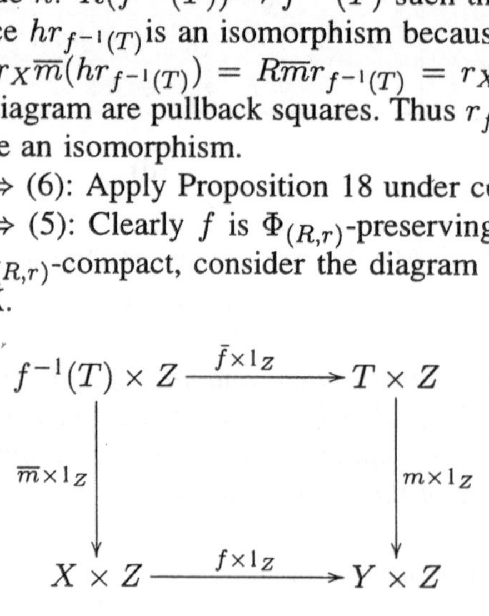

This commutes since $f\overline{m} = m\bar{f}$. By assumption, m is $\Phi_{(R,r)}$-closed, so its pullback $\overline{m}$ along f is $\Phi_{(R,r)}$-closed, and thus $\overline{m} \times 1_Z$ is also $\Phi_{(R,r)}$-closed. Since by Corollary 5 $\Phi_{(R,r)}$ is weakly hereditary, we conclude that $\overline{m} \times 1_Z$ is $\Phi_{(R,r)}$-preserving. But $f \times 1_Z$ is $\Phi_{(R,r)}$-preserving, thus $(f \times 1_Z)(\overline{m} \times 1_Z)$ is $\Phi_{(R,r)}$-preserving and hence $(m \times 1_Z)(\bar{f} \times 1_Z)$ is $\Phi_{(R,r)}$-preserving. So by Lemma 22, since $m \times 1_Z \in \mathcal{M}$, $\bar{f} \times 1_Z$ is $\Phi_{(R,r)}$-preserving. But $\bar{f} \times 1_Z \cong \pi_2\colon f^{-1}(T) \times Z \to Z$, so $f^{-1}(T)$ is $\Phi_{(R,r)}$-compact.
(4) ⇒ (5): Apply Proposition 19 since under (b), $\{(R,r)\text{-perfect}\} \subseteq \{\Phi_{(R,r)}\text{-preserving}\}$. □

EXAMPLE 24. (1) If (R,r) is induced by the Stone–Čech compactification β: TYCH → HCOMP then all the conditions in Theorem 23 are satisfied. As is well known, $\{\Phi_{(R,r)}\text{-compact}\} = \mathrm{Fix}(R,r) =$ HCOMP and all the notions of perfectness coincide.

(2) For the reflection $(-)_0$: TOP → TOP$_0$ and induced endofunctor (R,r) of Example 8(2), all conditions except (d) are satisfied. (Every $\{\bullet\} \xrightarrow{m} X$ is $\Phi_{(R,r)}$-closed iff $X \in$ TOP$_0$.) A continuous map $f\colon X \to Y$ is (R,r)-perfect iff f is $\Phi_{(R,r)}$-closed and preimages of points are T_0.

(3) Consider the torsion free reflection T: ABGRP → TFAB, and the induced endofunctor (R,r) and closure $\Phi_{(R,r)}$. Conditions (a), (b) and (c) hold. A homomorphism $f\colon G \to H$ is (R,r)-perfect iff f is $\Phi_{(R,r)}$-closed and $f^{-1}(e_H)$ is torsion free. Here e_G is the neutral element in a group G.

(4) For (R,r) and $\Phi_{(R,r)}$ induced by the Abelian reflection R: GRP → ABGRP conditions (b) and (c) are satisfied, but (R,r) is not direct. A homomorphism $f\colon G \to H$ is (R,r)-perfect iff f is $\Phi_{(R,r)}$-closed and $f^{-1}(e_H) \cap C = e_G$, where C is the commutator subgroup of G.

References

1. Adámek, J., Herrlich, H., and Strecker, G. E.: *Abstract and Concrete Categories*, Wiley, Inc., New York, 1990.
2. Brümmer, G. C. L. and Giuli. E.: Splitting operators, Unpublished manuscript, 1993.
3. Cassidy, C., Hébert, M., and Kelly, G. M.: Reflective subcategories, localisations and factorisation systems, *J. Austral. Math. Soc. Ser. A* **38** (1985), 287–329.
4. Castellini, G.: Compact objects, surjectivity of epimorphisms and compactifications, *Cahiers Topologie Géom. Différentielle Catégoriques* **31** (1990), 53–65.
5. Castellini, G.: Regular closure operators and compactness, *Cahiers Topologie Géom. Différentielle Catégoriques* **33** (1992), 21–31.
6. Clementino, M., Giuli, E., and Tholen, W.: Compact objects and perfect morphisms, Preprint, 1995.
7. Dikranjan, D.: Semiregular closure operators and epimorphisms in topological categories, *Suppl. Rend. Circolo Mat. Palermo (Serie II)* **29** (1992), 105–160.
8. Dikranjan, D. and Giuli, E.: Closure Operators I, *Topology Appl.* **27** (1987), 129–143.
9. Dikranjan, D. and Giuli, E.: C-perfect morphisms and C-compactness, Preprint, 1990.

10. Dikranjan, D., Giuli, E., and Tholen, W.: Closure Operators II, in *Categorical Topology and Its Relation to Analysis, Algebra and Combinatorics (Conference Proceedings, Prague 1988)* World Scientific, Singapore, 1988, pp. 297–335.
11. Dikranjan, D. and Tholen, W.: *Categorical Structure of Closure Operators*, Kluwer, Dordrecht, 1995.
12. Fay, T. H. and Walls, G.: Compact nilpotent groups, *Communications in Algebra* **17**(9) (1989), 2255–2268.
13. Herrlich, H.: A generalization of perfect maps, in *General Topology and its Relations to Modern Analysis and Algebra III (Prague Symp. 1971)* Academia, Prague, 1972, pp. 187–191.
14. Herrlich, H.: Perfect subcategories and factorisations, *Colloq. Math. Soc. János Bolyai* **8** (1974), 387–403.
15. Herrlich, H., Salicrup, G., and Strecker, G. E.: Factorisations, denseness, separation and relatively compact objects, *Topology Appl.* **27** (1987), 157–169.
16. Holgate, D.: Closure Operators in Categories, Thesis Reprints 015, University of Cape Town, 1992.
17. Nel, L.: Development classes: An approach to perfectness, reflectiveness and extension problems, in *TOPO 72 Proc. Conf. General Topology and Its Applications (Pittsburgh 1972)*, Lecture Notes in Math. 378, Springer, 1974, pp. 322–340.
18. Strecker, G.: Epireflection operators vs perfect morphisms and closed classes of epimorphisms, *Bull. Austral. Math. Soc.* **7** (1972), 359–366.
19. Strecker, G.: Perfect sources, in *Proc. Conf. Categorical Topology (Mannheim 1975)*, Lecture Notes in Math. 540, Springer, 1976, pp. 605–624.

Applied Categorical Structures **4**: 121–126, 1996.

A Remark on Fixed Points of Functors in Topological Categories

JIŘÍ ADÁMEK*
Technical University of Braunschweig, Postfach 33 29, 38092 Braunschweig, Germany

(Received: 6 December 1994; accepted: 6 July 1995)

Abstract. For a topological category $\mathcal{K}$ over **Set** we prove that if a functor $T\colon \mathcal{K} \to \mathcal{K}$ has a fixed cardinal α (i.e. for each object K with card $(UK) = \alpha$ we have card $(UTK) \leqslant \alpha$), then T has a least fixed point, and if T has a successive pair of fixed cardinals α and α^+, then T has a greatest fixed point. This extends results of Adámek and Koubek.

Mathematics Subject Classifications (1991). 18A99, 68Q65.

Key words: fixed point of a functor, topological category.

1. Least Fixed Point

Fixed points of endofunctors $T\colon \mathcal{K} \to \mathcal{K}$ play a fundamental role in the theory of domains for formal semantics of computer languages. We investigate here fixed points of endofunctors of concrete categories over **Set**, i.e., categories equipped with a faithful functor $U\colon \mathcal{K} \to \mathbf{Set}$, see [2].

A fixed point of a functor $T\colon \mathcal{K} \to \mathcal{K}$ is an object of $\mathcal{K}$ together with an isomorphism $k\colon TK \to K$. This is a special case of a *T-algebra* which is a pair (K, k) consisting of an object K of $\mathcal{K}$ and a morphism $k\colon TK \to K$. A *homomorphism* from a T-algebra (K, k) to a T-algebra (K', k') is a morphism $f\colon K \to K'$ of $\mathcal{K}$ such that $f.k = k'.Tf$.

This gives rise to the category of T-algebras and homomorphisms. If this category has an initial object (K, k), then k is an isomorphism, as proved by J. Lambek [5]; this fixed point is then called the *least fixed point* (LEP) of T. A transfinite iterative construction of the LFP has been introduced in [1]: let $\mathcal{K}$ be a category with an initial object 0 and and with colimits of chains. Given a functor $T\colon \mathcal{K} \to \mathcal{K}$ define the LFP-chain $W\colon \mathbf{Ord} \to \mathcal{K}$ as follows:

(a) $W_0 = 0$, $W_1 = T0$ and $W_{0,1}\colon 0 \to T0$ is the unique morphism;
(b) $W_{i+1} = TW_i$ and $W_{i+1,j+1} = TW_{i,j}$ for all $i \leqslant j$ in **Ord**;
(c) $W_j = \operatorname{colim} W_i$ with colimit morphisms $W_{i,j}$ $(i \leqslant j)$ for limit ordinals j.

* Partial financial support of the Grant Agency of the Czech Republic under Grant No. 201/93/0950 is gratefully acknowledged.

If the LFP-chain *stops,* i.e., if some $W_{i,i+1}$ is an isomorphism, then the pair $(W_i, W_{i,i+1}^{-1})$ is a least fixed point of T, see [1]. If this is the case, then we say that T has a *constructive* LFP.

Of the following obvious implications

$$T \text{ has constructive LFP} \Rightarrow T \text{ has an LFP} \Rightarrow T \text{ has a fixed point}$$

none is reversible, in general, see [4]; however, if T preserves monomorphisms and $\mathcal{K}$ satisfies some rather mild additional assumptions, then the above implications can be reversed, see [7]. In particular, one has:

COROLLARY. *For each fibre-small monotopological category $\mathcal{K}$ and each functor T: $\mathcal{K} \to \mathcal{K}$ preserving monomorphisms the existence of a fixed point guarantees the existence of a constructive LFP.*

We are now going to prove a closely related result for functors preserving bimorphisms. The methods of the following proof is a analogous to that presented by J. Reiterman [6].

DEFINITION. Let T: $\mathcal{K} \to \mathcal{K}$ be an endofunctor of a concrete category $(\mathcal{K}, U)$. A *fixed cardinal* of T is a cardinal α such that for each object K of $\mathcal{K}$ we have

$$\operatorname{card} UK = \alpha \quad \text{implies} \quad \operatorname{card} U(TK) \leqslant \alpha.$$

THEOREM 1. *For each fibre-small monotopological category $\mathcal{K}$ and each functor T: $\mathcal{K} \to \mathcal{K}$ preserving bimorphisms the existence of a fixed cardinal guarantees the existence of a constructive LFP.*

Proof. The forgetful functor has a left adjoint F: **Set** $\to \mathcal{K}$, viz., the discrete-object functor; let ε: $FU \to \mathrm{Id}$ denote the back adjuction (formed by bimorphisms). If α is a fixed cardinal of T, then the functor $\overline{T} = UTF$ fulfills $\operatorname{card} \overline{T}X \leqslant \operatorname{card} X$ for any set X of cardinality α; in fact, $\operatorname{card} UFX = \operatorname{card} X = \alpha$ implies $\operatorname{card} U(TFX) \leqslant \alpha$. By [7] it follows that $\overline{T}$ has a constructive LFP. Denote by W: Ord $\to \mathcal{K}$ and $\overline{W}$: Ord $\to$ **Set** the LFP-chains of T and $\overline{T}$, respectively. We know that for some cardinal i the chain W stops, i.e., $W_{i,j}$ are isomorphisms for all $j \geqslant i$. Consequently, $F\overline{W}$ also stops (after i steps). To prove that W stops, it is sufficient to find a natural transformation e: $F\overline{W} \to W$, formed by bimorphisms: since $\mathcal{K}$ is fibre-small, it follows that W stops.

We define e: $F\overline{W}_i \to W_i$ by transfinite induction:

e_0 is the canonical isomorphism (F preserves initial objects);
$e_{i+1} = Te_i \cdot \varepsilon_{TF\overline{W}_i}$: $F\overline{T}W_i \to W_i$;
$e_j = \operatorname*{colim}_{i<j} e_i$ for each limit ordinal j.

Since T preserves bimorphisms and a colimit of a chain of bimorphisms is a bimorphism, it follows by transfinite induction that e: $F\overline{W} \to W$ is a transformation formed by bimorphisms, as required. □

EXAMPLES.

(1) If $\mathcal{K}$ is a topological category and T: $\mathcal{K} \to \mathcal{K}$ a concrete functor, then the LFP-chain of T is formed on the level of sets. That is, if T_0: **Set** $\to$ **Set** denotes the underlying functor and W_0 is the LFP-construction of T_0, then W_0 carries the LFP-construction of T.

For example, let Σ be a finitary signature, i.e., a set together with prescribed arities $\text{ar}(\sigma) = 0, 1, 2, \ldots$, and let T_Σ be the functor defined on objects by $TX = \coprod X^{ar(\sigma)}$ and analogously on morphisms. The underlying set functor $(T_\Sigma)_0$ has the following LFP-construction: $(W_0)_i$ is the set of all Σ-labelled tress of depth less than i and $(W_0)_{ij}$ is the inclusion map. Thus, an LFP of $(T_\Sigma)_0$ is the algebra of all finite Σ-labelled trees, see [3]. Consequently, an LFP of T_Σ is carried by this algebra.

If, moreover, a product of finitely many discrete objects in $\mathcal{K}$ is discrete, then an LFP of T_Σ is the algebra of finite Σ-labelled trees with the discrete structure.

(2) A functor with a fixed point but without an LFP (see [4]): Let $\mathcal{K}$ be the category of graphs (i.e., sets with a binary relation) and homomorphisms. Let T: $\mathcal{K} \to \mathcal{K}$ be the functor which to each discrete graph (X, ϕ) assignes (PX, ϕ) and to each non-discrete graph assigns 1. The action on morphisms between discrete graphs is that of the power-set functor P, otherwise it is clear.

Although $\mathcal{K}$ is topological and T has essentially the unique fixed point 1, the LFP-chain never stops and T does not have an LFP. Observe that T preserves epimorphisms.

OPEN PROBLEM. Is the assumption that T preserve bimorphisms really necessary in the above theorem? That is, does there exist a functor on a monotopological category which has a fixed cardinal but does not have an LFP?

EXAMPLE (3). Theorem 1 does not hold for solid categories, in general: Consider the category $\mathcal{K}$ of 0-1-lattices and homomorphisms (preserving 0, 1). The functor T: $\mathcal{K} \to \mathcal{K}$ with $T1 = 1$ and $TX =$ free 0-1-lattice generated by $PX = \{M; M \leqslant X\}$ (for any $X \ncong 1$), which is defined on morphisms f: $X \to Y$ with $Y \ncong 1$ as the free homomorphism extending Pf, has a unique fixed point $T1 \to 1$. This is not an LFP because for any T-algebra $TK \to K$ with $K \ncong 1$ these exists no T-homomorphism from $T1 \to 1$ to $TK \to K$. Such T-algebras exist. (E.g., if $K = \{0, 1\}$, then TK is the free lattice on 4 generators, and k: $TK \to K$, defined by

$$k(0) = 0 \quad \text{and} \quad k(x) = 1 \quad \text{for all } x \neq 0$$

is, obviously, a morphism of K.)

2. Greatest Fixed Point

The dual concept to LFP is a greatest fixed point (GFP) of a functor $T\colon \mathcal{K} \to \mathcal{K}$, i.e., a terminal T-coalgebra (where a T-coalgebra is an arrow $k\colon K \to TK$). This can be obtained by the GFP-chain $W\colon \mathbf{Ord}^{op} \to \mathcal{K}$ which is obtained by iterating the unique morphism $W_{1,0}\colon T1 \to 1$: if that chain stops after i steps then $(W_i, W_{i+1,i}^{-1})$ is a GFP of T.

EXAMPLE (see [3]) of a functor T: Set $\to$ Set which has a fixed point, but does not have a GFP:

$$TX = \{M \not\subseteq X; \operatorname{card} M \neq \aleph_0\}$$

$$\text{and for} \quad f\colon X \to Y \quad \text{and} \quad M \in TX$$

$$Tf(M) = f[M] \quad \text{if } f \text{ restricted to } M \text{ is one-to-one; } \phi \text{ else.}$$

DEFINITION. Let $\mathcal{K}$ be a concrete category. A *fixed cardinal-pair* of a functor $T\colon \mathcal{K} \to \mathcal{K}$ is a cardinal α such that both α and its successor α^+ are fixed cardinals. (That is, for each object K with $\operatorname{card} UK = \alpha$ or $\operatorname{card} UK = \alpha^+$ we have $\operatorname{card} U(TK) \leqslant \operatorname{card} UK$.)

THEOREM 2. *Assuming the Generalized Continuum Hypothesis, for each endofunctor of a fibre-small topological category preserving monomorphisms the existence of a fixed cardinal-pair guarantees the existence of a constructive GFP.*

Proof. Let $(\mathcal{K}, U)$ be a topological category. The forgetful functor has a right adjoint G: $\mathbf{Set} \to \mathcal{K}$, viz., the indiscrete-object functor. Let μ: Id $\to GU$ be the front adjunction. If α, α^+ is a fixed cardinal-pair of $T\colon \mathcal{K} \to \mathcal{K}$, then the functor $\overline{T} = UTG$: $\mathbf{Set} \to \mathbf{Set}$ has the property that for each set X of cardinality α or α^+ we have

$$\operatorname{card} \overline{T}X \leqslant \operatorname{card} X.$$

As proved in [3], this implies, assuming the Generalized Continuum Hypothesis, that $\overline{T}$ has a constructive GFP. Let $W\colon \mathbf{Ord}^{op} \to \mathcal{K}$ and $\overline{W}\colon \mathbf{Ord}^{op} \to \mathbf{Set}$ denote the GFP-chains of T and $\overline{T}$, respectively. Since $\overline{W}$ stops, i.e., there is an ordinal i such that $W_{j,i}$ are isomorphisms for all $j \geqslant i$, the chain $G\overline{W}$ stops too. It is sufficient to find a monotransformation $m\colon W \to G\overline{W}$: since $\mathcal{K}$ is wellpowered, it follows that W stops as well.

We define m by transfinite induction as follows:

m_0: $1 \to G1 \cong 1$ is the canonical isomorphism;

$$m_{i+1} = \mu_{TG\overline{W}_i} \cdot Tm_i\colon TW_i \to GUTG\overline{W}_i = G\overline{T}\,\overline{W}_i;$$

$$m_j = \lim_{i<j} m_i \text{ for limit ordinal } j.$$

EXAMPLES (4). Let $\mathcal{K}$ be a topological category and Σ a finitary signature. The functor $T_\Sigma\colon \mathcal{K} \to \mathcal{K}$ has a greatest fixed point carried by the coalgebra of all (finite and infinite) Σ-labelled trees. If $\mathcal{K}$ is such that coproducts of indiscrete objects are indiscrete, then this coalgebra carries the indiscrete structure in $\mathcal{K}$.

(5) Theorem 2 does not hold for monotopological categories in general:
Let $\mathcal{K}$ be the category of irreflexive graphs with a terminal object 1 added. That is, objects of $\mathcal{K}$ are pairs (X, ρ) where X is set, ρ a subset of $X \times X$ and if $\operatorname{card} X \neq 1$, then ρ does not meet the diagonal of X. Morphisms are the usual homomorphisms. Equipped with the natural forgetful functor, $\mathcal{K}$ is obviously monotopological.

Given a cardinal γ, define

$$T_\gamma\colon \mathcal{K} \to \mathcal{K}$$

on objects as follows: $T_\gamma 1$ is the discrete object on the set $\gamma + \{\phi\}$ and if $(X, \varrho) \not\cong 1$, then $T_\gamma(X, \varrho)$ is the discrete object on the set

$$(\gamma \times \{M; M \not\subseteq X, \operatorname{card} M \geqslant \gamma\}) + \{\phi\}.$$

The definition of T_γ on morphisms $f\colon K \to K'$ is as follows: $T_\gamma f(\phi) = \phi$; if $K \not\cong 1 \not\cong K'$ then $T_\gamma f$ is the first projection; if $K \not\cong 1$ (thus, $K' \not\cong 1$), then for each $i \in \gamma$, and $M \not\subseteq X$ with $\operatorname{card} M \geqslant \gamma$

$$T_\gamma f(i, M) = \begin{cases} (i, f[M]) & \text{if } f \text{ restricted to } M \text{ is one-to-one,} \\ \phi & \text{else.} \end{cases}$$

It is easy to see that T_γ is a functor preserving monomorphisms and such that each cardinal α with $1 < \alpha < \gamma$ is fixed cardinal.

Nevertheless, T_γ does not have a constructive GFP: in the GFP-chain $W\colon \mathrm{Ord}^{op} \to \mathcal{K}$ the cardinality of W_i is larger than $\aleph_i$ for each ordinal i.

REMARK. Without the Generalized Continuum Hypothesis an analogous result to Theorem 2 holds: instead of fixed cardinals α and α^+ we have to assume that all cardinals between α and 2^α are fixed.

References

1. Adámek, J.: Free algebras and automata realizations in the language of categories, *Comment. Math. Univ. Carolinae* **15** (1974), 589–602.
2. Adámek, J., Herrlich, H. and Strecker, G. E.: *Abstracts and Concrete Categories*, Wiley, New York, 1991.

3. Adámek, J. and Koubek, V.: On the greatest fixed point of a set functors, *Theoret. Comput. Sci.*, to appear.
4. Adámek, J. and Trnková, V.: *Automata and Algebras in Category*, Kluwer, Dordrecht, 1990.
5. Lambek, J.: A fixpoint theorem in complete categories, *Math. Z.* **103** (1968), 151–161.
6. Reiterman, J.: A more categorical model of universal algebra, *Lecture Notes in Comput. Sci.* **56** (1977), 308–313.
7. Trnková, V., Adámek, J., Koubek, V., and Reiterman, J.: Free algebras, input processes and free monads, *Comment. Math. Univ. Carolinae* **16** (1975), 339–351.

Applied Categorical Structures **4:** 127–128, 1996.

Concrete Categories Are Concretely Equivalent iff Their Uniquely Transportable Modifications Are Strict Concretely Isomorphic

HANS-E. PORST
Fachbereich Mathematik und Informatik, Universität Bremen, 28334 Bremen, Germany
(e-mail: porst@informatik.uni-bremen.de)

(Received: 3 November 1994; accepted: 6 July 1995)

Abstract. A proof of the statement in the title is given.

Mathematics Subject Classifications (1991). Primary 18A05, Secondary 18C15.

Key words: concrete functor, concrete equivalence.

In [2] the following generalization of the notion of concrete functors was suggested: A concrete functor from the concrete category $(\mathcal{A}, U)$ into the concrete category $(\mathcal{B}, V)$ – both concrete over the same category $\mathcal{X}$ – is pair (F, φ) with $F\colon \mathcal{A} \to \mathcal{B}$ a functor and $\varphi\colon VF \overset{\sim}{\to} U$ a natural isomorphism. This allows for the following definition (c.f. [2]): a concrete functor $(F, \varphi)\colon (\mathcal{A}, U) \to (\mathcal{B}, V)$ is a concrete equivalence iff there exists a concrete functor $(G, \gamma)\colon (\mathcal{B}, V) \to (\mathcal{A}, U)$ and concrete natural isomorphisms $\eta\colon (F, \varphi) \circ (G, \gamma) \to (1_{\mathcal{B}}, 1_V)$ and $\varepsilon\colon (G, \gamma) \circ (F, \varphi) \to (1_{\mathcal{A}}, 1_U)$.

Having in mind that the importance of the notion of equivalence for abstract categories is given by the fact that categories $\mathcal{A}$ and $\mathcal{B}$ are equivalent iff their respective structural essence (i.e. their skeletons) are isomorphic, one is led to the question whether a similar statement holds w.r.t. concrete equivalence. Now the structural essence of a concrete category $(\mathcal{A}, U)$ is given by its uniquely transportable modification $(\mathcal{A}^\star, U^\star)$ as described in [1, 5.36]. Therefore I am going to prove the following theorem, thus complementing the paper [2]:

THEOREM 1. *Let $(\mathcal{A}, U)$ and $(\mathcal{B}, V)$ be concrete categories over $\mathcal{X}$. Then $(\mathcal{A}, U)$ and $(\mathcal{B}, V)$ are concretely equivalent iff their respective uniquely transportable modifications are strict concretely isomorphic.*

Proof. For any concrete category $(\mathcal{B}, V)$ there is a
– transportable modification $\widehat{E}_{(\mathcal{B},V)}\colon (\mathcal{B}, V) \to (\widehat{\mathcal{B}}, \widehat{V})$ and an
– amnestic modification $\overline{P}_{(\mathcal{B},V)}\colon (\mathcal{B}, V) \to (\overline{\mathcal{B}}, \overline{V})$,

both of which are strictly concrete equivalences (c.f. [1]). Recall also from [1] that, for any concrete category $(\mathcal{A}, U)$, its uniquely transportable modification $(\mathcal{A}^\star, U^\star)$ is determined up to a strictly concrete isomorphism by the property that there exists a strictly concrete equivalence $E^\star\colon (\mathcal{A},U) \to (\mathcal{A}^\star,U^\star)$ with $(\mathcal{A}^\star,U^\star)$ being uniquely transportable, and can be obtained as

$$E^\star_{(\mathcal{A},U)} = \overline{P}_{(\widehat{\mathcal{A}},\widehat{U})} \circ \widehat{E}_{(\mathcal{A},U)}\colon\ (\mathcal{A},U) \longrightarrow (\overline{\widehat{\mathcal{A}}},\overline{\widehat{U}}) =:\ (\mathcal{A}^\star,U^\star).$$

The following then is obvious: if $H\colon\ (\mathcal{A}^\star,U^\star) \longrightarrow (\mathcal{B}^\star,V^\star)$ is a strict concrete isomorphism and (P,ψ) is a concrete equivalence inverse of $E^\star_{(\mathcal{B},V)}$, then

$$(P,\psi)\circ(H,1)\circ(E^\star_{(\mathcal{A},U)},1_U)\colon\ (\mathcal{A},U) \longrightarrow (\mathcal{B},V)$$

is a concrete equivalence.

To prove the converse, one needs the following lemma:

LEMMA 2. *For every concrete functor* $(F,\varphi)\colon\ (\mathcal{A},U) \to (\mathcal{B},V)$ *there exists a strictly concrete functor* $\widehat{F}\colon\ (\widehat{\mathcal{A}},\widehat{U}) \to (\widehat{\mathcal{B}},\widehat{V})$ *together with a natural isomorphism* $\psi\colon \widehat{E}_{(\mathcal{B},V)}F \xrightarrow{\sim} \widehat{F}\widehat{E}_{(\mathcal{A},U)}$.

Hence, if $(F,\varphi)\colon\ (\mathcal{A},U) \to (\mathcal{B},V)$ is a concrete equivalence so is $(\widehat{F},1)$, and there results a strictly concrete equivalence

$$(\mathcal{A},U) \xrightarrow{\widehat{E}_{(\mathcal{A},U)}} (\widehat{\mathcal{A}},\widehat{U}) \xrightarrow{\widehat{F}} (\widehat{\mathcal{B}},\widehat{V}) \xrightarrow{\overline{P}_{(\widehat{\mathcal{B}},\widehat{V})}} (\mathcal{B}^\star,V^\star).$$

Uniqueness of $(\mathcal{A}^\star,U^\star)$ now shows that there exists a strictly concrete isomorphism $(\mathcal{A}^\star,U^\star) \to (\mathcal{B}^\star,V^\star)$. □

Proof of the lemma. Recall that $(\widehat{\mathcal{A}},\widehat{U})$ has objects (A,a,X) with A in $ob\mathcal{A}$, X in $ob\mathcal{X}$, and $a\colon\ UA \to X$ an $\mathcal{X}$–isomorphism, while morphisms are triples $(f,a,b)\colon\ (A,a,X) \to (B,b,Y)$ with $f\colon\ A \to B$ an $\mathcal{A}$-morphism. Define

$$\widehat{F}\Big((A,a,X) \xrightarrow{(f,a,b)} (B,b,Y)\Big) = (FA,a\varphi_A,X) \xrightarrow{(Ff,a\varphi_A,b\varphi_B)} (FB,b\varphi_B,Y).$$

Then, due to the definition of $\widehat{V}$ (c.f. [1]), $\widehat{F}$ is strictly concrete. Finally, by

$$(FA,1_{VFA},VFA) = \widehat{E}_{(\mathcal{B},V)}FA \xrightarrow{(1_{FA},1_{VFA},\varphi_A)} \widehat{F}\widehat{E}_{(\mathcal{A},U)}A = (FA,\varphi_A,UA).$$

is given a natural isomorphism $\psi\colon \widehat{E}_{(\mathcal{B},V)}F \to \widehat{F}\widehat{E}_{(\mathcal{A},U)}$. □

References

1. Adámek, J., Herrlich, H., and Strecker, G. E.: *Abstract and Concrete Categories*, Wiley Interscience, New York, 1990.
2. Porst, H.-E.: What is concrete equivalence? *J. Appl. Categorical Structures* **2** (1994), 57–70.

CONNECTEDNESS, DISCONNECTEDNESS AND CLOSURE OPERATORS, A MORE GENERAL APPROACH

G. Castellini*

Department of Mathematics, University of Puerto Rico, Mayagüez campus, P.O. Box 5000, Mayagüez, PR 00681-5000.

Abstract: Let $\mathcal{X}$ be an arbitrary category with an $(\mathbf{E}, \mathcal{M})$-factorization structure for sinks. A notion of constant morphism that depends on a chosen class of monomorphisms is introduced. This notion yields a Galois connection that can be seen as a generalization of the classical connectedness-disconnectedness correspondence (also called torsion-torsion free in algebraic contexts). It is shown that this Galois connection factors through the collection of all closure operators on $\mathcal{X}$ with respect to $\mathcal{M}$.

Mathematics Subject Classification (1991): 18A20, 18A32, 06A15.

Key Words: Closure operator, Galois connection, connectedness, disconnectedness.

Introduction

This paper presents in the setting of an arbitrary category some ideas that were introduced by D. Hajek and the author in a category whose objects were structured sets (cf. [3]).

Let $\mathcal{X}$ be an arbitrary category with an $(\mathbf{E}, \mathcal{M})$-factorization structure for sinks and let $\mathcal{N} \subseteq \mathcal{M}$. An $\mathcal{X}$-morphism $X \xrightarrow{f} Y$ is called $\mathcal{N}$-constant if its direct image is isomorphic to the direct image under f of every $\mathcal{N}$-subobject of X. So, if $S(\mathcal{X})$ denotes the class of all subclasses of objects of $\mathcal{X}$, ordered by inclusion, for every $\mathcal{N} \subseteq \mathcal{M}$, the relation: $X \mathcal{R}_{\mathcal{N}} Y$ if and only if every $\mathcal{X}$-morphism $X \xrightarrow{f} Y$ is $\mathcal{N}$-constant yields a Galois connection $S(\mathcal{X}) \underset{\nabla_{\mathcal{N}}}{\overset{\Delta_{\mathcal{N}}}{\rightleftarrows}} S(\mathcal{X})^{op}$. If $\mathcal{N}$ is closed under direct images, we have that this Galois connection factors through $CL(\mathcal{X}, \mathcal{M})$, i.e., the collection of all closure operators on $\mathcal{X}$ with respect to $\mathcal{M}$, via two Galois connections $CL(\mathcal{X}, \mathcal{M}) \underset{T_{\mathcal{N}}}{\overset{D_{\mathcal{N}}}{\rightleftarrows}} S(\mathcal{X})^{op}$ and $S(\mathcal{X}) \underset{I_{\mathcal{N}}}{\overset{J_{\mathcal{N}}}{\rightleftarrows}} CL(\mathcal{X}, \mathcal{M})$.

The development of a general theory of topological connectedness was started by Preuß (cf. [15-17]) and by Herrlich [10]. We recently became aware that an introduction to the study of connectedness and disconnectedness in an arbitrary category, using an approach similar to ours was made by Petz (cf. [14]).

We would like to point out that the definition of $\mathcal{N}$-constant morphism that

* Research supported by the University of Puerto Rico, Mayagüez Campus during a sabbatical visit at Kansas State University.

Eraldo Giuli (ed.), Categorical Topology, 129–138.

appears in this paper was not chosen with the intention of developing a general theory of connectedness and disconnectedness, but rather to support certain constructions with closure operators in an arbitrary setting. A general theory of connectedness and disconnectedness was recently presented by Clementino in [6], where she extends results in [12] to an arbitrary category.

Examples and further study of the Galois connections mentioned in this paper will appear in a subsequent paper (cf. [2]).

We use the terminology of [1] throughout the paper. We also acknowledge that Paul Taylor's commutative diagrams macro package was used to typeset most of the diagrams in this paper.

1. Preliminaries

Throughout we consider a category $\mathcal{X}$ together with a fixed class $\mathcal{M}$ of $\mathcal{X}$-monomorphisms and a class $\mathbf{E}$ of $\mathcal{X}$-sinks such that $\mathcal{X}$ is an $(E, \mathcal{M})$-*category for sinks*, (cf. [1] for the dual case), that is:

(a) each of $\mathbf{E}$ and $\mathcal{M}$ is closed under compositions with isomorphisms;

(b) $\mathcal{X}$ has $(E, \mathcal{M})$-factorizations (of sinks); i.e., each sink $\mathbf{s}$ in $\mathcal{X}$ has a factorization $s = m \circ e$ with $e \in E$ and $m \in \mathcal{M}$, and

(c) $\mathcal{X}$ has the unique $(E, \mathcal{M})$-diagonalization property; i.e., if $B \xrightarrow{g} D$ and $C \xrightarrow{m} D$ are $\mathcal{X}$-morphisms with $m \in \mathcal{M}$, and $e = (A_i \xrightarrow{e_i} B)_I$ and $s = (A_i \xrightarrow{s_i} C)_I$ are sinks in $\mathcal{X}$ with $e \in E$, such that $m \circ s = g \circ e$, then there exists a unique diagonal $B \xrightarrow{d} C$ such that $m \circ d = g$ and for every $i \in I$, $d \circ e_i = s_i$.

DEFINITION 1.1. A *closure operator* C on $\mathcal{X}$ (with respect to $\mathcal{M}$) is a family $\{(\ )_X^C\}_{X \in \mathcal{X}}$ of functions on the $\mathcal{M}$-subobject lattices of $\mathcal{X}$ with the following properties that hold for each $X \in \mathcal{X}$:

(a) [*expansiveness*] $m \leq (m)_X^C$, for every $\mathcal{M}$-subobject $M \xrightarrow{m} X$;

(b) [*order-preservation*] $m \leq n \Rightarrow (m)_X^C \leq (n)_X^C$ for every pair of $\mathcal{M}$-subobjects of X;

(c) [*morphism-consistency*] If p is the pullback of the $\mathcal{M}$-subobject $M \xrightarrow{m} Y$ along some $\mathcal{X}$-morphism $X \xrightarrow{f} Y$ and q is the pullback of $(m)_Y^C$ along f, then $(p)_X^C \leq q$, i.e., the closure of the inverse image of m is less than or equal to the inverse image of the closure of m.

Condition (a) implies that for every closure operator C on $\mathcal{X}$, every $\mathcal{M}$-subobject $M \xrightarrow{m} X$ has a canonical factorization $m = (m)_X^C \circ t$, where $(M)_X^C \xrightarrow{(m)_X^C} X$ is called the C-*closure* of the subobject (M, m).

When no confusion is likely we will write m^C rather than $(m)_X^C$ and for notational symmetry we will denote the morphism t by m_C.

If $X \xrightarrow{f} Y$ is an $\mathcal{X}$-morphism and $M \xrightarrow{m} X$ is an $\mathcal{M}$-subobject, then the $(E,\mathcal{M})$-factorization of $f \circ m$ will be denoted by $X \xrightarrow{e_{f \circ m}} M_f \xrightarrow{m_f} Y$. $M_f \xrightarrow{m_f} Y$ will be called the direct image of m along f. If $N \xrightarrow{n} Y$ is an $\mathcal{M}$-subobject, the pullback $f^{-1}(N) \xrightarrow{f^{-1}(n)} X$ of n along f will be called the inverse image of n along f. Whenever no confusion is likely to arise, to simplify the notation we will denote the morphism $e_{f \circ m}$ simply e_f.

REMARK 1.2. Notice that in the above definition, under condition (b), the morphism-consistency condition (c) is equivalent to the following statement concerning direct images: if $M \xrightarrow{m} X$ is an $\mathcal{M}$-subobject and $X \xrightarrow{f} Y$ is a morphism, then $((m)^C_Y)_f \leq (m_f)^C_Y$, i.e., the direct image of the closure of m is less than or equal to the closure of the direct image of m; (cf. [7]).

DEFINITION 1.3. Given a closure operator C, we say that $m \in \mathcal{M}$ is *C-closed* (*C-dense*) if m_C (m^C) is an isomorphism. We call C *idempotent* provided that m^C is C-closed for every $m \in \mathcal{M}$. C is called *weakly hereditary* if m_C is C-dense for every $m \in \mathcal{M}$.

Notice that Definition 1.1(c) implies that pullbacks of C-closed $\mathcal{M}$-subobjects are C-closed.

We denote the collection of all closure operators on $\mathcal{M}$ by $CL(\mathcal{X},\mathcal{M})$ preordered as follows: $C \sqsubseteq D$ if $m^C \leq m^D$ for all $m \in \mathcal{M}$ (where $\leq$ is the usual order on subobjects). Notice that arbitrary suprema and infima exist in $CL(\mathcal{X},\mathcal{M})$, they are formed pointwise in the $\mathcal{M}$-subobject fibers.

For more background on closure operators see, e.g., [4], [5], [7], [8] and [13]. For a detailed survey on the same topic one could check [11].

DEFINITION 1.4. For pre-ordered classes $\mathcal{X} = (X, \sqsubseteq)$ and $\mathcal{Y} = (Y, \sqsubseteq)$, a *Galois connection* $\mathcal{X} \underset{G}{\overset{F}{\rightleftarrows}} \mathcal{Y}$ consists of order preserving functions F and G that satisfy $F \dashv G$, i.e., $x \sqsubseteq GF(x)$ for every $x \in X$ and $FG(y) \sqsubseteq y$ for every $y \in Y$. (G is adjoint and has F as coadjoint).

If $x \in \mathcal{X}$ and $y \in \mathcal{Y}$ are such that $F(x) = y$ and $G(y) = x$, then x and y are said to be corresponding fixed points of the Galois connection $(\mathcal{X}, F, G, \mathcal{Y})$.

Properties and many examples of Galois connections can be found in [9].

2. General results

Throughout the paper we assume that $\mathcal{X}$ is an $(E,\mathcal{M})$-category for sinks.

Let $S(\mathcal{X})$ be the collection of all subcategories of $\mathcal{X}$, ordered by inclusion and let $\mathcal{N}$ be a fixed subclass of $\mathcal{M}$. For every $X \in \mathcal{X}$, we denote by $\mathcal{N}_X$ all the $\mathcal{N}$-subobjects that have X as codomain.

DEFINITION 2.1. Let $\mathcal{N} \subseteq \mathcal{M}$. An $\mathcal{X}$-morphism $X \xrightarrow{f} Y$ is called $\mathcal{N}$-constant if for every $\mathcal{N}$-subobject $N \xrightarrow{n} X$, we have that $n_f \simeq (id_X)_f$.

As a consequence we have the following

PROPOSITION 2.2. (cf. [10]) *Let $\mathcal{N} \subseteq \mathcal{M}$. Define $S(\mathcal{X}) \xrightarrow{\Delta_\mathcal{N}} S(\mathcal{X})^{op}$ and $S(\mathcal{X})^{op} \xrightarrow{\nabla_\mathcal{N}} S(\mathcal{X})$ as follows:*

$$\nabla_\mathcal{N}(\mathcal{A}) = \{X \in \mathcal{X} : \forall Y \in \mathcal{A}, X \xrightarrow{f} Y \text{ is } \mathcal{N}\text{-constant}\}$$

$$\Delta_\mathcal{N}(\mathcal{B}) = \{Y \in \mathcal{X} : \forall X \in \mathcal{B}, X \xrightarrow{f} Y \text{ is } \mathcal{N}\text{-constant}\}$$

Then, $S(\mathcal{X}) \underset{\nabla_\mathcal{N}}{\overset{\Delta_\mathcal{N}}{\rightleftarrows}} S(\mathcal{X})^{op}$ is a Galois connection. □

As in [3], with some minor modifications we have the following two Galois connections.

PROPOSITION 2.3. *Let $CL(\mathcal{X},\mathcal{M}) \xrightarrow{D_\mathcal{N}} S(\mathcal{X})^{op}$ and $S(\mathcal{X})^{op} \xrightarrow{T_\mathcal{N}} CL(\mathcal{X},\mathcal{M})$ be defined by:*

$$D_\mathcal{N}(C) = \{X \in \mathcal{X} : \text{ every } n \in \mathcal{N}_X \text{ is } C\text{-closed}\}$$

$$T_\mathcal{N}(\mathcal{A}) = Sup\{C \in CL(\mathcal{X},\mathcal{M}) : D_\mathcal{N}(C) \supseteq \mathcal{A}\}.$$

Then, $CL(\mathcal{X},\mathcal{M}) \underset{T_\mathcal{N}}{\overset{D_\mathcal{N}}{\rightleftarrows}} S(\mathcal{X})^{op}$ is a Galois connection. □

PROPOSITION 2.4. *Let $CL(\mathcal{X},\mathcal{M}) \xrightarrow{I_\mathcal{N}} S(\mathcal{X})$ and $S(\mathcal{X}) \xrightarrow{J_\mathcal{N}} CL(\mathcal{X},\mathcal{M})$ be defined by:*

$$I_\mathcal{N}(C) = \{X \in \mathcal{X} : \text{ every } n \in \mathcal{N}_X \text{ is } C\text{-dense}\}$$

$$J_\mathcal{N}(\mathcal{B}) = Inf\{C \in CL(\mathcal{X},\mathcal{M}) : I_\mathcal{N}(C) \supseteq \mathcal{B}\}.$$

Then, $S(\mathcal{X}) \underset{I_\mathcal{N}}{\overset{J_\mathcal{N}}{\rightleftarrows}} CL(\mathcal{X},\mathcal{M})$ is a Galois connection. □

Clearly, we have the following

COROLLARY 2.5. *The composition functions $D_\mathcal{N} \circ J_\mathcal{N}$ and $I_\mathcal{N} \circ T_\mathcal{N}$ give rise to a Galois connection between $S(\mathcal{X})$ and $S(\mathcal{X})^{op}$.* □

The following result provides a description of how to construct the closure operator $T_\mathcal{N}(\mathcal{A})$ defined in Proposition 2.3.

PROPOSITION 2.6. *Let $\mathcal{A} \in S(\mathcal{X})^{op}$ and let $\mathcal{N}$ be a subclass of $\mathcal{M}$. For every*

$X \in \mathcal{X}$ and for every $\mathcal{M}$-subobject $M \xrightarrow{m} X$, we define

$$_{\mathcal{A}}m = \cap\{f^{-1}(n) : Y \in \mathcal{A}, X \xrightarrow{f} Y, N \xrightarrow{n} Y \in \mathcal{N}_Y \text{ and } m \leq f^{-1}(n)\}.$$

For every $\mathcal{A} \in S(\mathcal{X})^{op}$ we have that the function $_{\mathcal{A}}(\,)$ that to every $\mathcal{M}$-subobject $M \xrightarrow{m} X$ associates $_{\mathcal{A}}m$ is an idempotent closure operator on $\mathcal{X}$ and $_{\mathcal{A}}m \simeq m^{T_{\mathcal{N}}(\mathcal{A})}$.

Proof: We first observe that as a consequence of $\mathcal{X}$ being an $(\mathbf{E}, \mathcal{M})$-category for sinks, $\mathcal{M}$ is closed under the formation of pullbacks and intersections and so $_{\mathcal{A}}m \in \mathcal{M}$.

We begin by showing that $_{\mathcal{A}}(\,)$ is a closure operator on $\mathcal{X}$. It is straightforward to show that $_{\mathcal{A}}(\,)$ is expansive and order-preserving. To show the remaining property, let us consider the following commutative diagram

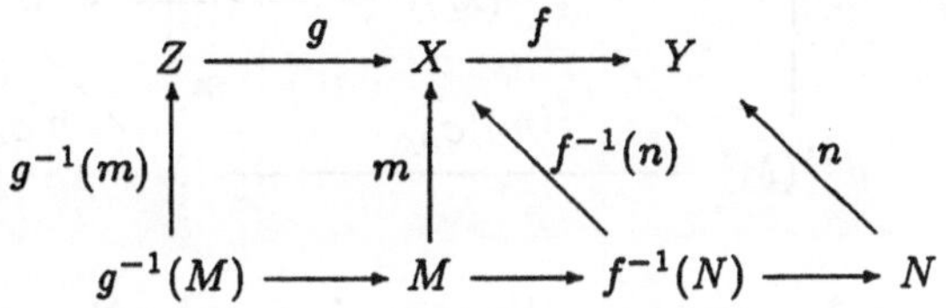

where $Y \in \mathcal{A}$, $n \in \mathcal{N}_Y$ and $m \leq f^{-1}(n)$.

Now, $g^{-1}(m) \leq g^{-1}(_{\mathcal{A}}m) = g^{-1}(\cap f^{-1}(n)) \simeq \cap g^{-1}(f^{-1}(n)) \simeq \cap(f \circ g)^{-1}(n)$, since pullbacks and intersections commute. Now since not all the morphisms from Z to Y are of the form $f \circ g$ we obtain that $_{\mathcal{A}}(g^{-1}(m)) \leq \cap(f \circ g)^{-1}(n) \simeq g^{-1}(f^{-1}(n)) \simeq g^{-1}(_{\mathcal{A}}m)$. Hence, $_{\mathcal{A}}(\,)$ is a closure operator.

To show idempotency it is enough to observe that if $n \in \mathcal{N}_Y$ and $_{\mathcal{A}}m \leq f^{-1}(n)$ then clearly $m \leq f^{-1}(n)$. On the other hand, if $m \leq f^{-1}(n)$, then by definition of $_{\mathcal{A}}m$ we also have that $_{\mathcal{A}}m \leq f^{-1}(n)$. Thus we obtain that $_{\mathcal{A}}m \simeq {_{\mathcal{A}}}(_{\mathcal{A}}m)$

Now, let $X \in \mathcal{A}$. The existence of $X \xrightarrow{id_X} X$ implies that for every $\mathcal{N}$-subobject $N \xrightarrow{n} X$ we have that $_{\mathcal{A}}n \simeq n$. Thus, $_{\mathcal{A}}(\,) \sqsubseteq T_{\mathcal{N}}(\mathcal{A})$.

Finally, let $X \in \mathcal{X}$, let $X \xrightarrow{f} Y$ be an $\mathcal{X}$-morphism with $Y \in \mathcal{A}$ and let $M \xrightarrow{m} X$ be an $\mathcal{M}$-subobject. Consider the $\mathcal{N}$-subobject $N \xrightarrow{n} Y$ with $m \leq f^{-1}(n)$. Then, we obtain that $m^{T_{\mathcal{N}}(\mathcal{A})} \leq (f^{-1}(n))^{T_{\mathcal{N}}(\mathcal{A})} \simeq f^{-1}(n)$, since n is $T_{\mathcal{N}}(\mathcal{A})$-closed and so is its pullback $f^{-1}(n)$. Therefore, by considering all morphisms $X \xrightarrow{f} Y$ with $Y \in \mathcal{A}$ and $n \in \mathcal{N}_Y$ with $m \leq f^{-1}(n)$, we obtain that $m^{T_{\mathcal{N}}(\mathcal{A})} \leq \cap f^{-1}(n) = {_{\mathcal{A}}m}$. Thus, $T_{\mathcal{N}}(\mathcal{A}) \sqsubseteq {_{\mathcal{A}}}(\,)$. Hence $_{\mathcal{A}}(\,) \simeq T_{\mathcal{N}}(\mathcal{A})$. □

Next we provide a description of how to construct the closure operator $J_{\mathcal{N}}(\mathcal{B})$ defined in Proposition 2.4.

PROPOSITION 2.7. *Let $\mathcal{B} \in S(\mathcal{X})$ and let $\mathcal{N} \subseteq \mathcal{M}$. For every $Y \in \mathcal{X}$ and for every $\mathcal{M}$-subobject $M \xrightarrow{m} Y$, we define*

$$C_{\mathcal{B}_{\mathcal{N}}}(m) = sup\left(\{m\} \cup \{(id_X)_f : X \in \mathcal{B}, X \xrightarrow{f} Y \text{ and } \exists n \in \mathcal{N}_X \text{ with } n_f \leq m\}\right).$$

For every $\mathcal{B} \in S(\mathcal{X})$, the function $C_{\mathcal{B}_{\mathcal{N}}}$ is a weakly hereditary closure operator on $\mathcal{X}$. Moreover, we have that $C_{\mathcal{B}_{\mathcal{N}}}(m) \simeq m^{J_{\mathcal{N}}(\mathcal{B})}$.

Proof: We first notice that due to the fact that $\mathcal{X}$ is an $(\mathbf{E}, \mathcal{M})$-category for sinks, suprema are formed via $(\mathbf{E}, \mathcal{M})$-factorizations. Therefore, $C_{\mathcal{B}_{\mathcal{N}}}(m) \in \mathcal{M}$.

It is easily seen that $C_{\mathcal{B}_{\mathcal{N}}}$ is expansive and order-preserving. Let us consider the following commutative diagram

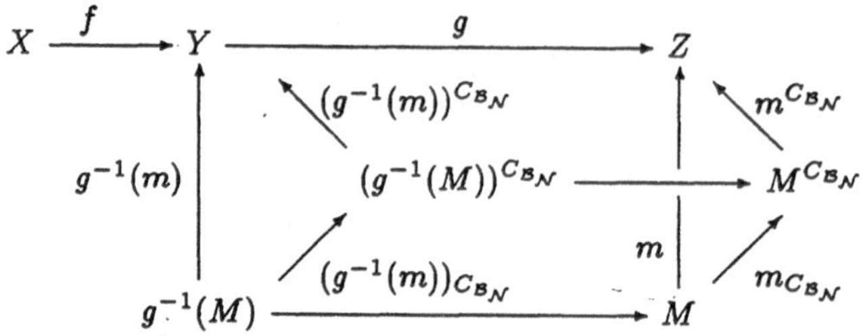

Let $X \xrightarrow{h} Z$ be an $\mathcal{X}$-morphism. We have that $(g^{-1}(m))^{C_{\mathcal{B}_{\mathcal{N}}}} =$

$$sup\Big(\{g^{-1}(m)\} \cup \{(id_X)_f : X \in \mathcal{B}, X \xrightarrow{f} Y \text{ and } \exists n \in \mathcal{N}_X \text{ with } n_f \leq g^{-1}(m)\}\Big) \simeq$$

$$sup\left(\{g^{-1}(m)\} \cup \{(id_X)_f : X \in \mathcal{B}, X \xrightarrow{f} Y \text{ and } \exists n \in \mathcal{N}_X \text{ with } n_{g\circ f} \leq m\}\right) \leq$$

$$sup\left(\{g^{-1}(m)\} \cup \{g^{-1}((id_X)_{g\circ f}) : X \in \mathcal{B}, X \xrightarrow{f} Y \text{ and } \exists n \in \mathcal{N}_X \text{ with } n_{g\circ f} \leq m\}\right) \simeq$$

$$g^{-1}\left(sup(\{m\} \cup \{(id_X)_{g\circ f} : X \in \mathcal{B}, X \xrightarrow{f} Y \text{ and } \exists n \in \mathcal{N}_X \text{ with } n_{g\circ f} \leq m\}\right) \leq$$

$$g^{-1}\left(sup(\{m\} \cup \{(id_X)_h : X \in \mathcal{B}, X \xrightarrow{h} Z \text{ and } \exists n \in \mathcal{N}_X \text{ with } n_h \leq m\}\right) = g^{-1}(m^{C_{\mathcal{B}_{\mathcal{N}}}}).$$

Notice that above we have used the fact that pullbacks and suprema commute.

This shows that, for every $\mathcal{B} \in S(\mathcal{X})$, $C_{\mathcal{B}_{\mathcal{N}}}$ is a closure operator.

To see that $C_{\mathcal{B}_{\mathcal{N}}}$ is weakly hereditary, let $M \xrightarrow{m} Y$ be an $\mathcal{M}$-subobject and let $X \xrightarrow{f} Y$, with $X \in \mathcal{B}$, be such that there exists $N \xrightarrow{n} X \in \mathcal{N}_X$ with $n_f \leq m$. Let us consider the following commutative diagram

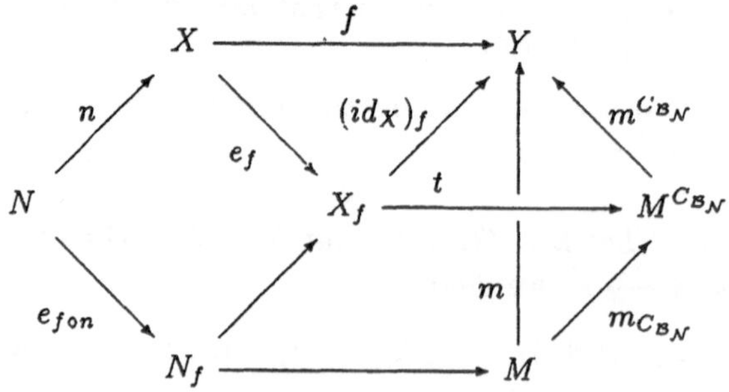

Notice that the morphism t exists by construction of $m^{C_{\mathcal{B}_\mathcal{N}}}$.

Consider the $(\mathbf{E},\mathcal{M})$-factorization of $t \circ e_f$, $(e_{t \circ e_f}, (id_X)_{t \circ e_f})$. Due to the $(\mathbf{E},\mathcal{M})$-diagonaliza-tion property, from the following commutative diagram

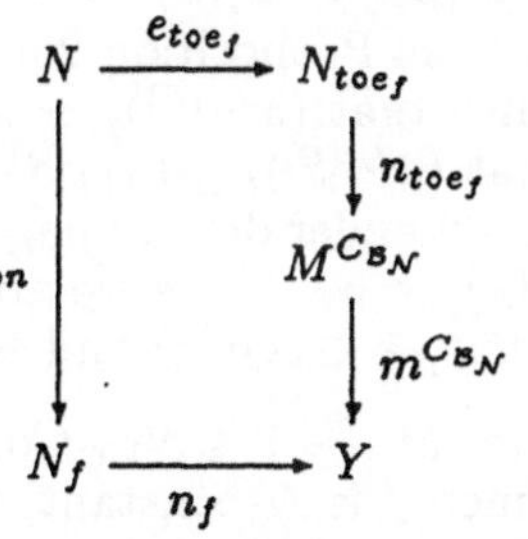

we obtain that $n_{t \circ e_f} \leq n_f \leq m$ and so $(id_X)_{t \circ e_f}$ occurs in the construction of $(m_{C_{\mathcal{B}_\mathcal{N}}})^{C_{\mathcal{B}_\mathcal{N}}}$. However, since $(id_X)_f$ and $C_{\mathcal{B}_\mathcal{N}}(m)$ both belong to $\mathcal{M}$, a property of $(\mathbf{E},\mathcal{M})$-categories implies that also $t \in \mathcal{M}$. Therefore, we have that $t \simeq (id_X)_{t \circ e_f}$. We conclude that $(id_X)_f \leq (m_{C_{\mathcal{B}_\mathcal{N}}})^{C_{\mathcal{B}_\mathcal{N}}}$, which implies $m^{C_{\mathcal{B}_\mathcal{N}}} \leq (m_{C_{\mathcal{B}_\mathcal{N}}})^{C_{\mathcal{B}_\mathcal{N}}}$. Consequently, $m^{C_{\mathcal{B}_\mathcal{N}}} \simeq (m_{C_{\mathcal{B}_\mathcal{N}}})^{C_{\mathcal{B}_\mathcal{N}}}$, i.e.; $C_{\mathcal{B}_\mathcal{N}}$ is weakly hereditary.

If $Y \in \mathcal{B}$ and $N \xrightarrow{n} Y \in \mathcal{N}_Y$, the identity morphism yields that $n^{C_{\mathcal{B}_\mathcal{N}}} \simeq id_Y$. Therefore we obtain that $J_\mathcal{N}(\mathcal{B}) \sqsubseteq C_{\mathcal{B}_\mathcal{N}}$.

Now, let $M \xrightarrow{m} Y$ be an $\mathcal{M}$-subobject of Y and let $X \xrightarrow{f} Y$ be an $\mathcal{X}$-morphism with $X \in \mathcal{B}$ and such that there exists $N \xrightarrow{n} X \in \mathcal{N}_X$ with $n_f \leq m$. By definition of $J_\mathcal{N}(\mathcal{B})$, we have that $n^{J_\mathcal{N}(\mathcal{B})} \simeq id_X$. This implies that $(id_X)_f \simeq (n^{J_\mathcal{N}(\mathcal{B})})_f \leq (n_f)^{J_\mathcal{N}(\mathcal{B})} \leq m^{J_\mathcal{N}(\mathcal{B})}$ (cf. Remark 1.2). Therefore, $C_{\mathcal{B}_\mathcal{N}}(m) \leq m^{J_\mathcal{N}(\mathcal{B})}$. Thus, $J_\mathcal{N}(\mathcal{B}) \simeq C_{\mathcal{B}_\mathcal{N}}$. □

THEOREM 2.8. *Let $\mathcal{N}$ be a subclass of $\mathcal{M}$ closed under the formation of direct images. Then the Galois connection $S(\mathcal{X}) \underset{\nabla_\mathcal{N}}{\overset{\Delta_\mathcal{N}}{\rightleftarrows}} S(\mathcal{X})^{op}$ factors through $CL(\mathcal{X},\mathcal{M})$ via the two Galois connections $S(\mathcal{X}) \underset{I_\mathcal{N}}{\overset{J_\mathcal{N}}{\rightleftarrows}} CL(\mathcal{X},\mathcal{M})$ and $CL(\mathcal{X},\mathcal{M}) \underset{T_\mathcal{N}}{\overset{D_\mathcal{N}}{\rightleftarrows}} S(\mathcal{X})^{op}$.*

Proof: Let $\mathcal{A} \in S(\mathcal{X})^{op}$ and let $X \in (I_\mathcal{N} \circ T_\mathcal{N})(\mathcal{A})$. Consider $X \xrightarrow{f} Y$ with $Y \in \mathcal{A}$ and let $N \xrightarrow{n} X$ belong to $\mathcal{N}_X$. From the properties of closure operators, we have that $(n^{T_\mathcal{N}(\mathcal{A})})_f \leq (n_f)^{T_\mathcal{N}(\mathcal{A})}$. Since n is $T_\mathcal{N}(\mathcal{A})$-dense, we have that $(n^{T_\mathcal{N}(\mathcal{A})})_f \simeq (id_X)_f$. From Proposition 2.3, we have that every $\mathcal{N}$-subobject of Y is $T_\mathcal{N}(\mathcal{A})$-closed, and since $\mathcal{N}$ is closed under direct images, we obtain that $(n_f)^{T_\mathcal{N}(\mathcal{A})} \simeq n_f$. Therefore we have that $(id_X)_f \leq n_f$. However, the reverse inequality always holds, thus we have that $n_f \simeq (id_X)_f$, i.e., $X \in \nabla_\mathcal{N}(\mathcal{A})$. Thus $(I_\mathcal{N} \circ T_\mathcal{N})(\mathcal{A}) \subseteq \nabla_\mathcal{N}(\mathcal{A})$.

Now, let $X \in \nabla_\mathcal{N}(\mathcal{A})$. Let us consider $X \xrightarrow{f} Y$ with $Y \in \mathcal{A}$ and an $\mathcal{N}$-subobject $N \xrightarrow{n} X$. By hypothesis we have that $n_f \simeq (id_X)_f$. Since $\mathcal{N}$ is closed under direct images, we have that n_f belongs to $\mathcal{N}_Y$. Consequently, $id_X \simeq f^{-1}((id_X)_f) \simeq$

$f^{-1}(n_f)$. Notice that, due to the existence of a Galois connection between direct and inverse images, we have that if $n \leq f^{-1}(n')$ then $n_f \leq n'$ and so $id_X \simeq f^{-1}(n_f) \leq f^{-1}(n')$, for every $n' \in \mathcal{N}_Y$ satisfying $n \leq f^{-1}(n')$. Therefore we obtain that $n^{T_\mathcal{N}(\mathcal{A})} \simeq id_X$, for every $N \xrightarrow{n} X$ in $\mathcal{N}_X$. So, $X \in (I_\mathcal{N} \circ T_\mathcal{N})(\mathcal{A})$. Thus $I_\mathcal{N} \circ T_\mathcal{N} = \nabla_\mathcal{N}$.

Let $\mathcal{B} \in S(\mathcal{X})$ and let $Y \in (D_\mathcal{N} \circ J_\mathcal{N})(\mathcal{B})$. Consider $X \xrightarrow{f} Y$ with $X \in \mathcal{B}$. If $N \xrightarrow{n} X$ belongs to $\mathcal{N}_X$, then from Proposition 2.4 we have that n is $J_\mathcal{N}(\mathcal{B})$-dense, i.e., $n^{J_\mathcal{N}(\mathcal{B})} \simeq id_X$. This implies that $(n^{J_\mathcal{N}(\mathcal{B})})_f \simeq (id_X)_f$. From the properties of closure operators, we have that $(n^{J_\mathcal{N}(\mathcal{B})})_f \leq (n_f)^{J_\mathcal{N}(\mathcal{B})}$. Since every $\mathcal{N}$-subobject of Y is $J_\mathcal{N}(\mathcal{B})$-closed and $\mathcal{N}$ is closed under direct images, we have that $(n_f)^{J_\mathcal{N}(\mathcal{B})} \simeq n_f$. Therefore, we obtain that $(id_X)_f \leq n_f$. Since we always have that $n_f \leq (id_X)_f$, we conclude that $n_f \simeq (id_X)_f$. Thus, $X \in \Delta_\mathcal{N}(\mathcal{B})$ and so $(D_\mathcal{N} \circ J_\mathcal{N})(\mathcal{B}) \subseteq \Delta_\mathcal{N}(\mathcal{B})$.

Now, let $Y \in \Delta_\mathcal{N}(\mathcal{B})$ and let $M \xrightarrow{m} Y \in \mathcal{N}_Y$. Consider $X \xrightarrow{f} Y$ with $X \in \mathcal{B}$ and $n \in \mathcal{N}_X$ such that $n_f \leq m$. Since f is $\mathcal{N}$-constant, we have that $(id_X)_f \simeq n_f \leq m$. Consequently, from Proposition 2.7 we have that $m^{J_\mathcal{N}(\mathcal{B})} \simeq m$ for every $M \xrightarrow{m} X$ in $\mathcal{N}_Y$. So, $X \in (D_\mathcal{N} \circ J_\mathcal{N})(\mathcal{B})$. Thus $D_\mathcal{N} \circ J_\mathcal{N} = \Delta_\mathcal{N}$. □

REMARK 2.9. (1) Notice that, since $(D_\mathcal{N} \circ J_\mathcal{N}, I_\mathcal{N} \circ T_\mathcal{N})$ is an adjoint situation, we could have proved either of the two equalities $I_\mathcal{N} \circ T_\mathcal{N} = \nabla_\mathcal{N}$ or $D_\mathcal{N} \circ J_\mathcal{N} = \Delta_\mathcal{N}$ and obtain the other one (up to isomorphism) by the uniqueness of adjoint situations (cf. [1, Proposition 19.9] and [9, Proposition 1.04]).

(2) Also notice that in the case that $\mathcal{X}$ is a construct satisfying all the assumptions of [3] and $\mathcal{N} = \mathcal{M}$, then our present notion of $\mathcal{N}$-constant morphism falls short of agreeing with the one of constant morphism in [3]. The only difference being the fact that in this case, morphisms with constant domain are $\mathcal{M}$-constant but they are not constant morphisms according to [3]. However, in the category **Top** of topological spaces (resp. **Ab** of abelian groups), by taking $\mathcal{M}$={ all extremal monomorphisms} and $\mathcal{N}$={ all extremal monomorphisms with nonempty domain} (obviously in **Ab** $\mathcal{N}$=$\mathcal{M}$), Proposition 2.2 produces connectedness and disconnectedness (resp. torsion theorics). More detailed examples that illustrate Theorem 2.8 can be found in [2].

We conclude with some results that show how the various constructions presented above interact with each other.

PROPOSITION 2.10. *Let $\mathcal{N} \subseteq \mathcal{M}$ and let $\mathcal{B} \in S(\mathcal{X})$ and $\mathcal{A} \in S(\mathcal{X})^{op}$ be two corresponding fixed points of the Galois connection $(S(\mathcal{X}), \Delta_\mathcal{N}, \nabla_\mathcal{N}, S(\mathcal{X})^{op})$. Then, the following hold:*

(a) If $Y \in \mathcal{A}$ and $m \in \mathcal{M}_Y$, then m is $J_\mathcal{N}(\mathcal{B})$-closed;

(b) Let $\mathcal{N}$ be closed under the formation of pullbacks. If $X \in \mathcal{B}$ and $m \in \mathcal{M}_X$, then m is $T_\mathcal{N}(\mathcal{A})$-dense;

(c) $J_\mathcal{N}(\mathcal{B}) \sqsubseteq T_\mathcal{N}(\mathcal{A})$;

(d) If $\mathcal{N}$ is closed under direct images, then for every $C \in CL(\mathcal{X}, \mathcal{M})$, we have that $D_\mathcal{N}(C) \subseteq (\Delta_\mathcal{N} \circ I_\mathcal{N})(C)$;

(e) *If $\mathcal{N}$ is closed under direct images, then for every $C \in CL(\mathcal{X}, \mathcal{M})$, we have that $I_{\mathcal{N}}(C) \subseteq (\nabla_{\mathcal{N}} \circ D_{\mathcal{N}})(C)$.*

Proof: (a). Let $Y \in \mathcal{A}$ and let $X \xrightarrow{f} Y$ be an $\mathcal{X}$-morphism. Let $M \xrightarrow{m} Y$ be an $\mathcal{M}$-subobject and let $N \xrightarrow{n} X$ be an $\mathcal{N}$-subobject such that $X \in \mathcal{B}$ and $n_f \leq m$. Then, f is $\mathcal{N}$-constant and therefore we have that $(id_X)_f \simeq n_f \leq m$. Thus, $m^{J_{\mathcal{N}}(\mathcal{B})} \simeq m$, i.e.; m is $J_{\mathcal{N}}(\mathcal{B})$-closed.

(b). Let $X \in \mathcal{B}$ and let $M \xrightarrow{m} X$ be an $\mathcal{M}$-subobject. Suppose that $X \xrightarrow{f} Y$ is an $\mathcal{X}$-morphism with $Y \in \mathcal{A}$ and let $N \xrightarrow{n} Y$ be an $\mathcal{N}$-subobject with $m \leq f^{-1}(n)$. Since $\mathcal{N}$ is closed under the formation of pullbacks, then $f^{-1}(n) \in \mathcal{N}_X$. Now, f is $\mathcal{N}$-constant and so we have that $(id_X)_f \simeq (f^{-1}(n))_f \leq n$. This implies that $id_X \leq f^{-1}((id_X)_f) \leq f^{-1}(n)$. Thus, $m^{T_{\mathcal{N}}(\mathcal{A})} \simeq id_X$. Hence, m is $T_{\mathcal{N}}(\mathcal{A})$-dense.

(c). It is a direct consequence of (a).

(d). Let $X \in I_{\mathcal{N}}(C)$, let $Y \in D_{\mathcal{N}}(C)$ and let $X \xrightarrow{f} Y$ be a morphism. From the properties of closure operators and the fact that $\mathcal{N}$ is closed under direct images, we have that for every $N \xrightarrow{n} X \in \mathcal{N}$, $(id_X)_f \simeq (n^C)_f \leq (n_f)^C \simeq n_f$. Thus we obtain that $Y \in (\Delta_{\mathcal{N}} \circ I_{\mathcal{N}})(C)$.

(e). Similarly to (d). □

References

1. J. Adamek, H. Herrlich, G.E. Strecker, *Abstract and Concrete Categories*, Wiley, New York, 1990.

2. G. Castellini, "Connectedness, disconnectedness and closure operators: further results," in progress.

3. G. Castellini, D. Hajek, "Closure operators and connectedness," *Topology and its Appl.*, 55 (1994), 29-45.

4. G. Castellini, J. Koslowski, G.E. Strecker, "Closure operators and polarities," Proccedings of the 1991 Summer Conference on General Topology and Applications in Honor of Mary Ellen Rudin and Her Work, *Annals of the New York Academy of Sciences*, Vol. 704 (1993), 38-52. .

5. G. Castellini, J. Koslowski, G.E. Strecker, "An approach to a dual of regular closure operators," *Cahiers Topologie Geom. Differentielle Categoriques*, 35(2) (1994), 219-244.

6. M. M. Clementino, "Constant morphisms and constant subcategories," preprint.

7. D. Dikranjan, E. Giuli, "Closure operators I," *Topology and its Appl.* 27 (1987), 129-143.

8. D. Dikranjan, E. Giuli, W. Tholen, "Closure operators II," *Proceedings of the Conference in Categorical Topology*, (Prague, 1988), World Scientific (1989), 297-335.

9. M. Erné, J. Koslowski, A. Melton, G. Strecker, "A primer on Galois connections," Proceedings of the 1991 Summer Conference on General Topology and Applications in Honor of Mary Ellen Rudin and Her Work, *Annals of the New York Academy of Sciences*, Vol. 704 (1993) 103-125.

10. H. Herrlich, "Topologische Reflexionen und Coreflexionen," L.N.M. 78, Sprin-ger, Berlin, 1968.

11. D. Holgate, *Closure operators in categories*, Master Thesis, University of Cape Town, 1992.

12. M. Hušek, D. Pumplün, "Disconnectednesses," *Quaestiones Mathematicae* 13 (1990), 449-459.

13. J. Koslowski, "Closure operators with prescribed properties," *Category Theory and its Applications* (Louvain-la-Neuve, 1987) Springer L.N.M. 1248 (1988), 208–220.

14. D. Petz, "Generalized connectednesses and disconnectednesses in topology," *Ann. Univ. Sci. Budapest Eötvös, Sect. Math.* 24 (1981), 247-252.

15. G. Preuss, "Eine Galois-Korrespondenz in der Topologie," *Monatsh. Math.* 75 (1971), 447-452.

16. G. Preuss, "Relative connectednesses and disconnectednesses in topological categories," *Quaestiones Mathematicae* 2 (1977), 297-306.

17. G. Preuss, "Connection properties in topological categories and related topics," Springer L.N.M. 719 (1979), 293-305.

Dold Type Theorems in Cubical Homotopy Theory

K.A. HARDIE
Department of Mathematics, University of Cape Town, Rondebosch 7700, South Africa

K.H. KAMPS
Fachbereich Mathematik, Fernuniversität, Postfach 940, D–58084 Hagen, Germany

and

R.W. KIEBOOM
Department of Mathematics, Vrije Universiteit Brussel, Pleinlaan 2, F10, B–1050 Brussels, Belgium

Abstract. It is shown how a generalised comparison theorem for topological spaces under a fixed space A and over a fixed space B due to the third author can be transferred to cubical homotopy theory.

Mathematics Subject Classifications (1991). 55P10, 55U35, 55P05, 55R05

Key words: Comparison theorem, homotopy equivalence, abstract homotopy, cubical homotopy, spaces under A and over B, fibration, cofibration, covering homotopy property, homotopy extension property, covering homotopy extension property

Introduction

In homotopy theory under a fixed space A and over a fixed space B comparison theorems generalising classical results of Dold ([4], 6.1, [5], 3.6) have been obtained amongst others by James [9], Heath [8], Lam [17], and Berrick [1,2]. A typical result is the following

THEOREM 0.1. *Let*

(0.2)
$$\begin{array}{ccc} & A & \\ {}^{\sigma}\swarrow & & \searrow^{\tau} \\ X & \xrightarrow{\;u\;} & Y \\ {}_{\rho}\searrow & & \swarrow_{\eta} \\ & B & \end{array}$$

be a commutative diagram of topological spaces where σ, τ are closed cofibrations and ρ, η are fibrations. Then, if u is a homotopy equivalence, u is a homotopy equivalence

Eraldo Giuli (ed.), Categorical Topology, 139–154.

under A and over B.

A categorical approach via cubical homotopy theory has been given in [3]. Recently, in [7] a new access to Theorem (0.1) based on the techniques of a certain type of coherent homotopy theory has been presented. In all these cases the arguments were based on the assumption that in diagram (0.2) pairs of maps such as $(\sigma, \eta), (\tau, \rho)$ etc. satisfy a suitable covering homotopy extension property (CHEP). A careful analysis and refinement of the arguments in [7] has lead the third author [16] to a substantial generalisation of Theorem (0.1) weakening the assumptions on the maps in diagram (0.2). In the present paper we show that the approach in [16] which has been given for topological spaces can be transferred to cubical homotopy theory.

1. Cubical homotopy theory

For convenience of the reader we recall the basic notions of cubical homotopy theory (see [11], [12], [13], [14], [15] where the terminology semicubical homotopy theory has been used).

A *cubical set* Q consists of a sequence of sets, Q_n, $n = 0, 1, 2, \cdots$, together with maps

$$\varepsilon_n^i : Q_n \longrightarrow Q_{n-1} \qquad (\varepsilon = 0, 1;\ 1 \le i \le n)$$

$$\zeta_n^j : Q_n \longrightarrow Q_{n+1} \qquad (1 \le j \le n+1)$$

satisfying the usual interchange rules (see [11], 2.1, [12], 1). The elements of Q_n are called *n–cubes*, the maps ε_n^i and ζ_n^j are called *face* resp. *degeneracy operators.*

For $\psi \in Q_n$ we define $\psi_\varepsilon^i = \varepsilon_n^i \psi$, for $\varphi \in Q_1$ we also use $\varphi_\varepsilon = \varepsilon_1^1 \varphi$. For $\psi \in Q_n$, we call

$$D\psi = (\psi_0^1, \psi_1^1, \cdots, \psi_0^n, \psi_1^n)$$

the *boundary* of ψ. For $n \ge 1$ let $\zeta_n : Q_0 \longrightarrow Q_n$ denote the map

$$\zeta_n = \zeta_{n-1}^1 \circ \cdots \circ \zeta_0^1.$$

A *cubical map* between cubical sets is a sequence of maps commuting with face and degeneracy operators. Thus we get the *category, Cub, of cubical sets.*

We will need two *Kan filler (extension) conditions* E(2), E(3) for cubical sets in low dimensions. For the definition we refer to [11], 2.2. The idea of E(2) is depicted as follows. If a cubical set Q satisfies E(2), then a *box*

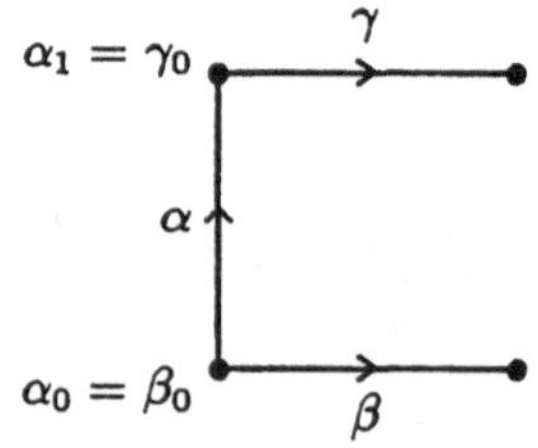

of three 1–cubes $\alpha, \beta, \gamma \in Q_1$ such that $\alpha_1 = \gamma_0$, $\alpha_0 = \beta_0$ can be filled with a 2–cube $\lambda \in Q_2$, i.e. there exists $\lambda \in Q_2$ such that

$$D\lambda = (\alpha, \lambda_1^1, \beta, \gamma).$$

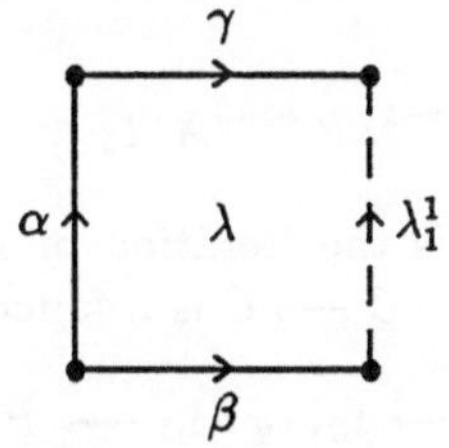

Clearly, E(3) involves dimension 3.

If a cubical set Q satisfies E(2) and E(3), then the *fundamental groupoid (track groupoid)* ΠQ of Q is defined as follows. The objects of ΠQ are the 0–cubes of Q. If f, g are 0–cubes of Q the morphisms $H : f \longrightarrow g$ in ΠQ are the classes $H = [\alpha]$, called *tracks*, of those elements $\alpha \in Q_1$ with $D\alpha = (f, g)$ with respect to the following relation $\sim$:

1.1. $\alpha \sim \beta$ if and only if there exists $\psi \in Q_2$ with

$$D\psi = (\alpha, \beta, \zeta_0^1 f, \zeta_0^1 g).$$

Composition in ΠQ which is written additively is given as follows: if $[\alpha] : f \longrightarrow g$, $[\beta] : g \longrightarrow h$ are morphisms of ΠQ and if $\lambda \in Q_2$ is such that

$$D\lambda = (\alpha, \lambda_1^1, \zeta_0^1 f, \beta)$$

(Note that λ exists by E(2).), then

$$[\beta] + [\alpha] = [\lambda_1^1].$$

A *cubical homotopy system* in a category $\mathcal{C}$ is a functor

$$Q : \mathcal{C}^{op} \times \mathcal{C} \longrightarrow \boldsymbol{Cub}$$

into the category of cubical sets such that in dimension 0 the induced functor

$$Q_0 : \mathcal{C}^{op} \times \mathcal{C} \longrightarrow \boldsymbol{Sets}$$

into the category of sets is the *hom*–functor $\mathcal{C}(-,-)$ of $\mathcal{C}$.

Note that the notion of a cubical homotopy system in a category is self dual.

If f is a morphism of $\mathcal{C}$, we use the notation $f_* = Q(1, f)$, $f^* = Q(f, 1)$.

EXAMPLE. Let $\mathcal{C}$ be a category, let (I, j_0, j_1, s) be a *cylinder* on $\mathcal{C}$, i.e. $I : \mathcal{C} \longrightarrow \mathcal{C}$ is a functor (*cylinder functor*) and

$$j_0, j_1 : 1_{\mathcal{C}} \longrightarrow I, \quad s : I \longrightarrow 1_{\mathcal{C}}$$

are natural transformations with $sj_0 = sj_1 = 1$. Then we have a cubical homotopy system Q in $\mathcal{C}$ with n–cubes

$$Q_n(X, Y) = \mathcal{C}(I^n X, Y),$$

face operators

$$\varepsilon_n^i = \mathcal{C}(I^{i-1} j_\varepsilon I^{n-i} X, 1_Y)$$

and degeneracy operators

$$\zeta_n^j = \mathcal{C}(I^{j-1}sI^{n+1-j}X, 1_Y),$$

where X, Y are objects of $\mathcal{C}$ and I^n is the iteration of I (see [12], 2.2). Dually, if (P, p_0, p_1, q) is a *cocylinder* on $\mathcal{C}$, i.e. $P : \mathcal{C} \longrightarrow \mathcal{C}$ is a functor (*cocylinder functor*) and

$$p_0, p_1 : P \longrightarrow 1_{\mathcal{C}}, \quad q : 1_{\mathcal{C}} \longrightarrow P$$

are natural transformations with $p_0 s = p_1 s = 1$, we have a cubical homotopy system Q in $\mathcal{C}$ with n-cubes

$$Q_n(X, Y) = \mathcal{C}(X, P^n Y).$$

In particular, there is a canonical homotopy system T in the category $\mathcal{T}\!\boldsymbol{op}$ of topological spaces with n-cubes

$$T_n(X, Y) = \mathcal{T}\!op(X \times [0,1]^n, Y) = \mathcal{T}\!op(X, Y^{[0,1]^n})$$

where $[0,1]$ denotes the unit interval of real numbers.

Let Q be a given cubical homotopy system in a category $\mathcal{C}$. We assume that Q satisfies the Kan conditions E(2) and E(3), i.e. for any objects X, Y of $\mathcal{C}$ the cubical set $Q(X, Y)$ satisfies E(2) and E(3). Then the basic notion of homotopy theory is induced as follows.

If $f, g : X \longrightarrow Y$ are morphisms of $\mathcal{C}$ then f is called *homotopic* to $g (f \simeq g)$ if there is an element $\varphi \in Q_1(X, Y)$ such that $D\varphi = (f, g)$. We write $\varphi : f \simeq g$; φ is called a *homotopy* from f to g. By E(2) the homotopy relation is an equivalence relation. The quotient category $\mathcal{C}/\simeq$ where each $\mathcal{C}(X, Y)$ is replaced by $\mathcal{C}(X, Y)/\simeq$, is called the *homotopy category* of $\mathcal{C}$. We shall write $\mathcal{C}h$ for $\mathcal{C}/\simeq$.

A morphism $f : X \longrightarrow Y$ of $\mathcal{C}$ is called a *homotopy equivalence* if its *homotopy class*

$$[f] \in \mathcal{C}h(X, Y)$$

is an isomorphism in $\mathcal{C}h$.

Furthermore we have an induced functor

$$\Pi : \mathcal{C}^{op} \times \mathcal{C} \longrightarrow \mathcal{G}\boldsymbol{rpd}$$

into the category of groupoids such that for any objects X, Y of $\mathcal{C}$, $\Pi(X, Y)$ is the fundamental groupoid of $Q(X, Y)$. The functor Π induces on $\mathcal{C}$ the structure of a groupoid enriched category if Q is assumed to satisfy the following axiom (A) (see [13], 2).

(A) For any objects X, Y, Z of $\mathcal{C}$ and any $\alpha \in Q_1(X, Y)$, $\beta \in Q_1(Y, Z)$ there exists $\lambda \in Q_2(X, Z)$ such that

$$D\lambda = (\beta_{0*}\alpha, \ \beta_{1*}\alpha, \ \alpha_0^*\beta, \ \alpha_1^*\beta).$$

Note that (A) holds if Q is induced either by a cylinder on $\mathcal{C}$ (choose $\lambda = \beta(I\alpha)$) or by a cocylinder on $\mathcal{C}$ (choose $\lambda = (P\beta)\alpha$).

If Q is a cubical homotopy system in a category $\mathcal{C}$, then for any objects A, B of $\mathcal{C}$ we have induced cubical homotopy systems Q^A in the category $\mathcal{C}^A$ of objects under A, Q_B in the category $\mathcal{C}_B$ of objects over B, and Q_B^A in the category $\mathcal{C}_B^A$ of objects under A and over B. Recall that the objects of $\mathcal{C}^A$ resp. $\mathcal{C}_B$ are the morphisms of $\mathcal{C}$ with domain

A resp. codomain B. An object $i: A \longrightarrow X$ of $\mathcal{C}^A$ will sometimes be denoted by (X, i). The objects of $\mathcal{C}^A_B$ are pairs of morphisms (σ, ρ), where

$$A \xrightarrow{\sigma} X \xrightarrow{\rho} B.$$

The morphisms in $\mathcal{C}^A, \mathcal{C}_B, \mathcal{C}^A_B$ are commutative diagrams in $\mathcal{C}$ of the form

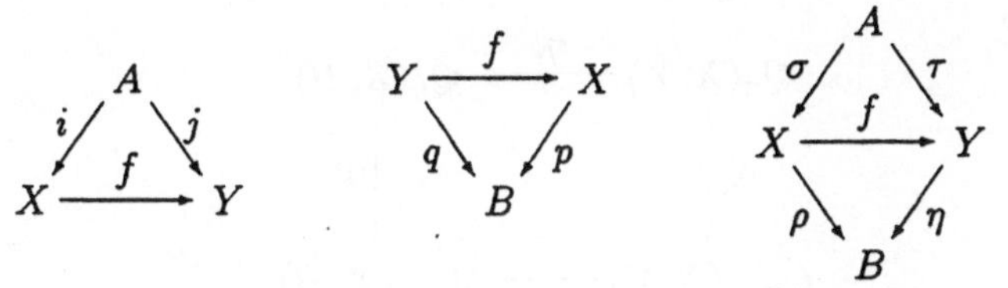

The induced cubical homotopy system are described by the formulae

$$Q^A_n(i,j) = \{\varphi \in Q_n(X,Y) \mid i^*\varphi = \zeta_n(j)\}$$

$$(Q_B)_n(q,p) = \{\psi \in Q_n(Y,X) \mid p_*\psi = \zeta_n(q)\}$$

$$(Q^A_B)_n((\sigma,\rho),(\tau,\eta)) = \{\gamma \in Q_n(X,Y) \mid \eta_*\gamma = \zeta_n(\rho), \sigma^*\gamma = \zeta_n(\tau)\}$$

where $i^* = Q(i, 1_Y)$, $p_* = Q(1_Y, p)$.

If Q satisfies the Kan conditions E(2) and E(3) there is no reason why these conditions should be inherited by Q^A, Q_B resp. Q^A_B. However, there is a variant of E(2) resp. E(3), denoted by DNE(2) resp. DNE(3) which does carry over to the induced cubical homotopy systems. For the definition of DNE(2) and DNE(3) we refer to [11], 3. Note that the 'N' in DNE means that fillers can be chosen naturally whereas the 'D' indicates that degenerate boxes give rise to degenerate fillers. Thus, if Q is assumed to satisfy DNE(2), DNE(3), we have induced notions of homotopy in $\mathcal{C}^A, \mathcal{C}_B$, and $\mathcal{C}^A_B$: homotopic under $A(\overset{A}{\simeq})$, over $B(\underset{B}{\simeq})$, under A and over B $(\overset{A}{\underset{B}{\simeq}})$, we have the corresponding induced homotopy categories under A, $\mathcal{C}^A h$, over B, $\mathcal{C}_B h$, resp. under A and over B, $\mathcal{C}^A_B h$, and the corresponding notions of homotopy equivalence.

If Q is induced by a cylinder on $\mathcal{C}$, then Q_B is automatically induced by a cylinder on $\mathcal{C}_B$ whereas Q^A resp. Q^A_B are induced by a cylinder on $\mathcal{C}^A$ resp. $\mathcal{C}^A_B$ if $\mathcal{C}$ has pushouts and pushouts are preserved by the cylinder functor (see [10], 2.4). Dually, if Q is induced by a cocylinder on $\mathcal{C}$, then Q^A is induced by a cocylinder on $\mathcal{C}^A$, whereas Q_B resp. Q^A_B are induced by a cocylinder on $\mathcal{C}_B$ resp. $\mathcal{C}^A_B$ if $\mathcal{C}$ has pullbacks and pullbacks are preserved by the cocylinder functor.

2. Fibration and cofibration concepts

In this section we present fibration and cofibration concepts in cubical homotopy theory. Some of them are cubical versions of well known topological concepts, others are new and have not been considered even in the topological case.

Throughout this section let

$$Q: \mathcal{C}^{op} \times \mathcal{C} \longrightarrow \mathcal{C}ub$$

be a cubical homotopy system in a category $\mathcal{C}$. We assume that Q satisfies the Kan conditions DNE(2) and DNE(3).

First, we consider fibration concepts and define various covering homotopy properties.

DEFINITION 2.1. Let $\eta : Y \longrightarrow B$ be a morphism of $\mathcal{C}$, let X be an object of $\mathcal{C}$, let $n \geq 1$ be an integer.

(a) η is said to satisfy the CHP(n) (CHP = *covering homotopy property*) w.r.t. X if the diagram

$$\begin{array}{ccc} Q_n(X,Y) & \xrightarrow{\eta_*} & Q_n(X,B) \\ {\scriptstyle 0_n^1}\downarrow & & \downarrow{\scriptstyle 0_n^1} \\ Q_{n-1}(X,Y) & \xrightarrow[\eta_*]{} & Q_{n-1}(X,B) \end{array} ,$$

where $\eta_* = Q(1_X, \eta)$ is a weak pullback in $\mathcal{S}ets$, i.e. for any

$$f \in Q_{n-1}(X,Y), \quad \psi \in Q_n(X,B)$$

such that $\psi_0^1 = \eta_* f$ there exists $\Psi \in Q_n(X,Y)$ such that

$$\Psi_0^1 = f \text{ and } \eta_*\Psi = \psi.$$

(b) η is said to satisfy the WCHP(n) (WCHP = *weak* CHP) w.r.t. X if for any

$$f \in Q_{n-1}(X,Y), \quad \psi \in Q_n(X,B)$$

such that $\psi_0^1 = \eta_* f$ there exist $\Psi, \Phi \in Q_n(X,Y)$ such that

$$\eta_*\Psi = \psi, \quad \Phi_0^1 = f, \quad \Phi_1^1 = \Psi_0^1 \text{ and } \eta_*\Phi = \zeta_{n-1}^1(\eta_* f).$$

(c) η is said to satisfy the CHP (resp. WCHP) w.r.t. X if it satisfies CHP(1) (resp. WCHP(1)) w.r.t. X, i.e. for any morphism $f : X \longrightarrow Y$ of $\mathcal{C}$ and any homotopy $\psi \in Q_1(X,B)$, $\psi : \eta f \simeq \psi_1$, there exists a homotopy $\Psi \in Q_1(X,Y)$ such that $\eta_*\Psi = \psi$ and $\Psi_0 = f$ (resp. $f \underset{B}{\simeq} \Psi_0$).

(d) η is said to satisfy the RWCHP (= *rather* WCHP) w.r.t. X, if for any morphism $f : X \longrightarrow Y$ of $\mathcal{C}$ and any homotopy $\psi \in Q_1(X,B)$, $\psi : \eta f \simeq \psi_1$, there exists a homotopy $\Psi \in Q_1(X,Y)$ such that $\Psi_0 = f$ and $\eta_*\Psi \sim \psi$ (see (1.1)).

Note that CHP(n) w.r.t. X implies WCHP(n) w.r.t. X (let $\Phi = \zeta_{n-1}^1 f$), (W)CHP(n) w.r.t. X implies (W)CHP(m) w.r.t. X for $m \leq n$. Furthermore WCHP w.r.t. X implies RWCHP w.r.t. X ([12], 3.6). If Q is induced by a cylinder on $\mathcal{C}$, then η satisfies the (W)CHP(n) w.r.t. X if and only if η satisfies the (W)CHP w.r.t. $I^{n-1}X$.

Cofibration concepts in cubical homotopy theory are obtained by dualisation of Definition (2.1). The following homotopy extension properties of a morphism of $\mathcal{C}$ w.r.t. an object of $\mathcal{C}$ are available:

HEP(n) (HEP = *homotopy extension property*)
WHEP(n) (WHEP = *weak* HEP)
HEP, WHEP, RWHEP (= *rather* WHEP).

We restrict ourselves to give the explicit definition of the HEP(n).

DEFINITION 2.2. Let $\sigma : A \longrightarrow X$ be a morphism of $\mathcal{C}$, let Y be an object of $\mathcal{C}$, let $n \geq 1$ be an integer. Then σ is said to satisfy the HEP(n) w.r.t Y if the diagram

$$\begin{array}{ccc} Q_n(X,Y) & \xrightarrow{\sigma^*} & Q_n(A,Y) \\ \downarrow 0_n^1 & & \downarrow 0_n^1 \\ Q_{n-1}(X,Y) & \xrightarrow[\sigma^*]{} & Q_{n-1}(A,Y) \end{array} ,$$

where $\sigma^* = Q(\sigma, 1_Y)$ is a weak pullback in $\mathcal{S}\boldsymbol{ets}$, i.e. for any $f \in Q_{n-1}(X,Y)$, $\varphi \in Q_n(A,Y)$ such that $\varphi_0^1 = \sigma^* f$ there exists $\Phi \in Q_n(X,Y)$ such that $\Phi_0^1 = f$ and $\sigma^*\Phi = \varphi$.

Let A be a fixed object of $\mathcal{C}$. If we apply the fibration resp. cofibration concepts to the induced cubical homotopy system Q^A in the category $\mathcal{C}^A$ of objects under A, we obtain the following fibration resp. cofibration properties under A of a morphism under A w.r.t. an object of $\mathcal{C}^A$:

$$\begin{array}{lllll} \mathrm{CHP}^A(n), & \mathrm{WCHP}^A(n), & \mathrm{CHP}^A, & \mathrm{WCHP}^A, & \mathrm{RWCHP}^A, \\ \mathrm{HEP}^A(n), & \mathrm{WHEP}^A(n), & \mathrm{HEP}^A, & \mathrm{WHEP}^A, & \mathrm{RWHEP}^A. \end{array}$$

We give an explicit description of $\mathrm{CHP}^A(n)$ and $\mathrm{WCHP}^A(n)$.

Definition 2.3. Let

$$\begin{array}{ccc} A & \xrightarrow{\tau} & Y \\ \sigma\downarrow & \searrow b & \downarrow\eta \\ X & & B \end{array}$$

be given morphisms of $\mathcal{C}$ with $\eta \circ \tau = b$, let $n \geq 1$ be an integer.

(a) $\eta : (Y,\tau) \longrightarrow (B,b)$ is said to have the $\mathrm{CHP}^A(n)$ w.r.t. (X,σ) if for any $f \in Q_{n-1}(X,Y)$, $\psi \in Q_n(X,B)$ such that

2.4. $$\sigma^* f = \zeta_{n-1}(\tau), \quad \sigma^*\psi = \zeta_n(b), \quad \psi_0^1 = \eta_* f$$

there exists $\Psi \in Q_n(X,Y)$ such that

$$\sigma^*\Psi = \zeta_n(\tau), \quad \Psi_0^1 = f \text{ and } \eta_*\Psi = \psi.$$

(b) $\eta : (Y,\tau) \longrightarrow (B,b)$ is said to have the $\mathrm{WCHP}^A(n)$ w.r.t. (X,σ) if for any $f \in Q_{n-1}(X,Y)$, $\psi \in Q_n(X,B)$ such that (2.4) holds there exist $\Psi, \Phi \in Q_n(X,Y)$ such that

$$\sigma^*\Psi = \zeta_n(\tau), \quad \sigma^*\Phi = \zeta_n(\tau), \quad \eta_*\Psi = \psi$$
$$\Phi_0^1 = f, \quad \Phi_1^1 = \Psi_0^1, \quad \eta_*\Phi = \zeta_{n-1}^1(\eta_* f).$$

In homotopy theory under A and over B, where A and B are objects of $\mathcal{C}$, there is a cubical version of the classical simultaneous covering homotopy extension property for topological spaces. We also present a weak covering homotopy extension property as a new concept.

Definition 2.5. Let (σ,η), $\sigma : A \longrightarrow X$, $\eta : Y \longrightarrow B$ be a pair of morphisms of $\mathcal{C}$, let $n \geq 1$ be an integer.

(a) (σ,η) is said to have the $\mathrm{CHEP}(n)$ (CHEP = *covering homotopy extension property*), if the diagram

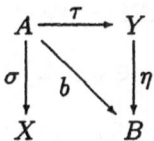

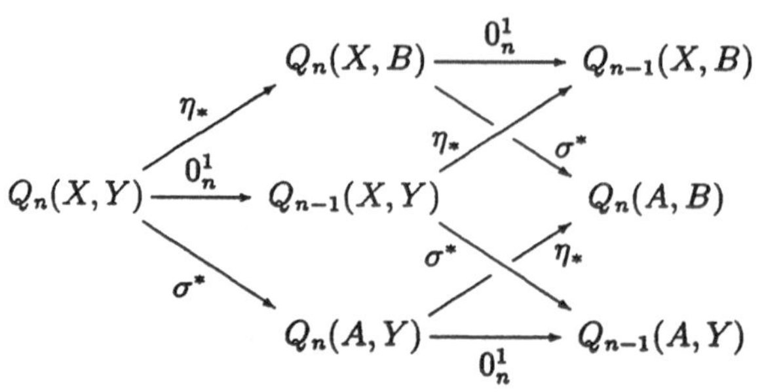

is a weak limit in *Sets*, i.e. for any

$$f \in Q_{n-1}(X,Y), \ \ \varphi \in Q_n(A,Y), \ \ \psi \in Q_n(X,B)$$

such that

2.6. $$\varphi^1_0 = \sigma^* f, \ \ \psi^1_0 = \eta_* f \ \text{ and } \ \eta_*\varphi = \sigma^*\psi$$

there exists $\Phi \in Q_n(X,Y)$ such that

$$\sigma^*\Phi = \varphi, \ \ \eta_*\Phi = \psi \ \text{ and } \ \Phi^1_0 = f.$$

(b) (σ,η) is said to have the WCHEP(n) (WCHEP = *weak* CHEP), if for any

$$f \in Q_{n-1}(X,Y), \ \ \varphi \in Q_n(A,Y), \ \ \psi \in Q_n(X,B)$$

such that (2.6) holds there exist $\Phi, \Psi \in Q_n(X,Y)$ such that

$$\sigma^*\Phi = \varphi, \eta_*\Phi = \psi, \ \ \Psi^1_0 = f, \ \ \Psi^1_1 = \Phi^1_0$$

$$\sigma^*\Psi = \zeta^1_{n-1}(\sigma^* f), \ \ \eta_*\Psi = \zeta^1_{n-1}(\eta_* f).$$

The following relation between some of the fibration / cofibration concepts is crucial.

PROPOSITION 2.7. *Let* (σ,η), $\sigma : A \longrightarrow X$, $\eta : Y \longrightarrow B$ *be a pair of morphisms of* C *which has the* (W)CHEP(n). *Then* $\eta : (Y,\tau) \longrightarrow (B, \eta\circ\tau)$ *has the* (W)CHP^A(n) w.r.t. (X,σ) *for all* $\tau : A \longrightarrow Y$.

Proof. Let $f \in Q_{n-1}(X,Y)$, $\psi \in Q_n(X,B)$ be given such that (2.4) holds. Now apply the (W)CHEP(n) to

$$f \in Q_{n-1}(X,Y), \ \ \zeta_n(\tau) \in Q_n(A,Y), \ \ \psi \in Q_n(X,B).$$

The result follows. □

3. Track homotopy

Let Q be a cubical homotopy system in a category C. We assume that Q satisfies the Kan conditions DNE(2) and DNE(3). Furthermore, we assume that axiom (A) holds. Then C has the structure of a groupoid enriched category. In particular, the notion of a homotopy commutative square in C is available and we have both a vertical and a horizontal composition of homotopy commutative squares.

DEFINITION 3.1. (a) A *homotopy commutative square* in C

$$X \xrightarrow{\ f\ } X'$$

$$p \downarrow \quad {}^{H)} \quad \downarrow p'$$

$$Y \xrightarrow{\ g\ } Y'$$

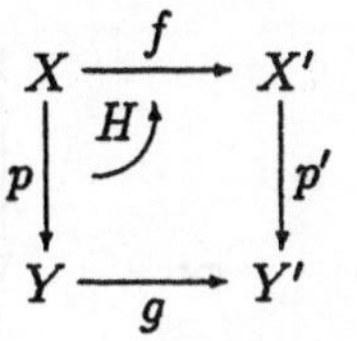

consists of morphisms of $\mathcal{C}$, p, p', f, g, and a track $H : gp \longrightarrow p'f$ in the fundamental groupoid $\Pi(X, Y')$.

(b) The vertical composition of homotopy commutative squares in $\mathcal{C}$

$$X \xrightarrow{\ f\ } X'$$

$$p \downarrow \quad {}^{H)} \quad \downarrow p'$$

$$Y \xrightarrow{\ g\ } Y'$$

$$q \downarrow \quad {}^{K)} \quad \downarrow q'$$

$$Z \xrightarrow{\ h\ } Z'$$

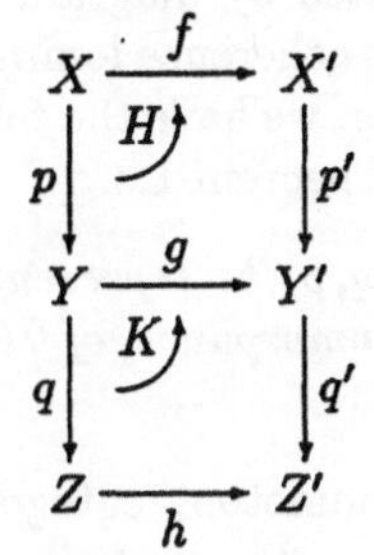

is defined as the homotopy commutative square

$$X \xrightarrow{\ f\ } X'$$

$$qp \downarrow \quad {}^{L)} \quad \downarrow q'p'$$

$$Z \xrightarrow{\ h\ } Z'$$

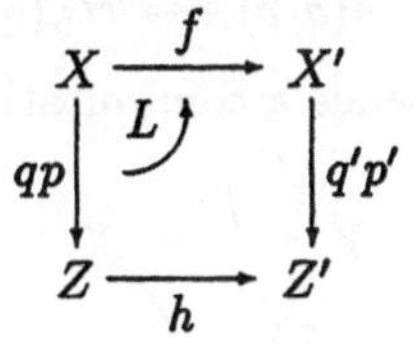

with

$$L = q'_* H + p^* K,$$

where $p^* : \Pi(Y, Z') \longrightarrow \Pi(X, Z')$, $q'_* : \Pi(X, Y') \longrightarrow \Pi(X, Z')$ are the induced morphisms of groupoids.

The definition of the horizontal composition is similar.

Let B be a fixed object of $\mathcal{C}$. Then we can define the track homotopy category $\mathcal{H}_B$ over B. The objects of $\mathcal{H}_B$ are the morphisms $p : X \longrightarrow B$ of $\mathcal{C}$ with codomain B. Thus $\mathcal{H}_B$ has the same objects as $\mathcal{C}_B$. The set $\mathcal{H}_B(q, p)$ of morphisms $q \longrightarrow p$ of $\mathcal{H}_B$ is obtained from the set of homotopy commutative squares of the form

(3.2)

$$Y \xrightarrow{\ f\ } X$$

$$q \downarrow \quad {}^{H)} \quad \downarrow p$$

$$B =\!=\!= B$$

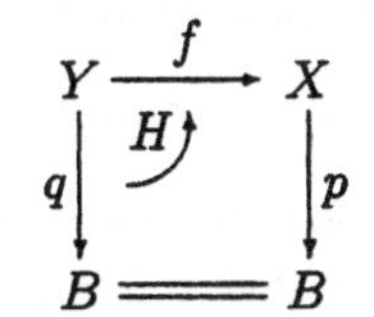

by factoring out by the equivalence relation

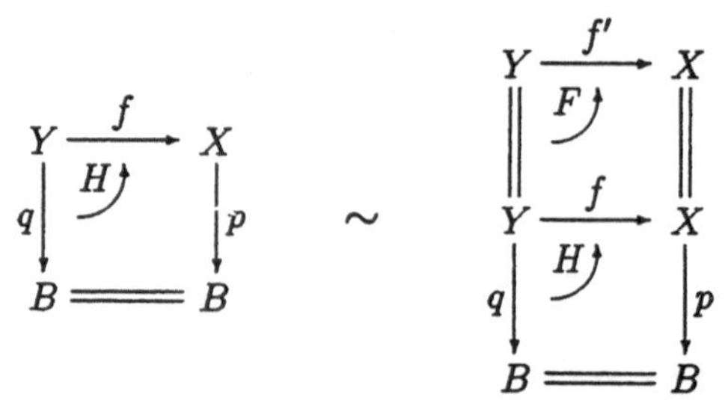

where $F : f \longrightarrow f'$ is a track in the fundamental groupoid $\Pi(Y, X)$.

The morphism $q \longrightarrow p$ represented by diagram (3.2) is denoted by $\{f, H\}$. Since the arguments of [6] including Vogt's coherence lemma for homotopy equivalences ([19]) remain valid in our abstract situation, we have the following characterisation of isomorphisms in $\mathcal{H}_B$ corresponding to [6] Theorem 1.3.

THEOREM 3.3. *Let* $\{f, H\} \in \mathcal{H}_B(q,p)$ *be represented by the homotopy commutative square* (3.2). *Then* $\{f, H\}$ *is an isomorphism of* $\mathcal{H}_B$ *if and only if* f *is a homotopy equivalence.*

We want to compare the track homotopy category $\mathcal{H}_B$ over B and the homotopy category over B, $\mathcal{C}_Bh$. Let $p : X \longrightarrow B$, $q : Y \longrightarrow B$ be morphisms of $\mathcal{C}$. Then we have a canonical map

$$\theta : \mathcal{C}_Bh(q,p) \longrightarrow \mathcal{H}_B(q,p)$$

induced by the assignment which sends a commutative triangle

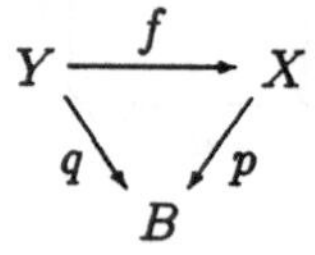

to the square

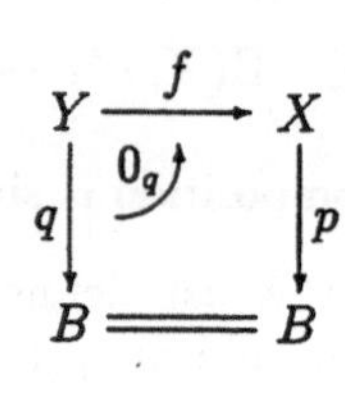

with the track 0_q of the constant homotopy $\zeta_0^1(q) : q \simeq q$. Thus we have

$$\theta[f]_B = \{f, 0_q\}$$

where $[f]_B$ denotes the homotopy class of f over B.

PROPOSITION 3.4. (a) *If p has the* RWCHP *w.r.t. Y, then θ is a surjection.*

(b) *If p has the* WCHP(2) *w.r.t. Y, then θ is an injection.*

Proof (a) Let $\{f, H\} \in \mathcal{H}_B(q,p)$ be represented by the homotopy commutative square (3.2). Choose a homotopy $\alpha : pf \simeq q$, $\alpha \in Q_1(Y, B)$ such that $[\alpha] = -H$. By the RWCHP of p w.r.t. Y there exists a homotopy $\Phi \in Q_1(Y, X)$ such that $\Phi_0 = f$ and $p_*\Phi \sim \alpha$. In particular, we have

$$p\Phi_1 = \alpha_1 = q.$$

Hence Φ_1 represents an element $[\Phi_1]_B \in \mathcal{C}_B h(q,p)$. Since

$$p_*(-[\Phi]) = -[p_*\Phi] = -[\alpha] = H,$$

it follows that

$$\theta[\Phi_1]_B = \{f, H\}.$$

(b) follows immediately from the definition of the track category $\mathcal{H}_B$ over B and [12], Lemma 3.7. □

Corollary 3.5. *If p has the* WCHP(2) *w.r.t. Y, then θ is a bijection.*

Proof. We have the implications WCHP(2) $\Longrightarrow$ WCHP(1) $\Longrightarrow$ RWCHP. □

4. Comparison theorems

We are now in a position to prove comparison theorems in cubical homotopy theory.

Theorem 4.1. *Let $\mathcal{C}$ be a category with a cubical homotopy system Q that satisfies* DNE(2), DNE(3) *and axiom* (A). *Let*

$$\begin{array}{ccc} E & \xrightarrow{\;f\;} & E' \\ & {\scriptstyle p}\searrow \quad \swarrow{\scriptstyle p'} & \\ & B & \end{array}$$

be a commutative diagram in $\mathcal{C}$. Assume that

(1) *p has the* WCHP(2) *w.r.t.* E
(2) *p has the* RWCHP *w.r.t.* E′
(3) *p' has the* WCHP(2) *w.r.t.* E′.

Then if f is a homotopy equivalence, f is a homotopy equivalence over B.

Proof. In order to show that f is a homotopy equivalence over B, i.e. $[f]_B : p \longrightarrow p'$ is an isomorphism in $\mathcal{C}_B h$, we have to show that

(a) the induced set map

$$f_* : \mathcal{C}_B h(p', p) \longrightarrow \mathcal{C}_B h(p', p')$$

is surjective,

(b) the induced set map

$$f_* : \mathcal{C}_B h(p, p) \longrightarrow \mathcal{C}_B h(p, p')$$

is injective.

Consider the diagrams

$$\begin{array}{ccc} \mathcal{C}_B h(p', p) & \xrightarrow{\;f_*\;} & \mathcal{C}_B h(p', p') \\ \Big\downarrow{\scriptstyle\theta} & & \Big\downarrow{\scriptstyle\theta} \\ \mathcal{H}_B(p', p) & \xrightarrow[\;f_\square\;]{} & \mathcal{H}_B(p', p') \end{array} \tag{4.2}$$

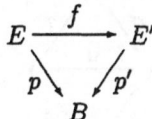

$$(4.3)\qquad \begin{array}{ccc} \mathcal{C}_B h(p,p) & \xrightarrow{f_*} & \mathcal{C}_B h(p,p') \\ \Big\downarrow{\scriptstyle\theta} & & \Big\downarrow{\scriptstyle\theta} \\ \mathcal{H}_B(p,p) & \xrightarrow[f_\square]{} & \mathcal{H}_B(p,p') \end{array}$$

where the arrows labelled $f_\square$ are induced by composition in $\mathcal{H}_B$. By Theorem (3.3) these arrows are bijections, since f is assumed to be a homotopy equivalence. It follows that (a) holds, if the left hand vertical arrow in (4.2) is surjective and the right hand vertical arrow in (4.2) is bijective while (b) holds if the left hand vertical arrow in (4.3) is injective. Application of Proposition (3.4) now completes the proof. □

We leave it to the reader to dualise Theorem (4.1).

THEOREM 4.4. *Let C be a category with a cubical homotopy system Q that satisfies* DNE(2) *and* DNE(3). *We assume that Q is induced by a cocylinder on C. Let*

$$\begin{array}{ccccc} & & A & & \\ & \sigma\swarrow & & \searrow\tau & \\ X & & \xrightarrow{u} & & Y \\ & \rho\searrow & & \swarrow\eta & \\ & & B & & \end{array}$$

be a commutative diagram in C. Let $b = \rho\circ\sigma = \eta\circ\tau$. Assume that

(1) *σ has the* WHEP(2) *w.r.t. X, τ has the* WHEP(2) *w.r.t. Y and the* RWHEP *w.r.t. X,*

(2) *$\rho : (X,\sigma) \longrightarrow (B,b)$ has the* WCHP^A(2) *w.r.t. (X,σ) and the* RWCHP^A *w.r.t. (Y,τ), and $\eta : (Y,\tau) \longrightarrow (B,b)$ has the* WCHP^A(2) *w.r.t. (Y,τ).*

Then if u is a homotopy equivalence, u is a homotopy equivalence under A and over B.

Proof. First, apply the dual of Theorem (4.2) to the commutative triangle in C

$$\begin{array}{ccccc} & & A & & \\ & \sigma\swarrow & & \searrow\tau & \\ X & & \xrightarrow[u]{} & & Y\ . \end{array}$$

It follows that u is a homotopy equivalence under A. Then apply Theorem (4.1) to the induced cubical homotopy system Q^A in C^A and the commutative triangle in C^A

$$\begin{array}{ccccc} (X,\sigma) & & \xrightarrow{u} & & (Y,\tau) \\ & \rho\searrow & & \swarrow\eta & \\ & & (B,b)\ . & & \end{array}$$

The result follows. □

We leave it to the reader to dualise Theorem (4.4).

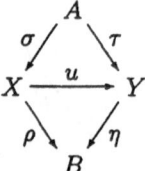

5. The topological case

In this section $\mathcal{C}$ will be the category $\boldsymbol{Top}$, equipped with the canonical homotopy system

$$T : \boldsymbol{Top}^{op} \times \boldsymbol{Top} \longrightarrow \boldsymbol{Cub}$$

introduced in section 1. In order to facilitate the interpretation of the (co–)fibration concepts of section 2 in the topological case we will describe the structure of the cubical set

$$T(X,Y) = (T_n(X,Y))_{n\in\mathbb{N}}$$

in greater detail.

We already know that the n–cubes $T_n(X,Y)$ are given by

$$T_n(X,Y) = \boldsymbol{Top}(X \times [0,1]^n, Y) = \boldsymbol{Top}(X, Y^{[0,1]^n})$$

for $n \geq 1$ and $T_0(X,Y) = \boldsymbol{Top}(X,Y.)$ The face operators

$$\varepsilon_n^i : T_n(X,Y) \longrightarrow T_{n-1}(X,Y) : \psi \longmapsto \psi_\varepsilon^i \quad (\varepsilon = 0,1,\ 1 \leq i \leq n)$$

are given by

$$\psi_\varepsilon^i(x, t_1, \cdots, t_{n-1}) = \psi(x, t_1, \cdots, t_{n-i}, \varepsilon, t_{n-i+1}, \cdots, t_{n-1})$$

and the degeneracy operators

$$\zeta_n^j : T_n(X,Y) \longrightarrow T_{n+1}(X,Y) \quad (1 \leq j \leq n+1)$$

by

$$\zeta_n^j \psi(x, t_1, \cdots, t_{n+1}) = \psi(x, t_1, \cdots, \hat{t}_{n+2-j}, \cdots, t_{n+1}).$$

The reader will verify that the usual interchange rules are satisfied.

Finally, T is defined on morphisms $f \in \boldsymbol{Top}^{op}(X,X') = \boldsymbol{Top}(X',X)$ and $g \in \boldsymbol{Top}(Y,Y')$ by

$$\begin{aligned} T_n(f,g) \ &: \ T_n(X,Y) \longrightarrow T_n(X',Y') \\ &\quad (\psi : X \times [0,1]^n \longrightarrow Y) \longmapsto (g \circ \psi \circ (f \times 1_{[0,1]^n}) : X' \times [0,1]^n \longrightarrow Y') \end{aligned}$$

(which, for $n = 0$, reduces to $T_0(f,g)(\psi) = g \circ \psi \circ f$).

Observe that T is induced by the cylinder (I, j_0, j_1, s) on $\boldsymbol{Top}$, where $I : \boldsymbol{Top} \longrightarrow \boldsymbol{Top}$ is the functor given by $IX = X \times [0,1]$ on objects and by $If = f \times 1_{[0,1]}$ on morphisms, and $j_\varepsilon : 1_{\boldsymbol{Top}} \longrightarrow I$, $s : I \longrightarrow 1_{\boldsymbol{Top}}$ are the natural transformations given by $j_\varepsilon X : X \longrightarrow IX : x \longmapsto (x,\varepsilon)$ $(\varepsilon = 0,1)$ and by $sX : IX \longrightarrow X : (x,t) \longmapsto x$.

T is also induced by the cocylinder (P, p_0, p_1, q) on $\boldsymbol{Top}$, where $P : \boldsymbol{Top} \longrightarrow \boldsymbol{Top}$ is the functor given by $PX = X^{[0,1]}$ on objects and by $Pf : X^{[0,1]} \longrightarrow Y^{[0,1]} : \alpha \longmapsto f \circ \alpha$ for a morphism $f : X \longrightarrow Y$. Here $p_\varepsilon : P \longrightarrow 1_{\boldsymbol{Top}}$, $q : 1_{\boldsymbol{Top}} \longrightarrow P$ are the natural transformations given by $p_\varepsilon X : PX \longrightarrow X : \alpha \longmapsto \alpha(\varepsilon)$ $(\varepsilon = 0,1)$ and by $qX : X \longrightarrow PX : x \longmapsto c_x$ (constant path at x). Moreover T satisfies DNE(2) and DNE(3) by [11](3.5) and axiom (A) by [13] 2.1.

Summary 5.1. *Top, equipped with the cubical homotopy system T, satisfies all the hypotheses of the comparison theorems* (4.1) *and* (4,4) (*and their duals*).

We will now consider the description in *Top* of the (co–)fibration concepts of section 2. We restrict ourselves to those concepts that appear in Theorem (4.4).

Let

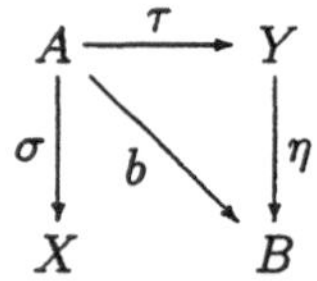

be a diagram of maps in *Top* with $\eta \circ \tau = b$.

Definition 5.2. $\eta : (Y, \tau) \longrightarrow (B, b)$ has the WCHP^A(2) with respect to (X, σ) if for any maps $F : X \times [0,1] \longrightarrow Y$, $\psi : X \times [0,1] \times [0,1] \longrightarrow B$ such that $F(\sigma(a), t) = \tau(a)$ for $a \in A$, $t \in [0,1]$, $\psi(\sigma(a), t, s) = b(a)$ for $a \in A$, $t, s \in [0,1]$ and $\psi(x, t, 0) = \eta F(x, t)$ for $x \in X$, $t \in [0,1]$, there exists a map $\Phi : X \times [0,1] \times [0,1] \longrightarrow Y$ such that

$$\Phi(\sigma(a), t, s) = \tau(a) \text{ for } a \in A, \ t, s \in [0,1], \ \eta \circ \Phi = \psi \text{ and } \Phi(\cdot, \cdot, 0) \overset{A \times [0,1]}{\underset{B}{\simeq}} F.$$

(Note: The reader will observe that $\eta : (Y, \tau) \longrightarrow (B, b)$ has the WCHP^A(2) w.r.t. (X, σ) if and only if $\eta : (Y, \tau \circ pr_1) \longrightarrow (B, b \circ pr_1)$ as a morphism of $\mathit{Top}^{A \times [0,1]}$ has the WCHP$^{A \times [0,1]}$ w.r.t. $X \times [0,1], \sigma \times 1_{[0,1]})$, where WCHP$^{A \times [0,1]}$ denotes the usual weak covering homotopy property in the category of spaces under $A \times [0,1]$.)

Definition 5.3. $\eta : (Y, \tau) \longrightarrow (B, b)$ has the RWCHP^A w.r.t. (X, σ) if for any map $f : X \longrightarrow Y$ 'under A' (i.e. $f \circ \sigma = \tau$) and any homotopy $\psi : X \times [0,1] \longrightarrow B$ of $\eta \circ f$ under A (i.e. $\psi \circ (\sigma \times 1_{[0,1]}) = b \circ pr_1$), there exists a homotopy $\Phi : X \times [0,1] \longrightarrow Y$ of f under A, such that $\eta \circ \Phi \simeq \psi$, rel $X \times \{0,1\} \cup A \times [0,1]$.

The reader will verify that WCHP^A(2) $\Longrightarrow$ WCHP^A $\Longrightarrow$ RWCHP^A.

By putting $A = \phi$ (the empty space) and identifying Top^{ϕ} with *Top* the properties WCHP^A(2), WCHP^A and RWCHP^A reduce to the WCHP(2), WCHP and RWCHP respectively. The corresponding dual notions are the WHEP(2), WHEP and RWHEP.

The topological interpretation of Theorem (4.4) is now clear. Notice that the previous comparison theorem (4.1) can be regarded (in the case $C = \mathit{Top}$) as a special case of (4.4), by taking $A = \phi$ and using the fact that $\phi \longrightarrow X$ has the HEP, hence also the WHEP(n) for all n.

Examples and Remarks 5.4. (i) = Example 2.9 of [7]. Let $A = \{*\}$ be a singleton, $X = [0,1] \times \{0\} \cup \{0\} \times [0,1]$, $Y = B = [0,1]$, $\sigma : A \longrightarrow X : x \longmapsto (0,1)$, $\tau : A \longrightarrow Y : x \longmapsto 0$, $u = \rho : X \longrightarrow [0,1] : (s,t) \longmapsto s$, $\eta = 1_{[0,1]}$. The reader will verify that the homotopy equivalence u is not a homotopy equivalence under A and over B. The reason for this is not so much the fact that ρ is only a weak fibration and no fibration, but the fact that $\rho : (X, \sigma) \longrightarrow (B, b)$ does not have the WCHP^A(2) w.r.t. (Y, τ). (It suffices to verify that ρ does not have the WCHP^A w.r.t. (Y, τ).)

(ii) = (i) with σ replaced by $\sigma' : A \longrightarrow X : * \longmapsto (0,0)$. Now all the hypotheses are satisfied. (Use the existence of a cross section $\lambda : B \longrightarrow X : s \longmapsto (s,0)$ of ρ such that $\lambda \circ \rho \overset{A}{\underset{B}{\simeq}} 1_X$.) The map u is now a homotopy equivalence under A and over B.

(iii) The pair of maps (τ, ρ) in the previous examples does not have the WCHEP (in ***Top***). (Use Proposition (2.7) in combination with Example (i).) This observation shows that the present comparison theorem (based on the notion of the WCHP$^A(n)$) is sharper than the obvious generalization of Theorem (0.1) using the WCHEP. Also note that

$$\tau \text{ a closed cofibration} + \rho \text{ a weak fibration} \not\Longrightarrow (\tau, \rho) \text{ has the WCHEP},$$

in sharp contrast with the well known result of Strøm [18] on the CHEP used in [7] p. 206.

References

1. Berrick, A.J.: 1976, 'The Samelson ex–product', *Quart. J. Math. Oxford* **(2) 27**, 173–180.

2. Berrick, A.J.: 1983, 'Ex–homotopy comparison theorems', *J. Singapore Nat. Acad. Sci.* **10–12**, 52–55.

3. Berrick, A.J., Kamps, K.H.: 1985, 'Comparison theorems in homotopy theory via operations on homotopy sets', *Math. Nachr.* **121**, 25–31.

4. Dold, A.: 1963, 'Partitions of unity in the theory of fibrations', *Ann. of Math.* **78**, 223–255.

5. Dold, A.: 1966, *Halbexakte Homotopiefunktoren*, Springer Lecture Notes Math. **12**.

6. Hardie, K.A., Kamps K.H.: 1989, 'Track homotopy over a fixed space', *Glas. Mat.* **24(44)**, 161–179.

7. Hardie, K.A., Kamps, K.H.: 1992, 'Variations on a theme of Dold', *Category Theory* 1991: *Proceedings, Montreal* 1991, CMS Conference Proceedings **13**, 201–209.

8. Heath, P.R.: 1977, 'Homotopy equivalence of a cofibre fibre composite', *Canad. J. Math.* **29**, 1152–1156.

9. James, I.M.: 1969, 'Bundles with special structure I', *Ann. of Math.* **69**, 359–390.

10. Kamps, K.H.: 1968, *Faserungen und Cofaserungen in Kategorien mit Homotopiesystem*, Dissertation, Saarbrücken.

11. Kamps, K.H.: 1972, 'Kan–Bedingungen und abstrakte Homotopietheorie', *Math. Z.* **124**, 215–236.

12. Kamps, K.H.: 1972, 'Zur Homotopietheorie von Gruppoiden', *Arch. Math.* **23**, 610–618.

13. Kamps, K.H.: 1973, 'Fundamentalgruppoid und Homotopien', *Arch. Math.* **24**, 456–460.

14. Kamps, K.H.: 1978, 'On a sequence of K.A. Hardie', *Cahiers Top. et Géom. Diff.* **19**, 147–154.

15. Kamps, K.H., Porter, T.: 1990, 'Note on bridging theorems for semisimplicial sets', *Quaestiones Math.* **13**, 349–360.

16. Kieboom, R.W.: 1993, Unpublished notes.

17. Lam, S.P.: 1980, 'A note on ex–homotopy equivalence', *Indag. Math.* **42**, 33–37.

18. Strøm, A.: 1966, 'Note on cofibrations', *Math. Scand.* **19**, 11–14.

19. Vogt, R.M.: 1972, 'A note on homotopy equivalence', *Proc. Amer. Math. Soc.* **32**, 627–629.

A TOPOLOGICAL BANACH SPACE MODEL OF LINEAR LOGIC

HEINRICH KLEISLI[1], HANS-PETER KÜNZI[1] and JIŘÍ ROSICKÝ[1]

ABSTRACT. We will show that the category $\mathcal{V}$ of topological Banach balls (introduced by M. Barr [1] - [3]) is a model of the full linear logic. The cotripel ! on $\mathcal{V}$ is constructed from the adjointness between $\mathcal{V}$ and the cartesian closed category of Hausdorff topological spaces and k-continuous maps.

1. Introduction

M. Barr [4] introduced a $*$-autonomous category as a symmetric monoidal closed category $\mathcal{K}$ with a dualizing object $\perp$. It means that $- \multimap \perp$ is the duality of $\mathcal{K}$ (where $\multimap$ denotes the internal hom). The importance of $*$-autonomous categories lies in the fact that they model linear logic (created by J.-Y. Girard in [8]). More precisely, they model the multiplicative fragment of linear logic. The additive part is given by finite products and, as shown by R. A. G. Seely [11], the exponential ! corresponds to a suitable cotriple on $\mathcal{K}$.

M. Barr also built up a $*$-autonomous category $\mathcal{V}$ starting from Banach spaces (see [1] - [3]). It consists of suitable topological Banach balls, i.e., of unit balls of complex Banach spaces endowed with a topology. The category $\mathcal{V}$ has finite products and our aim is to show that it bears a cotriple corresponding to !. We construct this cotriple from an adjoint situation

$$(*) \qquad \mathcal{V} \underset{M}{\overset{U}{\rightleftarrows}} \mathcal{H} \qquad\qquad M \dashv U$$

between $\mathcal{V}$ and the category $\mathcal{H}$ of Hausdorff topological spaces and k-continuous maps. Here, U is the forgetful functor. The category $\mathcal{H}$ is cartesian closed (see [6]) and, from the point of view of logic, $(*)$ connects the linear world $\mathcal{V}$ with the intuitionistic world $\mathcal{H}$. We will also show that M is a tensor functor, i.e.,

$$M(X \times Y) \cong M(X) \otimes M(Y),$$

which makes M suitable for defining group algebras for arbitrary Hausdorff topological groups (in the style of Dorofeev, Kleisli [9]). We are going to detail it in a subsequent paper.

AMS Classification: 46A70, 18D15.

Key words: Saks space, $*$-autonomous category, linear logic.

[1] This paper was written while the second author was supported by the Swiss National Science Foundation under grant 21-30585.91 and by the Spanish Ministry of Education and Sciences under the DGICYT grant SAB94-0120. The first author acknowledges partial support under grant 21-36185.92 by the Swiss National Science Foundation. The third author acknowledges support under grant 201/93/0950 of the Grant Agency of the Czech Republic.

Eraldo Giuli (ed.), Categorical Topology, 155–162.

2. Preliminaries

By a *topological Banach ball*, briefly called a *ball*, we understand a unit ball B of a complex Banach space endowed with a separated locally convex topology which is coarser than its norm topology and which satisfies the following condition: For each x in B,

$$\|x\| = \sup\{\varphi(x);\ \varphi \text{ is a continuous semi-norm such that } \varphi \leq \|\ \|\}.$$

In other words, a ball can be regarded as a unit ball of a Saks space (see [7, Chapt. I, Definition 3.2.]).

If no topology is specified we endow the unit ball with the norm topology and speak of a *discrete ball*.

There are two kinds of structure on a ball B:

(i) *An absolute convex structure*, given by a binary operation $(x, y) \mapsto \alpha x + \beta y$ on B for each pair (α, β) of complex numbers which satisfy $|\alpha| + |\beta| \leq 1$;
(ii) *a uniform structure*, given by a family of semi-norms $\varphi : B \to [0,1]$ defining a family of pseudo-metrics $d(x,y) = 2\varphi(\frac{1}{2}x - \frac{1}{2}y)$ on B.

Given two balls B and C, a *morphism* $f : B \to C$ is a map which preserves those two structures, i.e., which satisfies the following two conditions:

(i) $f(\alpha x + \beta y) = \alpha f(x) + \beta f(y)$ for all $(x, y) \in B^2$ and all $(\alpha, \beta) \in \mathbb{C}^2$ with $|\alpha| + |\beta| \leq 1$;
(ii) f is uniformly continuous.

Observe that it suffices to require that f is continuous in 0.

The balls together with their morphisms form a category which will be denoted by $\mathcal{B}$ (instead of by $\mathcal{TB}$ as in [10]) and is called the *category of balls.*

If B is a ball, a subset A of B which is closed under the absolute convex operations and is complete in the Minkowski norm $\|a\| = \inf\{\lambda > 0;\ a \in \lambda A\}$ is called a *subball*. We do not require the inclusion to be an isometry.

Products, equalizers, coproducts and coequalizers in $\mathcal{B}$ are constructed in a straightforward manner. For instance, the product of a family $(B_i)_{i \in I}$ of balls is given by the cartesian product $\prod_{i \in I} B_i$ equipped with the product topology. For a detailed discussion, see [1, §2].

3. Duality

Given two balls B and C, we denote by $[B, C]$ the set of morphisms $f : B \to C$ and endow it with the topology of compact convergence, i.e., the topology generated by the semi-norms ψ of the form

$$\psi(f) = \sup_{x \in K} \varphi(f(x)) \quad \text{for all} \quad f \in [B, C],$$

where K is a compact subset of B and φ a continuous semi-norm on C. It is easily seen that $[B, C]$ is again a ball.

In [1], a slightly different topology is given on the set $[B, C]$, namely the topology of uniform convergence on compact subballs of B. The two topologies coincide if B is a ζ-ball. By a *ζ-ball* (see [3, §2]) we understand a ball where every closed precompact subball is compact. Hence, in a ζ-ball each closed subball generated by a compact subset is compact (see [10, Proposition 4.13.1]) so that uniform convergence on compact subsets and on compact subballs coincide.

Let B, C and D be balls, and $f : B \to C$ a morphism. Then the maps

$$[f, D] : [C, D] \to [B, D], \quad \text{given by} \quad [f, D](g) = g \circ f \quad \text{for every} \quad g \in [C, D],$$

and

$$[D, f] : [D, B] \to [D, C], \quad \text{given by} \quad [D, f](g) = f \circ g \quad \text{for every} \quad g \in [D, B],$$

are again morphisms.

Let $O\mathbb{C}$ denote the unit ball of $\mathbb{C}$ (endowed with the norm topology). For a ball B, we denote the ball $[B, O\mathbb{C}]$ by B' and call it the *dual ball* of B. Likewise, for a morphism $f : B \to C$, we denote the morphism $[f, O\mathbb{C}]$ by $f' : C' \to B'$ and speak of the *dual morphism* of f. By iteration we obtain the *bidual ball* B'' and the *bidual morphism* $f'' : B'' \to C''$.

If B is a ball, the *evaluation map* $e_B : B \to B''$ is given by $e_B(x)(f) = f(x)$ for all $x \in B$ and $f \in B'$. It preserves the absolute convex structures on B and on B'' but is, in general, not continuous. If it is an isomorphism in the category $\mathcal{B}$, we shall call B a *reflexive ball*.

The following facts are well known (see [1] or [10]):

3.1. *Every discrete ball is reflexive.*

3.2. *The dual of a discrete ball is a compact ball.*

3.3. *Every compact ball is reflexive.*

3.4. *The dual of a compact ball is a discrete ball.*

It follows that the restriction of the duality functor to the full subcategory $\mathcal{B}an$ of $\mathcal{B}$ consisting of the discrete balls yields an equivalence between $\mathcal{B}an^{op}$ and the full subcategory $\mathcal{C}$ of $\mathcal{B}$ consisting of the compact balls.

4. The $*$-autonomous category $\mathcal{V}$

The concept of a *$*$-autonomous category* is due to M. Barr. It is defined in [4] as a closed symmetric monoidal category with tensor product $\otimes$, unit object $\top$ and internal hom $\multimap$ together with a *dualizing object* $\bot$. We shall assume here that $\bot = \top$.

The following example of a $*$-autonomous category which is a full subcategory of $\mathcal{B}$ containing the dual categories $\mathcal{B}an$ and $\mathcal{C}$, is likewise due to M. Barr.

Recall that a ζ-ball is a ball where every closed precompact subset is compact. By a *δ-ball* we understand a ball B where B and B' are reflexive ζ-balls. The full subcategory of $\mathcal{B}$ consisting of δ-balls is denoted by $\mathcal{V}$. It can be made into a $*$-autonomous category. We shall explain the construction of the tensor product $B \otimes C$ and of the internal hom object $B \multimap C$. For most of the proofs, we refer to [2] and [3].

The full subcategory of $\mathcal{B}$ consisting of the ζ-balls is denoted by $\zeta\mathcal{B}$. If B is a ball, it has a uniform completion $\hat{B}$, and we denote by ζB the intersection of all ζ-subballs of $\hat{B}$ which contain B.

4.1. *The object function $B \mapsto \zeta B$ yields a functor $\zeta : \mathcal{B} \to \zeta\mathcal{B}$ which is left adjoint to the inclusion $\zeta\mathcal{B} \to \mathcal{B}$, the unit being given by the dense embedding $u_B : B \to \zeta B$.* (See [2, Proposition 2.4].)

We have denoted the full subcategory of $\mathcal{B}$ consisting of the δ-balls by $\mathcal{V}$. If B is a ζ-ball, we denote by δB the ball $(\zeta(B'))'$.

4.2. *The object function $B \mapsto \delta B$ yields a functor $\delta : \zeta\mathcal{B} \to \mathcal{V}$ which is right adjoint to the inclusion $\mathcal{V} \to \zeta\mathcal{B}$, the counit being given by a bijective morphism $c_B : \delta B \to B$.* (See [2, Proposition 4.1].)

4.3. *Suppose B' and C are reflexive δ-balls. Then $[B, C]$ is a ζ-ball.* (See [3, Lemma 3.3].)

We define now, for B and C in $\mathcal{V}$, the *internal hom object* $B \multimap C$ by setting $B \multimap C = \delta[B, C]$, where $[B, C]$ is the ζ-ball of all morphisms $B \to C$, endowed with the topology of compact convergence. When $C = O\mathbb{C}$, we get $B' = \delta(B') = \delta[B, O\mathbb{C}] = B \multimap O\mathbb{C}$ so that the dual is unchanged.

For B and C in $\mathcal{V}$, the *tensor product* $B \otimes C$ is defined by setting $B \otimes C = (C \multimap B')'$. The *unit object* $\top$ is given by $O\mathbb{C}$.

4.4. *The category $\mathcal{V}$ equipped with $\otimes$, $\multimap$ and $\top$ is a closed symmetric monoidal category in which every object is reflexive.* (See [3, Theorem 4.10].)

In other words, $(\mathcal{V}, \otimes, \top, \multimap)$ is a $*$-autonomous category with *dualizing object* $\perp$ given by $O\mathbb{C}$.

The following two results will be needed later.

Lemma 4.5. *If B and C are objects of $\mathcal{V}$, then the map $t : (B \multimap C) \to (C' \multimap B')$, given by $t(f) = f'$ for all $f \in [B, C]$, is an isomorphism in $\mathcal{V}$.*

Proof. Consider the following sequence of isomorphisms: $\mathcal{V}(A, B \multimap C) \cong \mathcal{V}(B \otimes A, C) \cong \mathcal{V}(B \otimes A, C'') \cong \mathcal{V}(C' \otimes (B \otimes A), O\mathbb{C}) \cong \mathcal{V}(B \otimes (C' \otimes A), O\mathbb{C}) \cong \mathcal{V}(C' \otimes A, B') \cong \mathcal{V}(A, C' \multimap B')$, and verify the naturality in A. □

If B is a ζ-ball with locally convex topology τ, we denote by κB the underlying discrete ball of B endowed with the topology τ^κ generated by the *k-continuous* semi-norms on B, i.e., by the semi-norms on B which need not be continuous but whose restriction to each compact subset of B is continuous. It is not difficult to see that κB is a ζ-ball. Moreover, it follows from 4.2 that δB can be regarded as the underlying discrete ball of B with the coarsest topology τ^δ containing τ such that (B, τ^δ) is a δ-ball. Hence, the counit $c_B : \delta B \to B$ is given by the identity map of B.

Lemma 4.6. *For every ζ-ball B, κB is a δ-ball, and therefore the map $c_B^{-1} : B \to \delta B$ is k-continuous.*

Proof. Since κB is a ζ-ball, the evaluation $e_{\kappa B} : \kappa B \to (\kappa B)''$ is an open bijection (see the argument in [1] leading to Theorem 3.1). The continuity of $e_{\kappa B}$ is a direct consequence of the Ascoli Theorem. Hence, the ball κB is reflexive. It follows that the dual ball $(\kappa B)'$ is likewise reflexive. Finally, $(\kappa B)'$ is complete and therefore a ζ-ball. Hence, κB is a δ-ball.

By the characterization of δB given above, the identity map $d_B : \kappa B \to \delta B$ is continuous. Let φ be a continuous semi-norm on δB and K a compact subset of B. Then $\varphi \circ d_B$ is a continuous semi-norm on κB. It follows that the restriction of $\varphi \circ c_B^{-1}$ to K is continuous, i.e., $\varphi \circ c_B^{-1}$ is k-continuous. □

5. The Girard category $\mathcal{V}$

The concept of a *Girard category* was introduced by R. A. G. Seely (see [11]) as a $*$-autonomous category with finite products equipped with one more structure to make it a model of full linear logic. The additional structure is a cotriple $(\,!\,, c, d)$ such that there are isomorphisms $!B \otimes !C \cong !(B \times C)$ and $\top \cong !1$, where 1 denotes the terminal object.

The category $\mathcal{V}$ of δ-balls obviously has finite products. In particular, there is a terminal object 1 given by the 1-point ball $\{0\}$. We shall show that $\mathcal{V}$ is a Girard category by defining a cotriple $(\,!\,, c, d)$ via a pair of adjoint functors $M \dashv U : \mathcal{V} \to \mathcal{H}$. The latter is given as follows.

We denote by $\mathcal{H}$ the category of topological Hausdorff spaces and k-continuous maps. Recall that a map $\varphi : X \to Y$ between topological spaces is k-*continuous* if the restriction of φ to each compact subset of X is continuous. It is well known that the category $\mathcal{H}$ is cartesian closed for the internal hom set $[X, Y]_k = \{k\text{-continuous } f : X \to Y\}$, endowed with the compact-open topology (see [6]).

Let $U : \mathcal{V} \to \mathcal{H}$ be the forgetful functor which forgets the absolute convex structure of each ball.

Theorem 5.1. *The forgetful functor U admits a left adjoint $M : \mathcal{H} \to \mathcal{V}$.*

Proof. We shall construct, for each Hausdorff space X, a δ-ball MX and a k-continuous map $\mu : X \to U(MX)$ satisfying the following universal mapping property: For every object B of $\mathcal{V}$ and each k-continuous map $\varphi : X \to UB$, there exists a unique morphism $f : MX \to B$ such that $(Uf) \circ \mu = \varphi$.

Put $C_k(X) = [X, O\mathbb{C}]_k$. It is the unit ball of the Banach space of all complex-valued k-continuous bounded functions on X for the supremum norm, endowed with the topology of compact convergence. Moreover, it is easily seen to be a complete uniform space, and hence a ζ-ball. We may therefore define MX and μ by setting $MX = (\delta C_k(X))'$ and $\mu(x)(\varphi) = \varphi(x)$ for all x in X and φ in $C_k(X)$, and we perform the necessary verifications in three steps:

(i) By construction MX is a δ-ball, and the k-continuity of μ is a direct consequence of the Ascoli Theorem.

(ii) *Uniqueness of f*: Assume that a morphism $f : MX \to B$ satisfying $(Uf) \circ \mu = \varphi$ exists. Since B is reflexive, we may assume $B = C'$ for another object C of $\mathcal{V}$. Thus, $f : (\delta C_k(X))' \to C'$ is the dual morphism of a morphism $g : C \to \delta C_k(X)$. The latter must satisfy the following relation: For all points x in X and all y in C,

$$g(y)(x) = \mu(x)(g(y)) = g'(\mu(x))(y) = f(\mu(x))(y) = \varphi(x)(y) .$$

Hence, the map g is uniquely determined by the given k-continuous map φ, and therefore also the dual morphism $f = g'$.

(iii) *Existence of f*: We define a map $g : C \to \delta C_k(X)$ by setting $g(y)(x) = \varphi(x)(y)$ for all points x in X and all y in C. Clearly, it is well defined. Moreover,

g preserves the absolute convex structures. It remains to prove that it is continuous, because then the dual morphism g' is the desired morphism f. Since C is an object of $\mathcal{V}$, we have $C = \delta C$. It therefore suffices to verify the continuity of g considered as a map into $C_k(X)$. Since C is reflexive, the topology on C is generated by the semi-norms ζ_L of the form

$$\zeta_L(y) = \sup_{h\in L} |h(y)| \quad \text{for all} \quad y \in C\,,$$

where L is a compact subset of $C' = B$. Let K be a compact subset of X. Then

$$\sup_{x\in K} |g(y)(x)| = \sup_{x\in K} |\varphi(x)(y)| = \sup_{h\in\varphi(K)} |h(y)|\,,$$

and $\varphi(K)$ is a compact subset of B. That implies the desired continuity. □

The next theorem is crucial for the verification that the cotripel generated by the pair of adjoint functors $M \dashv U$ defines on $\mathcal{V}$ the structure of a Girard category.

Theorem 5.2. *If Y is a topological Hausdorff space and C a δ-ball, then there is an isomorphism*

$$U(MY \multimap C) \cong [Y, UC]_k \quad \textit{in} \quad \mathcal{H}.$$

Proof. We define a bijection $\beta : U[MY, C] \to [Y, UC]_k$ by setting

$$\beta(f) = (Uf) \circ \mu \quad \text{for all} \quad f \in [MY, C]\,,$$

where $\mu : Y \to U(MY)$ is the unit of the adjunction $M \dashv U$; i.e., $\mu(y)(\psi) = \psi(y)$ for all y in Y and ψ in $C_k(Y)$. We shall first show that β is continuous and β^{-1} (only) k-continuous.

(i) *Continuity of β*: Since UC is a uniform space, the compact-open topology on $[Y, UC]_k$ coincides with the topology of compact convergence. Let K be a compact subset of Y and η a continuous semi-norm on C. Then

$$\sup_{y\in K} \eta(\beta(f)(y)) = \sup_{y\in K} \eta(f(\mu(y))) = \sup_{\nu\in\mu(K)} \eta(f(\nu)) \quad \text{for all} \quad f \in [MY, C]\,.$$

Since $\mu(K)$ is a compact subset of MY, the continuity of β follows.

(ii) *k-continuity of β^{-1}*: Recall that, for every φ in $[Y, UC]_k$, the morphism $\beta^{-1}(\varphi)$ was constructed as the dual g' of the morphism $g : D \to \delta C_k(Y)$, given by $g(z)(y) = \varphi(y)(z)$ for all $y \in Y$ and $z \in D$, where $C = D'$. Consider the commutative diagram

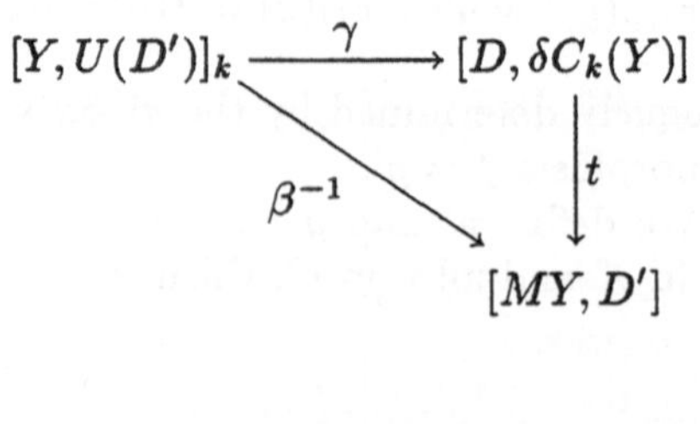

where γ is defined by setting

$$\gamma(\varphi)(z)(y) = \varphi(y)(z) \quad \text{for all} \quad \varphi \in [Y, U(D')]_k, \quad z \in D \text{ and } y \in Y,$$

and t is the isomorphism of Lemma 4.5. Since $U[D, \delta C_k(Y)]$ is a subspace of $[UD, U(\delta C_k(Y))]_k$, it suffices to verify the k-continuity of γ considered as a map into $[UD, U(\delta C_k(Y))]_k$. By 4.6., the identity map $c : U(\delta C_k(Y)) \to U(C_k(Y)) = [Y, O\mathbb{C}]_k$ is an isomorphism in $\mathcal{H}$. Hence, the induced map $[UD, c]_k : [UD, U(\delta C_k(Y))]_k \to [UD, [Y, O\mathbb{C}]_k]_k$ is again an isomorphism in $\mathcal{H}$ (see [6, Proposition 3.4]). Since $U(D') = U[D, O\mathbb{C}]$ is a subspace of $[UD, O\mathbb{C}]_k$, the embedding $j : U(D') \to [UD, O\mathbb{C}]_k$ induces a k-continuous map $[Y, j]_k : [Y, U(D')]_k \to [Y, [UD, O\mathbb{C}]_k]_k$ [6, Proposition 3.4], and we obtain a commutative diagram

$$\begin{array}{ccc} [Y, U(D')]_k & \xrightarrow{\gamma} & [UD, U(\delta C_k(Y))]_k \\ {\scriptstyle [Y,j]_k}\downarrow & & \downarrow{\scriptstyle [UD,c]_k} \\ [Y, [UD, O\mathbb{C}]_k]_k & \xrightarrow[\gamma_1]{} & [UD, [Y, O\mathbb{C}]_k]_k \end{array}$$

where the map γ_1 is given by the same formula as γ. It follows from the cartesian closedness of $\mathcal{H}$ [6, Theorem 3.3] that γ_1 is a well defined isomorphism in $\mathcal{H}$. Hence, γ is k-continuous.

Finally, we invoke again Lemma 4.6 which tells us that $U(MY \multimap C) = U(\delta[MY, C])$ and $U[MY, C]$ are isomorphic objects in $\mathcal{H}$. That yields the desired isomorphism. □

Corollary 5.3. *If X and Y are topological Hausdorff spaces, then there is an isomorphism*

$$MX \otimes MY \cong M(X \times Y) \quad \textit{in } \mathcal{V}.$$

Proof. Consider the following sequence of isomorphisms: $\mathcal{V}(MX \otimes MY, C) \cong \mathcal{V}(MX, MY \multimap C) \cong \mathcal{H}(X, U(MY \multimap C)) \cong \mathcal{H}(X, [Y, UC]_k) \cong \mathcal{H}(X \times Y, UC) \cong \mathcal{V}(M(X \times Y), C)$, and verify the naturality in C. □

Corollary 5.4. *The category $\mathcal{V}$ equipped with the cotriple $(\,!, c, d)$, generated by the pair of adjoint functors $M \dashv U : \mathcal{V} \to \mathcal{H}$, is a Girard category.*

Proof. Using Corollary 5.3 and the fact that the forgetful functor U is right adjoint and therefore preserves products, we have for any two objects B and C of $\mathcal{V}$: $!B \otimes\, !C = M(UB) \otimes M(UC) \cong M(UB \times UC) \cong M(U(B \times C)) = \,!(B \times C)$. Finally, the verification that $O\mathbb{C} \cong\, !(\{0\})$ is straightforward. □

References

[1] M. Barr, *Duality of Banach spaces*, Cahiers de Topologie et Géométrie Différentielle, Vol. XVII-1 (1976), 15–32.

[2] M. Barr, *Closed categories and topological vector spaces*, Cahiers de Topologie et Géométrie Différentielle, Vol. XVII-3 (1976), 223–234.

[3] M. Barr, *Closed categories of Banach spaces*, Cahiers de Topologie et Géométrie Différentielle, Vol. XVII-4 (1976), 335–342.

[4] M. Barr, **-Autonomous Categories*, Lect. Notes in Math. 752, Springer-Verlag, Berlin 1979.

[5] M. Barr, **-Autonomous categories and linear logic*, Math. Struct. in Comp. Science 1 (1991), 159–178.

[6] R. Brown, *Function spaces and product topologies*, Quat. Journal Math. Oxford 15 (1964), 238–250.

[7] J. B. Cooper, Saks Spaces and Applications to Functional Analysis, 2nd edition. North-Holland-Mathematics Studies 139, Amsterdam 1987.

[8] J.-Y. Girard, *Linear Logic*, Theor. Comp. Sci. 50 (1987), 1-102.

[9] S. Dorofeev and H. Kleisli, *Functorial methods in the theory of group representations*, Preprint Université de Fribourg, Institut de Mathématiques, March 1994.

[10] H. Kleisli and H.-P. Künzi, *Topological totally convex spaces II.* Cahiers de Topologie et Géométrie Différentielle Catégorique 36 (1995), 11–52.

[11] R. A. G. Seely, *Linear logic, *-autonomous categories and cofree coalgebras*, Contemporary Mathematics 92 (1989), 371–382.

Heinrich Kleisli
Math. Institute
University of Fribourg
1700 Fribourg (Switzerland)

Hans-Peter Künzi
Math. Institute
University of Berne
3012 Berne (Switzerland)

Jiří Rosický
Dept. of Algebra and Geometry
Masaryk University
662 95 Brno (Czech Republic)

PARAMETRIZING THE THEORY OF CLOSURE OPERATORS

Extended Version

Jürgen Koslowski
Institut für Theoretische Informatik, TU Braunschweig, < *koslowj@iti.cs.tu-bs.de* >

September 22, 1995

Abstract. We present a streamlined treatment of $\mathcal{Z}$-modal closure operators. Originally introduced to facilitate the integration of modal closure operators (induced, e.g., by Grothendieck topologies) into the classical theory of closure operators, this notion and its dual also provides insight into the structure of the lattice of all closure operators. A suitable adjustment of the notion of orthogonality between composable pairs enables us to develop the theory to a large extent parallel to the theory of all closure operators.

Key words: $\langle \boldsymbol{E}, \mathcal{M}\rangle$-category for sinks, orthogonality relation, closure operator

0 Introduction

This paper extends previous joint work with G. Castellini and G. E. Strecker, e.g., [2], [3], and [4]. For a category $\mathcal{X}$ with a factorization structure $\langle \boldsymbol{E}, \mathcal{M}\rangle$ for sinks two of the main achievements of this collaboration were:

- we made precise the connection between closure operators on $\mathcal{M}$ and the orthogonality relation *at its proper level of generality*, namely viewed in the category $\mathcal{M} \diamond \mathcal{M}$ of composable pairs of arrows in $\mathcal{M}$;
- we identified the crucial role the composition functor $W : \mathcal{M} \diamond \mathcal{M} \to \mathcal{M}$[1] plays in the theory.

Essentially, we identified the lattice of all closure operators on $\mathcal{M}$ up to equivalence as the lattice of fixed points for the polarity induced by the orthogonality relation. This lead to a very satisfactory picture in [2] of Galois connections interconnecting various complete lattices, briefly recalled in Section 2.

Even though for special applications it turned out to be necessary to make adjustments to the basic orthogonality relation (cf., e.g., the notion of *regular closure operator* in [3]), the consequences for the "big picture" of [2] were not studied systematically. The modifications to the orthogonality relation we consider in Section 3 may be thought of as "breaking the symmetry". Section 4 then shows that the outer square of our previous "big picture" may be viewed as a slice in a diagram of functors and natural transformations, parametrized by suitable collections of $\mathcal{M}$-morphisms. Under different hypotheses a similar result is obtained for the inner square in Section 5. This result, and its dual, clarify the fine structure of

[1] Diagrams and morphisms were typeset with XY-pic, version 3.2, by Kristoffer H. Rose.

Eraldo Giuli (ed.), Categorical Topology, 163–174.

the lattice of all closure operators. Finally in Section 6 we ask, when both squares can be combined in the fashion of the "big picture". Not surprisingly, a categorical property of the composition functor W plays a crucial role here.

1 The General Set-Up

We wish to keep this section short and to the point. For additional details and motivation the reader may consult the preliminary sections of [2] or [3].

As usual, we consider a category $\mathcal{X}$ with pullbacks and a factorization system $\langle \boldsymbol{E}, \mathcal{M}\rangle$ for sinks. In particular, $\mathcal{M}$ consists of monos, is stable by pullbacks, satisfies the cancellation property

$$p \circ n^2 \in \mathcal{M} \quad \text{and} \quad p \in \mathcal{M} \quad \text{implies} \quad n \in \mathcal{M}$$

and the $\mathcal{M}$-subobjects of any $\mathcal{X}$-object form a complete lattice.

We view $\mathcal{M}$ as a full subcategory of the comma category $\mathcal{X}/\mathcal{X}$, with domain functor U and codomain functor V into $\mathcal{X}$. Pulling V back along U induces the category $\mathcal{M} \diamond \mathcal{M}$ of composable pairs $\langle n, p\rangle \in \mathcal{M} \times \mathcal{M}$. (Arrows are triples of $\mathcal{X}$-arrows which make the appropriate squares commute.) The obvious composition functor W from $\mathcal{M}\diamond\mathcal{M}$ to $\mathcal{M}$ is a *bifibration*: W*-inverse* and W*-direct images* of pairs along $\mathcal{M}$-morphisms are formed via suitable pullbacks resp. via appropriate $\langle \boldsymbol{E}, \mathcal{M}\rangle$-factorizations. The W*-fibre* over $m \in \mathcal{M}$ consists of all factorizations of m into two factors from $\mathcal{M}$; it is just the comma category W/m, which in fact is pre-ordered. We write $\ll$ for this pre-order. Given $\langle f, g\rangle : m \to n$, the W-direct resp. W-inverse image functors between the fibres W/m and W/n are denoted by $\langle f, g\rangle_{\rightarrow}$ and $\langle f, g\rangle^{\leftarrow}$, respectively.

A key notion for the theory of closure operators on $\mathcal{M}$ is the orthogonality of composable pairs, which slightly generalizes the concept introduced by Tholen (cf. [10]). For an even more general version, cf. [7], and [3].

DEFINITION 1.0. We call a pair $\langle q, r\rangle \in W/m$ **left-orthogonal** to $\langle s, t\rangle \in W/n$, written as $\langle q, r\rangle \perp \langle s, t\rangle$, iff for every $\mathcal{M}$-morphism $\langle f, g\rangle : m \to n$ there exists a (unique) horizontal fill-in d such that $d \circ q = s \circ f$ and $t \circ d = g \circ r$.

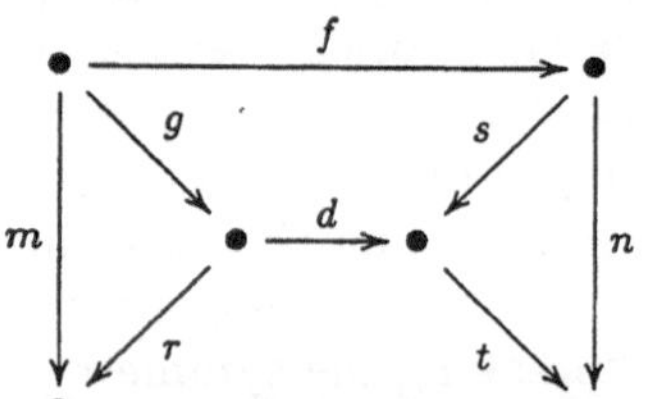

(1-0)

[2] A version of this document that uses the logical left-to-right order of composition is available as a PostScript file from the author upon request.

These tools allow us to give a very simple definition of a closure operator on $\mathcal{M}$. For other possible approaches, see [5], or [7].

DEFINITION 1.1. A **closure operator** on $\mathcal{M}$ is a functor $F : \mathcal{M} \to \mathcal{M} \diamond \mathcal{M}$ that is left inverse to W, i.e., $WF = \mathcal{M}$, and satisfies $F(m) \perp F(n)$ for all $m, n \in \mathcal{M}$. We write $\langle m_F, m^F \rangle$ for $F(m)$.

If F is a closure operator, $m \in \mathcal{M}$ is called **F-dense**, if m^F is iso, and **F-closed**, if m_F is iso. The corresponding subclasses of $\mathcal{M}$ are denoted by $\Delta^*(F)$ and $\nabla_*(F)$, respectively.

A closure operator F is called **idempotent**, if m^F is always F-closed, and **weakly hereditary**, if m_F is always F-dense.

The collection $\boldsymbol{CL}_{\mathcal{M}}$ of all closure operators on $\mathcal{M}$, pre-ordered by $F \sqsubseteq G$ iff $m^F \leq m^G$ for all $m \in \mathcal{M}$, (up to equivalence) is a complete lattice. The weakly hereditary closure operators form a coreflective sublattice $\boldsymbol{wCL}_{\mathcal{M}}$, while the idempotent ones form a reflective sublattice $\boldsymbol{iCL}_{\mathcal{M}}$. We write $\boldsymbol{iwCL}_{\mathcal{M}}$ for their intersection. F_{w} denotes the *weakly hereditary core*, and F^{i} the *idempotent hull* of a closure operator F.

Next we recall some facts about Galois connections between pre-ordered classes, especially between power collections, and introduce some convenient notation.

DEFINITION 1.2. For pre-ordered classes $\mathcal{A} = \langle A, \leq \rangle$ and $\mathcal{B} = \langle B, \sqsubseteq \rangle$ a **Galois connection** $\pi = \langle \pi_*, \pi^* \rangle : \mathcal{A} \multimap \mathcal{B}$ consists of order-preserving functions $\pi_* : \mathcal{A} \to \mathcal{B}$ and $\pi^* : \mathcal{B} \to \mathcal{A}$ that satisfy $id_{\mathcal{A}} \leq \pi^* \circ \pi_*$ and $\pi_* \circ \pi^* \sqsubseteq id_{\mathcal{B}}$, i.e., π_* is left adjoint to π^*. We call π a **(co)reflection** iff π^* (resp. π_*) is a one-to-one function, and an **equivalence** iff $\langle \pi^*, \pi_* \rangle$ is a Galois connection from $\mathcal{B}$ to $\mathcal{A}$.

The **composite** of two Galois connections $\pi : \mathcal{A} \multimap \mathcal{B}$ and $\rho : \mathcal{B} \multimap \mathcal{C}$ is defined as $\rho \circ \pi = \langle \rho_* \circ \pi_*, \pi^* \circ \rho^* \rangle$. Thus we obtain a quasi-category $\mathcal{GAL}$.

Any Galois connection $\pi : \langle A, \leq \rangle \multimap \langle B, \sqsubseteq \rangle$ restricts to an equivalence between the classes of **left fixed points** $\{ a \in A \mid a \cong \pi^* \pi_*(a) \}$ and **right fixed points** $\{ b \in B \mid \pi_* \pi^*(b) \cong b \}$ with the induced orders, respectively (cf., e.g., [6]).

DEFINITION 1.3. We call a factorization $\ddot{\varphi} \circ \dot{\varphi}$ of $\varphi : \mathcal{A} \multimap \mathcal{C}$ through $\mathcal{B}$ **essentially canonical with center** $\mathcal{B}$, if $\mathcal{B}$ is equivalent to $\mathcal{A}^{\varphi}$ and $\mathcal{C}_{\varphi}$. The dot notation throughout indicates an essentially canonical factorization (e.c.f.).

DEFINITION 1.4 (cf. [6]). Any relation $R \subseteq A \times B$ between classes defines at least two Galois connections between the corresponding power collections. The **axiality** $\varphi : \boldsymbol{P}(A) \multimap \boldsymbol{P}(B)$ and the **polarity** $\psi : \boldsymbol{P}(A) \multimap \boldsymbol{P}^{\mathrm{op}}(B)$ are defined by setting for $U \subseteq A$ and $V \subseteq B$

$$\psi_*(U) := \{ b \in B \mid \exists_{a \in A} \ \langle a, b \rangle \in R \text{ and } a \in U \}$$
$$\psi^*(V) := \{ a \in A \mid \forall_{b \in B} \ \langle a, b \rangle \in R \text{ implies } b \in V \}$$
$$\varphi_*(U) := \{ b \in B \mid \forall_{a \in A} \ a \in U \text{ implies } \langle a, b \rangle \in R \}$$
$$\varphi^*(V) := \{ a \in A \mid \forall_{b \in B} \ b \in V \text{ implies } \langle a, b \rangle \in R \}$$

2 The Fundamental Polarity

The polarity $\omega : \boldsymbol{P}(\mathcal{M} \diamond \mathcal{M}) \longrightarrow\!\bullet\, \boldsymbol{P}^{\text{op}}(\mathcal{M} \diamond \mathcal{M})$ induced by the orthogonality relation $\perp$ has been shown in [2] to be of central importance for the theory of closure operators. In particular, essentially canonical factorizations of the form $\varphi = \ddot{\varphi} \circ \dot{\varphi}$ were established for

- ω itself, with $\boldsymbol{CL}_{\mathcal{M}}$ as center;
- the composition $\upsilon := \curlywedge \circ \omega \circ \Upsilon$ with $\boldsymbol{iwCL}_{\mathcal{M}}$ as center; the first resp. second projection from $\mathcal{M} \diamond \mathcal{M}$ to $\mathcal{M}$ induces the axiality $\Upsilon : \boldsymbol{P}(\mathcal{M}) \longrightarrow\!\bullet\, \boldsymbol{P}(\mathcal{M} \diamond \mathcal{M})$ resp. $\curlywedge : \boldsymbol{P}^{\text{op}}(\mathcal{M} \diamond \mathcal{M}) \longrightarrow\!\bullet\, \boldsymbol{P}^{\text{op}}(\mathcal{M})$, and υ is the polarity induced by the orthogonality relation restricted to $\mathcal{M}$;
- the composite $\Delta = \dot{\omega} \circ \Upsilon$ with $\boldsymbol{wCL}_{\mathcal{M}}$ as center;
- the composite $\nabla = \curlywedge \circ \ddot{\omega}$ with $\boldsymbol{iCL}_{\mathcal{M}}$ as center;
- the composite $\epsilon = \dot{\nabla} \circ \ddot{\Delta}$ with $\boldsymbol{iwCL}_{\mathcal{M}}$ as center.

We recall the corresponding "big picture" from [2]

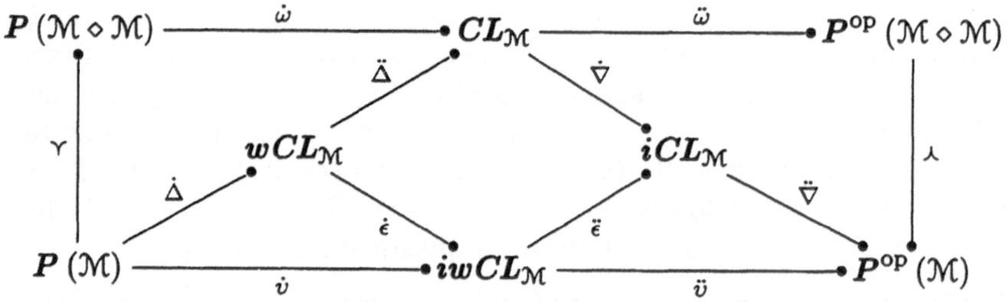

(2-0)

3 The Parameter $\mathcal{Z}$

We would like to know, to what extent the "big picture" (2-0) can be preserved, if the fundamental relation $\perp$ is modified. The idea for such a modified orthogonality relation dates back to [7], Definition 3.07. Besides studying certain orthogonality conditions on a closure operator F, there we considered so-called "naturality conditions", which required F to preserve certain (co-) cartesian $\mathcal{M}$-morphisms. That notion did not yet capture the fundamental similarity between hereditary and modal closure operators, and later [3] mutated into the notion of $\mathcal{Z}$-modality. However, there and also in [4] $\mathcal{Z}$-modality was treated as just an additional property closure operators might have. We did not ask whether $\mathcal{Z}$-modal closure operators might themselves arise as the fixed points of a modified polarity, even though we had shown certain partial results in that direction.

Let $\mathcal{Z}$ be an arbitrary collection of $\mathcal{M}$-morphisms. Parts (1) and (2) of the following definition can already be found in [3], Definition 2.00. Part (0) constitutes the missing link. In [4] we used a notion of "$\mathcal{Z}$-stable relation" instead, which had its roots in Definition 3.03 of [3].

DEFINITION 3.0. (0) A pair $\langle q,r\rangle \in W/m$ is called **left-$\mathcal{Z}$-orthogonal** to $\langle s,t\rangle \in W/n$, written as $\langle q,r\rangle \perp^{\mathcal{Z}} \langle s,t\rangle$, if any $\langle f,g\rangle \in \mathcal{Z}$ with codomain m satisfies $\langle f,g\rangle^{\leftarrow}(\langle q,r\rangle) \perp \langle s,t\rangle$.

(1) A collection $C \subseteq \mathcal{M} \diamond \mathcal{M}$ is called **$\mathcal{Z}$-stable**, if for each $\langle f,g\rangle : m \rightarrow n$ in $\mathcal{Z}$ the W-inverse image functor $\langle f,g\rangle^{\leftarrow}$ maps pairs in $C \cap W/n$ to pairs in $C \cap W/m$. A collection $A \subseteq \mathcal{M}$ is called **$\mathcal{Z}$-stable**, if $\Upsilon_*(A) = \{\langle n,p\rangle \in \mathcal{M} \diamond \mathcal{M} \mid n \in A\}$ has this property. $\boldsymbol{STAB}(\mathcal{Z})$ and $\boldsymbol{stab}(\mathcal{Z})$ denote the corresponding sublattices of $\boldsymbol{P}(\mathcal{M} \diamond \mathcal{M})$ and $\boldsymbol{P}(\mathcal{M})$, respectively, and we write $\Upsilon^{\mathcal{Z}}$ for the restriction of the axiality $\Upsilon : \boldsymbol{P}(\mathcal{M}) \longrightarrow \boldsymbol{P}(\mathcal{M} \diamond \mathcal{M})$.

(2) A closure operator F on $\mathcal{M}$ is called **$\mathcal{Z}$-modal** if F commutes with W-inverse images along members of $\mathcal{Z}$; i.e., if $\langle f,g\rangle : m \rightarrow n$ belongs to $\mathcal{Z}$, then $F(m) \cong \langle f,g\rangle^{\leftarrow}(F(n))$. We write $\boldsymbol{CL}(\mathcal{Z})$ for the collection of all $\mathcal{Z}$-modal closure operators on $\mathcal{M}$, and $\boldsymbol{wCL}(\mathcal{Z})$, $\boldsymbol{iCL}(\mathcal{Z})$, $\boldsymbol{iwCL}(\mathcal{Z})$ for the corresponding collections of $\mathcal{Z}$-modal closure operators that are either weakly hereditary, or idempotent, or both.

The orthogonality relation as specified in Definition 1.0 displays a particular kind of symmetry:

$$\langle q,r\rangle \perp \langle s,t\rangle \quad \text{in} \quad \mathcal{M} \diamond \mathcal{M} \quad \text{iff} \quad \langle s,t\rangle \perp \langle q,r\rangle \quad \text{in} \quad (\mathcal{M} \diamond \mathcal{M})^{\text{op}}$$

Our modification above breaks this symmetry. The relation $\perp^{\mathcal{Z}}$ no longer connects just single pairs on both sides, but rather "$\mathcal{Z}$-sieves" of pairs on the left with single pairs on the right.

We invite the reader to explore the dual notions of **right-$\mathcal{Z}$-orthogonality**, **$\mathcal{Z}$-costability**, and **$\mathcal{Z}$-comodality**, where W-direct images rather than W-inverse images are used. In this setting we denote the relevant sublattices of $\boldsymbol{P}^{\text{op}}(\mathcal{M} \diamond \mathcal{M})$ and $\boldsymbol{P}^{\text{op}}(\mathcal{M})$ by $\boldsymbol{COST}(\mathcal{Z})$ and $\boldsymbol{cost}(\mathcal{Z})$, respectively.

Remark 3.1. (0) For $\mathcal{Z} = \mathcal{I} := \boldsymbol{Iso}(\mathcal{M})$ we recover ordinary closure operators. Setting $\mathcal{Z} = \mathcal{H} := \{\langle f,g\rangle \mid f \text{ iso and } g \in \mathcal{M}\}$ yields the familiar notion of hereditary closure operators. Note that all $\mathcal{M}$-morphisms in $\mathcal{H}$ are in fact cartesian with respect to the codomain functor V, i.e., they form pullbacks in $\mathcal{X}$. If $\mathcal{Z}$ is the collection $\mathcal{C}$ of all cartesian $\mathcal{M}$-morphisms, the corresponding $\mathcal{Z}$-modal closure operators are just the modal ones.

(1) $\mathcal{Z}$-stability of a collection $A \subseteq \mathcal{M}$ is a pullback stability condition; the pullbacks in question are those that arise in forming W-inverse images along morphisms in $\mathcal{Z}$. But even if $\mathcal{Z}$ consists of cartesian $\mathcal{M}$-morphisms, or if $\mathcal{Z}$ is stable under pullbacks in $\mathcal{M}$, the condition that whenever the codomain of $\langle f,g\rangle \in \mathcal{Z}$ belongs to A so does the domain in general is weaker than $\mathcal{Z}$-stability, contrary to the assertion after Definition 2.00 in [3].

(2) Since for any $\mathcal{M}$-morphism $\langle f,g\rangle$ the W-inverse image functor $\langle f,g\rangle^{\leftarrow}$ right adjoint and hence preserves infima, $CL(\mathcal{Z})$ is closed under the forma tion of infima in $CL(\mathcal{I}) = CL_{\mathcal{M}}$. In particular, every closure operator has a **$\mathcal{Z}$-modal hull** $F^{\mathcal{Z}}$. Since $iCL(\mathcal{I}) = iCL_{\mathcal{M}}$ also is closed under th formation of infima in $CL(\mathcal{I})$, this is true for $iCL(\mathcal{Z}) = CL(\mathcal{Z}) \cap iCL(\mathcal{I}$ as well. Hence every closure operator F also has an **idempotent $\mathcal{Z}$-moda hull** $F^{(i\mathcal{Z})}$. A dual argument shows that F also has a **$\mathcal{Z}$-comodal cor** $F_{\mathcal{Z}}$ and a **weakly hereditary $\mathcal{Z}$-comodal core** $F_{(w\mathcal{Z})}$.

EXAMPLE 3.2. (0) If $\mathcal{Z}$ is the collection for all $\mathcal{M}$-morphisms $\langle f,g\rangle$ suc that $f \in \mathcal{M}$ and g is iso (the dual of $\mathcal{H}$ in Remark 3.1(0)), then the onl $\mathcal{Z}$-modal closure operator is the indiscrete one: for $m \in \mathcal{M}$ with codomain just consider $\langle m, id_X\rangle : m \twoheadrightarrow id_X$ in $\mathcal{Z}$. Dually, the only $\mathcal{H}$-comodal closu operator is the discrete one.

(1) Let $\mathcal{Z}$ be as above. Then a $\mathcal{Z}$-comodal closure operator F (which we a tempted to call *strongly idempotent* in order to restore some terminologic symmetry) is already determined (up to equivalence) by its value on the lea $\mathcal{M}$-subobject 0_X of each $\mathcal{X}$-object X. Any $\mathcal{M}$-subobject n of X satisfie $n^F = n \vee (0_X)^F$, i.e., the closure is given by the supremum (= union) with fixed $\mathcal{M}$-subobject.

(2) If $\mathcal{M}$, and hence $\mathcal{X}$, has binary products, consider the collection of cartesia $\mathcal{M}$-morphisms

$$\mathcal{P} := \{\, \langle f,g\rangle : X \times n \twoheadrightarrow n \mid \ f \text{ and } g \text{ project onto the second factor} \,\}$$

(cf. [3], before Proposition 2.03). $\mathcal{P}$-modal closure operators are **finitely pro ductive**. Given m and n in $\mathcal{M}$, the universal property of products implie $F(m \times n) \ll F(m) \times F(n)$. Since

$$F(m) \times id = F(m \times id) \perp F(m \times n) \quad \text{and} \quad id \times F(n) = F(id \times n) \perp F(m \times n$$

the pair $F(m \times n)$ is an upper bound in $W/(m \times n)$ for both $\langle m_F \times id, m^F \times n$ and $\langle id \times n_F, m \times n^F\rangle$. But $F(m) \times F(n)$ is the supremum of these last tw pairs, which shows that $F(m) \times F(n) \ll F(m \times n)$.

The usual closure operator on the class of embeddings in $\mathcal{X} = \textbf{\textit{Top}}$ clearl is $\mathcal{P}$-modal, but the regular closure operator induced by Hausdorff spac by means of the Salbany construction [8] does not have this property (cf. [3 Example 3.08). [

4 The Outer Square

Our goal is to interpret the "big picture" as the instance at $\mathcal{Z} = \mathcal{I}$ of a diagra of functors and natural transformations. We begin by studying the simplest par the outer square.

Clearly, $\boldsymbol{CL}(\mathcal{Z})$ remains unchanged if we close $\mathcal{Z}$ under composition, or add isomorphisms of $\mathcal{M}$. Hence without loss of generality we may restrict our attention for the remainder of this section to those collections $\mathcal{Z}$ that are closed under composition and contain $\boldsymbol{Iso}(\mathcal{M})$.

PROPOSITION 4.0. *Let $\mathcal{Y}$ and $\mathcal{Z}$ be collections of $\mathcal{M}$-morphisms that satisfy $\mathcal{Y} \subseteq \mathcal{Z}$.*

(0) *Both inclusions $\boldsymbol{STAB}(\mathcal{Z}) \hookrightarrow \boldsymbol{STAB}(\mathcal{Y})$ and $\boldsymbol{stab}(\mathcal{Z}) \hookrightarrow \boldsymbol{stab}(\mathcal{Y})$ have left adjoints that map a $\mathcal{Y}$-stable collection $C \subseteq \mathcal{M} \diamond \mathcal{M}$ resp. $A \subseteq \mathcal{M}$ to its $\mathcal{Z}$-stable hull, i.e., the smallest $\mathcal{Z}$-stable supercollection in $\mathcal{M} \diamond \mathcal{M}$ resp. $\mathcal{M}$. We denote the corresponding Galois connections by $\boldsymbol{STAB}\langle \mathcal{Y}, \mathcal{Z} \rangle$ and $\boldsymbol{stab}\langle \mathcal{Y}, \mathcal{Z} \rangle$, respectively.*

If in addition $\mathcal{Z}$ and $\mathcal{Y}$ are compositive and contain $\boldsymbol{Iso}(\mathcal{M})$, then

(1) *the polarity $\boldsymbol{P}(\mathcal{M} \diamond \mathcal{M}) \multimap \boldsymbol{P}^{\mathrm{op}}(\mathcal{M} \diamond \mathcal{M})$ induced by the relation $\perp^{\mathcal{Z}}$ restricts to a Galois connection $\omega^{\mathcal{Z}} : \boldsymbol{STAB}(\mathcal{Z}) \multimap \boldsymbol{P}^{\mathrm{op}}(\mathcal{M} \diamond \mathcal{M})$;*

(2) *the following diagrams commute*

$$\begin{array}{ccc} \boldsymbol{STAB}(\mathcal{Y}) & \xrightarrow{\boldsymbol{STAB}\langle \mathcal{Y},\mathcal{Z}\rangle} & \boldsymbol{STAB}(\mathcal{Z}) \\ {\scriptstyle \Upsilon^{\mathcal{Y}}} \Big| & & \Big| {\scriptstyle \Upsilon^{\mathcal{Z}}} \\ \boldsymbol{stab}(\mathcal{Y}) & \xrightarrow[\boldsymbol{stab}\langle \mathcal{Y},\mathcal{Z}\rangle]{} & \boldsymbol{stab}(\mathcal{Z}) \end{array} \quad \text{and} \quad \begin{array}{ccc} & \boldsymbol{P}^{\mathrm{op}}(\mathcal{M} \diamond \mathcal{M}) & \\ {\scriptstyle \omega^{\mathcal{Y}}} \diagup & & \diagdown {\scriptstyle \omega^{\mathcal{Z}}} \\ \boldsymbol{STAB}(\mathcal{Y}) & \xrightarrow[\boldsymbol{STAB}\langle \mathcal{Y},\mathcal{Z}\rangle]{} & \boldsymbol{STAB}(\mathcal{Z}) \end{array}$$

Proof. Straightforward. □

Let $\mathbb{Z}$ be the order category of all collections of $\mathcal{M}$-morphisms that are compositive and contain $\boldsymbol{Iso}(\mathcal{M})$, ordered by inclusion.

THEOREM 4.1. *There exist functors $\boldsymbol{STAB}$, $\boldsymbol{stab}$, $\boldsymbol{P}^{\mathrm{op}}(\mathcal{M} \diamond \mathcal{M})$ and $\boldsymbol{P}^{\mathrm{op}}(\mathcal{M})$ (the latter two constant) from $\mathbb{Z}$ to $\mathcal{GAL}$ that map inclusions to reflections, and there exist natural transformations ω, υ, Υ, and $\curlywedge$ (the latter constant) such that the following diagram commutes*

$$\begin{array}{ccc} \boldsymbol{STAB} & \xrightarrow{\omega} & \boldsymbol{P}^{\mathrm{op}}(\mathcal{M} \diamond \mathcal{M}) \\ {\scriptstyle \Upsilon} \Big\uparrow & & \Big\downarrow {\scriptstyle \curlywedge} \\ \boldsymbol{stab} & \xrightarrow[\upsilon]{} & \boldsymbol{P}^{\mathrm{op}}(\mathcal{M}) \end{array} \tag{4-0}$$

□

Notice that the dual theorem involves the constant functors $\boldsymbol{P}(\mathcal{M} \diamond \mathcal{M})$ and $\boldsymbol{P}(\mathcal{M})$, and the functors $\boldsymbol{COST}$ and $\boldsymbol{cost}$ that map inclusions to coreflections.

5 The Inner Square

The interaction of $\mathcal{Z}$-modal closure operators with weakly hereditary cores is crucial for handling the inner square of the "big picture". In fact, we need to impose a mild restriction on the collections $\mathcal{Z}$ in order to obtain useful results.

DEFINITION 5.0. We say that $\mathcal{Z}$ has property (T) if for every instance of Diagram (1-0) where d is a pullback of g we have

$$\langle f, g\rangle \in \mathcal{Z} \quad \text{implies} \quad \langle f, d\rangle \in \mathcal{Z}$$

Clearly, the collections $\mathcal{I}$, $\mathcal{H}$, $\mathcal{C}$, and if $\mathcal{M}$ has products of pairs also $\mathcal{P}$, have this property.

PROPOSITION 5.1. *If* $\mathcal{Z}$ *satisfies* (T), *then*

(0) *weakly hereditary cores of* $\mathcal{Z}$*-modal closure operators are* $\mathcal{Z}$*-modal;*

(1) *(idempotent)* $\mathcal{Z}$*-modal hulls of weakly hereditary closure operators are weakly hereditary.*

Proof. This was proved in Lemma 2.07 of [4] under the assumption that $\mathcal{Z}$ is pullback-stable in $\mathcal{M}$. But we only used this assumption to show that whenever $\langle f, g\rangle$ in Diagram 1-0 belongs to $\mathcal{Z}$, and d is a pullback of g along r, then $\langle f, d\rangle$ belongs to $\mathcal{Z}$ as well, which is just what (T) guarantees. □

Let $\mathcal{Z}'$ be the order category of all collections $\mathcal{Z}$ of $\mathcal{M}$-morphisms that satisfy (T), ordered by inclusion. Here we do not require $\mathcal{Z}$ to be closed under composition, or to contain $\boldsymbol{Iso}(\mathcal{M})$.

THEOREM 5.2. *There exist functors* $\boldsymbol{CL}$, $\boldsymbol{wCL}$, $\boldsymbol{iCL}$, *and* $\boldsymbol{iwCL}$ *from* $\mathcal{Z}'$ *to* $\mathcal{GAL}$ *that map inclusions to reflections. Furthermore, there exist natural transformations* $\ddot{\Delta}$ *and* $\ddot{\epsilon}$ *that are point-wise coreflections, as well as natural transformations* $\dot{\nabla}$ *and* $\dot{\epsilon}$ *that are point-wise reflections such that*

$$\begin{array}{ccccc} & & \boldsymbol{CL} & & \\ & \overset{\ddot{\Delta}}{\nearrow} & & \overset{\dot{\nabla}}{\searrow} & \\ \boldsymbol{wCL} & & & & \boldsymbol{iCL} \\ & \underset{\dot{\epsilon}}{\searrow} & & \underset{\ddot{\epsilon}}{\nearrow} & \\ & & \boldsymbol{iwCL} & & \end{array} \qquad (5\text{-}0)$$

where $\ddot{\epsilon} \circ \dot{\epsilon}$ *is a point-wise e.c.f. of* $\dot{\nabla} \circ \ddot{\Delta}$.

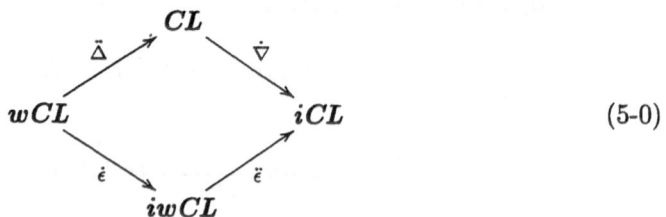

Proof. In all cases, the left adjoints of the reflections take the appropriate hulls ($\mathcal{Z}$-modal ones or idempotent $\mathcal{Z}$-modal ones) of their arguments, and the right adjoints of the coreflections take the weakly hereditary cores. This yields the four functors and the naturality of $\ddot{\Delta}$ and $\ddot{\epsilon}$.

To show the naturality of $\dot{\nabla}$ and $\dot{\epsilon}$, observe that for $\mathcal{Y} \subseteq \mathcal{Z}$ and any closure operator F we have $F \sqsubseteq F^{(\mathrm{i}\mathcal{Y})} \sqsubseteq F^{(\mathrm{i}\mathcal{Z})}$ and $F \sqsubseteq F^{\mathcal{Z}} \sqsubseteq F^{(\mathrm{i}\mathcal{Z})}$, and hence $(F^{(\mathrm{i}\mathcal{Y})})^{(\mathrm{i}\mathcal{Z})} = F^{(\mathrm{i}\mathcal{Z})} = (F^{\mathcal{Z}})^{(\mathrm{i}\mathcal{Z})}$. This establishes that the left adjoint arrows in the appropriate naturality diagrams commute.

That $\ddot{\epsilon} \circ \dot{\epsilon}$ constitutes point-wise an e.c.f. of $\dot{\nabla} \circ \ddot{\Delta}$ is clear. □

This result sheds new light on the structure of the lattice $\boldsymbol{CL}_{\mathcal{M}}$ of all closure operators on $\mathcal{M}$. There exists a "reflective tower" of sublattice-squares, the instances of (5-0), parametrized by the collections $\mathcal{Z}$ of $\mathcal{M}$-morphisms that satisfy (T). At least from level $\mathcal{H}$ onwards the squares collapse to an arrow, since hereditary closure operators are weakly hereditary. Dually, there is a corresponding "coreflective tower" of sublattice squares, parametrized by those collections $\mathcal{Z}$ which satisfy the dual of (T), namely that for every instance of Diagram 1-0 where $\langle\langle d, s\rangle, r\rangle$ is an $\langle \boldsymbol{E}, \mathcal{M}\rangle$-factorization of the 2-sink $\langle g \circ r, n\rangle$ we have

$$\langle f, g\rangle \in \mathcal{Z} \quad \text{implies} \quad \langle d, g\rangle \in \mathcal{Z}$$

The collapse of squares to arrows happens at least from the dual level of $\mathcal{H}$ (cf. Example 3.2(0)) onwards.

6 Linking Both Squares

The question remains, whether Diagrams (4-0) and (5-0), suitably restricted, can be linked by identifying $\boldsymbol{CL}$ and $\boldsymbol{iwCL}$ point-wise as centers of essentially canonical factorizations of ω and υ, respectively. Diagram (2-0) displays this for $\mathcal{Z} = \mathcal{I}$.

Given a compositive collection of pairs $\mathcal{Z}$ that contains $\boldsymbol{Iso}(\mathcal{M})$, notice that a fixed point $\langle C, D\rangle$ of $\omega^{\mathcal{Z}}$ (i.e., $(\omega^{\mathcal{Z}})_*(C) = D$ and $(\omega^{\mathcal{Z}})^*(D) = C$) is also a fixed point of $\omega^{\mathcal{I}}$, and therefore corresponds to any closure operator F that satisfies $(\dot{\omega}^{\mathcal{I}})_*(C) \cong F \cong (\ddot{\omega}^{\mathcal{I}})^*(D)$. Hence up to equivalence the fixed points of $\omega^{\mathcal{Z}}$ can be realized by some complete lattice of closure operators. But these closure operators may not all be $\mathcal{Z}$-modal! And if they are not, it is not obvious at all that the formation of $\mathcal{Z}$-modal hulls will result in an equivalent complete lattice, preferably $\boldsymbol{CL}(\mathcal{Z})$. Hence to obtainin an e.c.f. of $\omega^{\mathcal{Z}}$ with center $\boldsymbol{CL}(\mathcal{Z})$ we must guarantee that all fixed points of $\omega^{\mathcal{Z}}$ correspond to $\mathcal{Z}$-modal closure operators.

Recall from [3] that $(\dot{\omega}^{\mathcal{I}})^*$ maps a closure operator F to

$$\{\langle q, r\rangle \mid \forall n \in \mathcal{M}.\, \langle q, r\rangle \perp F(n)\} = \bigcup_{m \in \mathcal{M}} \{\langle s, t\rangle \in W/m \mid \langle s, t\rangle \ll F(m)\}$$

It is immediately clear that $(\dot{\omega}^{\mathcal{I}})^*$ maps $\mathcal{Z}$-modal closure operators to $\mathcal{Z}$-stable collections of pairs, cf. [3], Proposition 2.02.

The question, under which conditions $(\dot{\omega}^{\mathcal{I}})_*$ maps $\mathcal{Z}$-stable subcollections of $\mathcal{M} \diamond \mathcal{M}$ to $\mathcal{Z}$-modal closure operators, also was addressed in [3], Proposition 3.02, albeit under the tacit assumption that $\mathcal{Z}$ consists of cartesian $\mathcal{M}$-morphisms. A corrected version can be found in [4], Proposition 2.09, but without proof. When formulated for all appropriate $\mathcal{Z}$, i.e., for functors and natural transformations, the odd hypothesis used there turns into the familiar Beck-Chevalley condition, which lends further creedence to the belief that the functorial version is the right one.

DEFINITION 6.0 (cf. [1]). The functor W satisfies the **Beck-Chevalley condition** if for any pullback $\langle a', b'\rangle : o \twoheadrightarrow m$ and $\langle f', g'\rangle : o \twoheadrightarrow p$ of $\langle f, g\rangle : m \twoheadrightarrow n$ and $\langle a, b\rangle : p \twoheadrightarrow n$ in $\mathcal{M}$ and for every $\langle q, r\rangle \in W/m$ we have

$$\langle a, b\rangle^{\leftarrow} \langle f, g\rangle_{\rightarrow} (\langle q, r\rangle) \cong \langle f', g'\rangle_{\rightarrow} \langle a', b'\rangle^{\leftarrow} (\langle q, r\rangle)$$

I.e., it doesn't matter whether we first take the W-direct image of $\langle q, r\rangle$ along $\langle f, g\rangle$, and then the W-inverse image along $\langle a, b\rangle$, or first the W-inverse image along $\langle a', b'\rangle$ and then the W-direct image along $\langle f', g'\rangle$.

Let $\mathfrak{Z}''$ be the order-category of all those compositive subcollections of $\mathcal{M}$-morphisms that contain ***Iso*** $(\mathcal{M})$ and are stable under pullbacks in $\mathcal{M}$ (these are formed componentwise). Notice that stability under pullbacks in $\mathcal{M}$ implies property (T), hence $\mathfrak{Z}''$ is contained in $\mathfrak{Z} \cap \mathfrak{Z}'$.

THEOREM 6.1. *Suppose that W satisfies the Beck-Chevalley condition, and that for every $\mathcal{M}$-morphism the corresponding W-inverse image functor is left adjoint, i.e., preserves suprema. If we restrict Diagrams (4-0) and (5-0) to $\mathfrak{Z}''$, then*

(0) *ω admits a point-wise e.c.f. $\omega = \ddot{\omega} \circ \dot{\omega}$ with center* ***CL***;

(1) *v admits a point-wise e.c.f. with center* ***iwCL***;

(2) *$\Delta := \dot{\omega} \circ \curlyvee$ admits a point-wise e.c.f. with center* ***wCL***;

(3) *$\nabla := \curlywedge \circ \ddot{\omega}$ admits a point-wise e.c.f. with center* ***wCL***.

Proof.

(0) Consider a $\mathcal{Z}$-stable collection $C \subseteq \mathcal{M} \diamond \mathcal{M}$ and $\langle f, g\rangle \in \mathcal{Z}$. According to [2], Proposition 8, $F(n)$ is the supremum in W/n of all possible W-direct images $\langle s, t\rangle$ of C-elements. The Beck-Chevalley condition together with the pullback-stability of $\mathcal{Z}$ in $\mathcal{M} \diamond \mathcal{M}$ and the $\mathcal{Z}$-stability of C now implies that the W-inverse image of any such $\langle s, t\rangle$ along $\langle f, g\rangle$ itself is a W-direct image of some C-element. Since $\langle f, g\rangle^{\leftarrow}$ preserves suprema, it follows that

$\langle f,g\rangle^{\leftarrow}(F(n)) \ll F(m)$, and hence that $(\omega^{\mathcal{I}})^*(C)$ is indeed $\mathcal{Z}$-stable. This allows us to define $\dot{\omega}^{\mathcal{Z}}$ and $\ddot{\omega}^{\mathcal{Z}}$ as the appropriate restrictions of $\dot{\omega}^{\mathcal{I}}$ and $\ddot{\omega}^{\mathcal{I}}$, respectively.

(1) Analogous.

(2) and (3) follow from the known results at the instance $\mathcal{Z} = \mathcal{I}$. □

As has been observed before ([4], Proposition 2.10), the hypotheses of Theorem 6.1 are satisfied if we restrict the parameter $\mathcal{Z}$ to collections of cartesian $\mathcal{M}$-morphisms (most relevant examples seem to be of that form, and pullback-stability in $\mathcal{M}$ then is automatic), provided $\boldsymbol{E}$ is well-behaved in the following sense: 2-sinks in $\boldsymbol{E}$ with one component in $\mathcal{M}$ and supremum-sinks (with all components in $\mathcal{M}$) are stable under point-wise pullbacks. This is the case in every familially regular category of [9], hence in particular in every Grothendieck topos.

It seems far less likely, though, to find $\langle \boldsymbol{E}, \mathcal{M}\rangle$-categories $\mathcal{X}$ such that every W-direct image functor preserves infima. Hence the dual of Theorem 6.1 may not be very useful.

If all W-inverse image functors preserve suprema, (**weakly hereditary**) **$\mathcal{Z}$-modal cores** of closure operators exist. The inclusions $\boldsymbol{STAB}(\mathcal{Z}) \hookrightarrow \boldsymbol{STAB}(\mathcal{Y})$ and $\boldsymbol{stab}(\mathcal{Z}) \hookrightarrow \boldsymbol{stab}(\mathcal{Y})$ both have right adjoints that map a $\mathcal{Y}$-stable collection $C \subseteq \mathcal{M} \diamond \mathcal{M}$ resp. $A \subseteq \mathcal{M}$ to its $\mathcal{Z}$-stable core, i.e., its largest $\mathcal{Z}$-stable subcollection. Hence it is clear how to form the $\mathcal{Z}$-modal core of F: take the $\mathcal{Z}$-stable core of $(\dot{\omega}^{\mathcal{I}})^*(F)$ and apply $(\dot{\omega}^{\mathcal{Z}})_*$ to it. For the weakly hereditary $\mathcal{Z}$-modal core use $\Delta^{\mathcal{Z}}$ instead of $\dot{\omega}^{\mathcal{Z}}$. More details can be found in [3], Section 3.

Although it may be tempting, one cannot "dualize" Proposition 5.1(0) in order to show that $\mathcal{Z}$-modal cores of idempotent closure operators are idempotent. This statement is equivalent to the assertion that idempotent hulls of $\mathcal{Z}$-modal closure operators are $\mathcal{Z}$-modal, cf. [3], Theorem 4.00. Unless $\mathcal{Z}$ in addition has the rather strong cancellation property that for every instance of Diagram 1-0 where d is a pullback of g we have

$$\langle f,g\rangle \in \mathcal{Z} \quad \text{implies} \quad \langle d,g\rangle \in \mathcal{Z}$$

these assertions can fail, as is shown in Example 4.03 of [3]. The correct dualization of Proposition 5.1(0) states that idempotent hulls of $\mathcal{Z}$-comodal closure operators are $\mathcal{Z}$-comodal, provided $\mathcal{Z}$ satisfies the dual of (T) as described at the end of Section 5.

References

1. Jean Benabou and Jaques Roubaud. Monades et descente. *C. R. Acad. Sci. Paris Sér. A*, 270:96–98, 1970.
2. Gabriele Castellini, Jürgen Koslowski, and George E. Strecker. Closure operators and polarities. In Aaron R. Todd, editor, *Papers on General Topology and Applications*, volume 704

of *Annals of the New York Academy of Sciences*, pages 38–52, Madison, WI, June 1991. New York Academy of Science, 1993.
3. Gabriele Castellini, Jürgen Koslowski, and George E. Strecker. Hereditary and modal closure operators. In R. A. G. Seely, editor, *Category Theory 1991, Proceedings of the Summer International Meeting*, number 13 in C.M.S. Conference Proceedings, pages 111–132, Montreal, Canada, June 1991. American Mathematical Society, 1992.
4. Gabriele Castellini, Jürgen Koslowski, and George E. Strecker. Regular closure operators. *Applied Categorical Structures*, 2(3):219–244, September 1994.
5. D. Dikranjan and E. Giuli. Closure operators I. *Topology Appl.*, 27:129–143, 1987.
6. Marcel Erné, Jürgen Koslowski, Austin Melton, and George E. Strecker. A primer on Galois connections. In Aaron R. Todd, editor, *Papers on General Topology and Applications*, volume 704 of *Annals of the New York Academy of Sciences*, pages 103–125, Madison, WI, June 1991. New York Academy of Science, 1993.
7. Jürgen Koslowski. Closure operators with prescribed properties. In Francis Borceux, editor, *Category Theory and its Applications*, number 1248 in Lecture Notes in Mathematics, pages 208–220, Louvain-la-Neuve, 1987. Springer-Verlag, Berlin – New York, 1988.
8. S. Salbany. Reflective subcategories and closure operators. In E. Binz and H. Herrlich, editors, *Categorical Topology*, number 540 in Lecture Notes in Mathematics, pages 548–565, Mannheim, 1975. Springer-Verlag, Berlin – New York, 1976.
9. Ross Street. The family approach to total cocompleteness and toposes. *Trans. Amer. Math. Soc.*, 284:355–369, 1984.
10. Walter Tholen. Factorizations, localizations and the orthogonal subcategory problem. *Math. Nachr.*, 114:63–85, 1983.

Tychonoff compactifications and R-completions of mappings and rings of continuous functions

Hans-Peter A. Künzi and Boris A. Pasynkov

Abstract

All perfect extensions (with Tychonoff domains) of a continuous mapping $f : X \to Y$ between two Tychonoff spaces X and Y (in the category $\mathbf{Top}_Y$) are described by means of presheaves of subrings of the rings $C^*(f^{-1}U)$ where U is open in Y. In fact, a general description of all Tychonoff compactifications of a Tychonoff mapping $f : X \to Y$ is obtained. Our methods yield even a characterization of all Tychonoff compactifications of Tychonoff continuous images of f in the category $\mathbf{Top}_Y$.

With the help of a similar technique the well-known theorem (due to Hewitt) that two $\mathbf{R}$-complete spaces X and T are homeomorphic provided that the corresponding rings $C(X)$ and $C(T)$ are isomorphic is generalized to the category $\mathbf{Top}_Y$.

Introduction

Throughout this article a space means a topological space and a continuous mapping means a continuous mapping between spaces.

One of the most useful methods to investigate a Tychonoff space X is to study the subrings of $C(X)$ and $C^*(X)$. (Here $C(X)$ (resp. $C^*(X)$) denotes the ring of all continuous (resp. all bounded continuous) real-valued functions defined on X.) This

[0]AMS (1991) Subject Classifications: 54B10, 54C10, 54C20, 54D35, 54D45, 54D60

Key Words and Phrases: partial topological products, extensions of maps, compactifications of continuous maps, realcompactifications of maps, local compactness of maps, Stone-Čech compactifications of maps, Alexandroff compactifications of maps, rings of continuous functions.

This paper was written while the first author was supported by the Swiss National Science Foundation under grant 21-30585.91 and by the Spanish Ministry of Education and Sciences under DGICYT grant SAB94-0120.

During his visit to the University of Berne the second author was supported by the first author's grant 21-32382.91 from the Swiss National Science Foundation. He would like to thank his Bernese colleagues for their hospitality.

Eraldo Giuli (ed.), Categorical Topology, 175–201.

of X and any point $x \in X \setminus F$ there exist an open neighborhood O of fx and a continuous function $g : f^{-1}O \to [0,1]$ such that $g(x) = 0$ and $g(F \cap f^{-1}O) \subseteq \{1\}$.

A completely regular mapping is called a *Tychonoff mapping* if it is a T_0-*mapping* (i.e. for any $x, x' \in X$ such that $fx = fx'$ and $x \neq x'$, either $x \notin \mathrm{cl}\{x'\}$ or $x' \notin \mathrm{cl}\{x\}$).

It is readily checked that any continuous mapping with a Tychonoff domain is Tychonoff. More generally, if a continuous mapping $f : X \to Y$ is parallel to a Tychonoff space Z (i.e. there exists a topological embedding $e : X \to Y \times Z$ such that $f = pr \circ e$, where pr denotes the projection from the product $Y \times Z$ onto its factor space Y), then f is Tychonoff. It is essential to observe that a continuous mapping $f : X \to Y$ to a Tychonoff space Y is Tychonoff if and only if X is Tychonoff.

Let us also note that every Tychonoff mapping $f : X \to Y$ is *Hausdorff* ($= T_2$). This means that for any $x, x' \in X$ such that $fx = fx'$ and $x \neq x'$, the points x and x' have disjoint neighborhoods in X. Every submapping of a Tychonoff mapping $f : X \to Y$ is Tychonoff (i.e. for any $C \subseteq X$ the restriction $f|_C$ is Tychonoff).

In the following we fix a topological space Y with topology θ. Moreover for any $y \in Y$ let $\mathcal{N}(y)$ denote the system of all open neighborhoods of y.

Next we recall the concept of a partial topological product [15,16].

Given a space Z_α and an open set O_α in Y, we consider the disjoint union D of the sets $Y \setminus O_\alpha$ and $O_\alpha \times Z_\alpha$ and define a mapping $p_\alpha : D \to Y$ by setting $p_\alpha(y) = y$ if $y \in Y \setminus O_\alpha$, and $p_\alpha(y, z) = y$ if $(y, z) \in O_\alpha \times Z_\alpha$.

The *elementary partial topological product* (briefly, EPTP) $P(Y, Z_\alpha, O_\alpha)$ (with *base* Y, *fiber* Z_α *and open set* O_α) consists of D endowed with the topology generated by the base $p_\alpha^{-1}\theta \cup \{V : V \text{ is open in the topological product } O_\alpha \times Z_\alpha\}$. The continuous mapping $p_\alpha : P(Y, Z_\alpha, O_\alpha) \to Y$ is called the *projection* of the EPTP $P(Y, Z_\alpha, O_\alpha)$ (to its base Y). The projection q_α of the product $O_\alpha \times Z_\alpha \subseteq P(Y, Z_\alpha, O_\alpha)$ to its factor space Z_α is called the *side projection* of the EPTP $P(Y, Z_\alpha, O_\alpha)$.

Given a family of spaces $\{Z_\alpha\}_{\alpha \in A}$ and open sets $\{O_\alpha\}_{\alpha \in A}$ in Y, for any $\alpha \in A$ we have the EPTP $P_\alpha = P(Y, Z_\alpha, O_\alpha)$ and its projection $p_\alpha : P_\alpha \to Y$. Let Π be the Tychonoff product $\prod\{P_\alpha : \alpha \in A\}$ and for any $\alpha \in A$ let $pr_\alpha : \Pi \to P_\alpha$ be the projection onto its factor space P_α. The subspace

$$P = \{t = \{t_\alpha\}_{\alpha \in A} \in \Pi : p_\alpha t_\alpha = p_\beta t_\beta; \alpha, \beta \in A\}$$

of Π is called the *partial topological product* (briefly, PTP) $P(Y, \{Z_\alpha\}, \{O_\alpha\}; A)$ (with *base* Y, *fibers* Z_α *and open sets* O_α $(\alpha \in A)$).

For any $\alpha \in A$ the restriction $\pi_\alpha = pr_\alpha|_P : P \to P_\alpha$ is called the *short projection* of the PTP P. Obviously, $p_\alpha \circ \pi_\alpha = p_\beta \circ \pi_\beta$ for any $\alpha, \beta \in A$. The mapping

$$p = p_\alpha \circ \pi_\alpha : P \to Y \quad (\alpha \in A) \tag{1}$$

is called the *long projection* of the PTP P (to its base Y). Note that p is the so-called fiberwise product of the mappings p_α $(\alpha \in A)$.

Let mappings $f : X \to Y$ and $\varphi_\alpha : f^{-1}O_\alpha \to Z_\alpha$ where O_α is open in Y for some fixed $\alpha \in A$ be given. The so-called *diagonal product* $\Delta_\alpha = \Delta(f, \varphi_\alpha) : X \to P_\alpha = P(Y, Z_\alpha, O_\alpha)$ *of* f *and* φ_α is defined in the following way: $\Delta_\alpha(x)$ is equal to fx if $x \in X \setminus f^{-1}O_\alpha$, and equal to $(fx, \varphi_\alpha x)$ if $x \in f^{-1}O_\alpha$.

If both f and φ_α are continuous, then $\Delta(f, \varphi_\alpha)$ is continuous, too. Obviously,

$$p_\alpha \circ \Delta_\alpha = f \tag{2}$$

and

$$q_\alpha \circ \Delta_\alpha|_{f^{-1}O_\alpha} = \varphi_\alpha. \tag{3}$$

Remark 1.1 Strictly speaking we cannot take the composition of the mappings $\Delta_\alpha|_{f^{-1}O_\alpha}$ and q_α in (3). But here and in the following (see for example (5),(11),(17)), we shall often not distinguish between a mapping and its corestrictions.

Suppose now that such mappings $f : X \to Y$ and $\varphi_\alpha : f^{-1}O_\alpha \to Z_\alpha$ for *any* $\alpha \in A$ be given. It is easy to prove that for the (usual) diagonal product

$$\overline{\Delta} = \Delta\{\Delta_\alpha = \Delta(f, \varphi_\alpha) : \alpha \in A\} : X \to \Pi = \prod\{P_\alpha : \alpha \in A\}$$

we have $\overline{\Delta}X \subseteq P = P(Y, \{Z_\alpha\}, \{O_\alpha\}; A)$. Therefore the *diagonal product*

$$\Delta = \Delta(f, \{\varphi_\alpha : \alpha \in A\}) : X \to P$$

of the mappings f *and* $\{\varphi_\alpha\}_{\alpha\in A}$ can be defined as the corestricted mapping $\overline{\Delta} : X \to P$. Clearly, if f and all φ_α are continuous, then their diagonal product is continuous, too.

It is not difficult to prove the following relations, which will turn out to be very useful in the present article:

$$p \circ \Delta = f \tag{4}$$

and

$$q_\alpha \circ \pi_\alpha \circ \Delta|_{f^{-1}O_\alpha} = q_\alpha \circ \Delta_\alpha|_{f^{-1}O_\alpha} = \varphi_\alpha \quad (\alpha \in A). \tag{5}$$

Furthermore it is readily checked that the following assertion is true:

(#) If a mapping $\nabla : X \to P$ satisfies the relations $p \circ \nabla = f$ and $q_\alpha \circ \pi_\alpha \circ \nabla|_{f^{-1}O_\alpha} = \varphi_\alpha$ whenever $\alpha \in A$, then $\nabla = \Delta(f, \{\varphi_\alpha : \alpha \in A\})$.

In the situation just discussed we say that the system $\{\varphi_\alpha : \alpha \in A\}$ *separates*

(1) *points of* f if for any $x, x' \in X$ such that $fx = fx'$ and $x \neq x'$, there exists $\alpha \in A$ such that $\varphi_\alpha x \neq \varphi_\alpha x'$;

(2) *points and closed sets of* f if for any $x \in X$ and closed set F of X such that $x \notin F$ and $fx \in \mathrm{cl} fF$ there exists $\alpha \in A$ such that $x \in f^{-1}O_\alpha$ and $\varphi_\alpha x \notin \mathrm{cl}\,\varphi_\alpha(F \cap f^{-1}O_\alpha)$.

Let us note that for any Tychonoff (even for any T_0-)mapping f condition (2) implies condition (1). The following basic result is proved in [16].

Main Lemma *If the system* $\{\varphi_\alpha : \alpha \in A\}$ *separates points of* f *as well as points and closed sets of* f, *then the diagonal product* $\Delta(f, \{\varphi_\alpha : \alpha \in A\})$ *is a topological embedding.*

We shall say that a continuous mapping $\lambda : X \to T$ is a *morphism* $\lambda : f \to g$ (in the category $\mathbf{Top_Y}$) from a continuous mapping $f : X \to Y$ to a continuous mapping $g : T \to Y$ if $g \circ \lambda = f$. Such a morphism $\lambda : f \to g$ is called *dense (injective, surjective, bijective, a topological embedding (of f into g), a closed topological embedding (of f into g), a homeomorphism)* if $\lambda : X \to T$ is dense (i.e. $\mathrm{cl}\lambda X = T$) (injective, surjective, bijective, a topological embedding, a topological embedding and λX is closed in T, a homeomorphism).

In this terminology, (2) (resp. (4)) means that Δ_α (resp. Δ) is a morphism from f to p_α (resp. f to p) and, under the conditions of the Main Lemma, Δ is a topological embedding of f into the long projection p.

A continuous mapping $f : X \to Y$ is called *compact* if it is perfect, i.e. f is a closed mapping and all fibers $f^{-1}y$, $y \in Y$, are compact. It is straightforward to check that

(1) if a submapping $f|_C$ of a compact mapping $f : X \to Y$ is closed (i.e. C is closed in X), then $f|_C$ is compact, too;

(2) if a mapping $f : X \to Y$ is Hausdorff and the submapping $f|_C$ of f is a compact mapping, then $f|_C$ is a closed submapping of f (i.e. C is closed in X).

It is proved in [16,17] that the long projection of a PTP is compact (Tychonoff) if all fibers of the PTP are compact (Tychonoff).

For a continuous mapping $f : X \to Y$ a pair (λ, g) consisting of a continuous mapping $g : T \to Y$ and a morphism $\lambda : f \to g$ is called an *extension of f* if λ is a dense topological embedding. Usually in this situation we shall identify X and λX by means of λ. Then $T = \mathrm{cl}\, X$, $f = g|_X$ and we shall say that g (not (λ, g)) is an extension of f. An extension g of f is a *compactification* ($\equiv$ *compact extension*) of f if g is compact. For two extensions $c : cX \to Y$ and $d : dX \to Y$ of f, a morphism $\lambda : d \to c$ is called *canonical* if $\lambda(x) = x$ for any $x \in X$. For extensions c and d of $f : X \to Y$ we put $c < d$ (and say that d *follows* c) if there exists a canonical morphism $\lambda : d \to c$. Extensions c and d of $f : X \to Y$ are called *equivalent* (and we write $c \sim d$) if $c < d$ and $d < c$. It is not difficult to prove (see, for example [17]) that two Hausdorff extensions c and d of a mapping $f : X \to Y$ are equivalent if and only if there exists a canonical homeomorphism $\lambda : d \to c$. Usually we shall identify equivalent Tychonoff extensions of a Tychonoff mapping $f : X \to Y$ by means of some canonical homeomorphism. (In fact, it is possible to prove that such a homeomorphism is necessarily unique.)

It is shown in [17] that for any Tychonoff mapping $f : X \to Y$ there exists a unique (up to equivalence) maximal compactification $\beta f : \beta_f X \to Y$ among all Tychonoff compactifications of f. The following characterization of βf is obtained in [17].

For a Tychonoff compactification bf of a Tychonoff mapping $f : X \to Y$ the following conditions are equivalent:

(β1) $bf \equiv \beta f$ (i.e. $bf \sim \beta f$);

(β2) for any $O \in \theta$ and any bounded continuous real-valued function φ with domain $f^{-1}O$ there exists a (unique) continuous extension of φ with domain $(bf)^{-1}O$;

(β3) for any Tychonoff compact mapping $c : T \to Y$ and any morphism $\lambda : f \to c$ there exists a morphism $\overline{\lambda} : bf \to c$ extending λ, i.e. $\overline{\lambda}|_X = \lambda$.

It is evident that if $|Y| = 1$, then X is a Tychonoff space and $\beta_f X \equiv \beta X$.

If there exists a surjective morphism $\lambda : f \to g$, then g is called *the image of* f (*under the morphism* λ). For images g and h of f under the morphisms λ and μ, respectively, let us put $g > h$ (or $h < g$) if there exists a morphism $\eta : g \to h$ such that $\eta \circ \lambda = \mu$. Clearly such a morphism is unique. It will be called *f-canonical.* Images g and h of f will be called *f-equivalent* if $g > h$ and $h > g$. It is clear that in this case the f-canonical morphisms $\eta : g \to h$ and $\zeta : h \to g$ are homeomorphisms that are inverse to each other. Either of them will be called *an f-canonical homeomorphism* (either from g onto h or from h onto g). Usually, f-equivalent images of a continuous mapping f will be identified by means of the f-canonical homeomorphisms. Let us note that if g, g', h, h' are images of f; $g > h$; and g and g' are f-equivalent as well as h and h' are f-equivalent, then $g' > h'$.

If $g : Z \to Y$ and $h : T \to Y$ are images of a continuous mapping f under morphisms λ and μ, respectively, $\eta : g \to h$ is the f-canonical morphism, and cg, ch are compactifications of g and h such that there exists a morphism $c\eta : cg \to ch$ with the property that $\eta = c\eta : g \to h$ (i.e. $c\eta(z) = \eta(z)$ for any $z \in Z$), then $c\eta$ will be called an *f-canonical morphism* (*from the compactification cg to the compactification ch*). In this case we shall write $cg > ch$ (or $ch < cg$). It is not difficult to prove that an f-canonical morphism $c\eta : cg \to ch$ is unique if the mapping ch is Hausdorff. Two compactifications cg and ch of images g and h of a continuous mapping f are called *f-equivalent* if $cg > ch$ and $ch > cg$. Let $c\eta : cg \to ch$ and $c\zeta : ch \to cg$ be f-canonical morphisms between the compactifications cg and ch. It is possible to prove that $c\eta$ and $c\zeta$ are homeomorphisms that are inverse to each other if the mappings cg and ch are Hausdorff. An f-canonical morphism from a compactification cg of an image g of f to a compactification ch of an image h of f which is a homeomorphism is called an *f-canonical homeomorphism* (*from cg onto ch.*) Usually, f-equivalent Hausdorff (in particular, Tychonoff) compactifications of images of f will be identified by means of the corresponding f-canonical homeomorphisms. Evidently, if h and g are images of f; cg, dg and ch, dh are compactifications of g and h, respectively; $cg > ch$; and cg and dg are f-equivalent as well as ch and dh are f-equivalent, then $dg > dh$.

The last part of this introductory section deals with the Tychonoff functor τ. It is well-known (see, for example, [13]) that for any space X there exist a Tychonoff space τX and a continuous mapping $\tau_X : X \to \tau X$ such that for any continuous mapping $\varphi : X \to T$ there exists a continuous mapping $\tau\varphi : \tau X \to \tau T$ with the property

$$\tau_T \circ \varphi = (\tau\varphi) \circ \tau_X. \tag{6}$$

For a continuous mapping $\varphi : X \to T$ the mapping $\varphi^*(f) = f \circ \varphi$ where $f \in C(T)$ (or $C^*(T)$) is a ring homomorphism from $C(T)$ to $C(X)$ (or from $C^*(T)$ to $C^*(X)$) and

$$\tau_X^* \circ (\tau\varphi)^* = \varphi^* \circ \tau_T^*. \tag{7}$$

If we identify $C(\tau X)$ with $C(X)$ by means of τ_X^* and $C(\tau T)$ with $C(T)$ by means of τ_T^*, then instead of (7) we have that

$$(\tau\varphi)^* \equiv \varphi^*. \tag{8}$$

2 Tychonoff compactifications of mappings

In this section, for any bounded real-valued function $\varphi : T \to \mathbf{R}$, we set

$$\|\varphi\| = \sup\{|\varphi(t)| : t \in T\}.$$

Fix a Tychonoff mapping $f : X \to Y$.

Let $X(U)$ (more exactly, $X(U, f)$) be $f^{-1}U$, C_U^* (more exactly, C_{Uf}^*) be $C^*(X(U))$ whenever $U \in \theta$. If $U, V \in \theta$ and $V \subseteq U$, then i_{VU} (more exactly, i_{VUf}) is the identical embedding of $X(V)$ into $X(U)$ and j_{UV} (more exactly, j_{UVf}) is i_{VU}^*. The system

$$C^*(f) = \{C_{Uf}^*, j_{UV}; U \in \theta\}$$

forms a presheaf (and, in fact, a sheaf in the sense of [21, ch. 6, §7]) of rings on Y. It will be called *the ring of all bounded continuous real-valued functions on f*.

Any system of subrings R_U of the rings C_{Uf}^*, $U \in \theta$, such that $j_{UV} R_U \subseteq R_V$ whenever $U, V \in \theta$ and $V \subseteq U$, also forms a presheaf of rings on Y. It will be called *a subring of the ring* $C^*(f)$ and will be denoted by $\{R_U; U \in \theta\}$.

For two subrings $R = \{R_U; U \in \theta\}$ and $S = \{S_U; U \in \theta\}$ of $C^*(f)$ we put $S < R$ (and shall say that *R follows S*) if the following condition is fulfilled:

($\star$) for any $U \in \theta, \psi \in S_U, y \in U$ and $\epsilon > 0$ there exist $V \in \mathcal{N}(y)$ and $\varphi \in R_V$ such that $V \subseteq U$ and $\|\psi|_{X(V)} - \varphi\| < \epsilon$.

Subrings R and S of $C^*(f)$ are called *equivalent* if $R < S$ and $S < R$.

Consider now a fixed subring $R = \{R_U; U \in \theta\}$ of $C^*(f)$. We are going to define the image of f (and the compactification of the image of f) generated by R.

Set $I_\varphi = [\inf\varphi, \sup\varphi]$ whenever $\varphi \in R_U$; $Z_{RU} = \prod\{I_\varphi : \varphi \in R_U\}$ and $\Delta R_U = \Delta\{\varphi : \varphi \in R_U\} : X(U) \to Z_{RU}$ whenever $U \in \theta$. Then we can build the PTP

$$P_R = P(Y, \{Z_{RU}\}, \theta; \theta)$$

and take the diagonal product

$$\Delta_R = \Delta(f, \{\Delta R_U : U \in \theta\}) : X \to P_R.$$

Let p_R be the long projection of P_R and $\pi_{RU} : P_R \to P_{RU} = P(Y, Z_{RU}, U)$, $U \in \theta$, the short projections and $q_{RU} : U \times Z_{RU} \to Z_{RU}$, $U \in \theta$, the side projections. The mapping p_R is compact and Tychonoff because of the compactness and Tychonoff property of all fibers Z_{RU}. Therefore the mappings

$$f_R = p_R|_{\Delta_R X} \text{ and } cf_R = p_R|_{\text{cl}\Delta_R X}$$

are Tychonoff and cf_R is compact. Let

$$\delta_R = \Delta_R : X \to \Delta_R X \text{ and } c\delta_R = \Delta_R : X \to \text{cl}\Delta_R X.$$

The mapping f_R will be called the *image of f (under the morphism δ_R) generated by R*. The mapping cf_R is a Tychonoff compactification of f_R. It will be called *the compactification of the image f_R of f generated by R.*

Lemma 2.1 *For any $U \in \theta$ and $\varphi \in R_U$ there exists a unique continuous function $R\varphi : (cf_R)^{-1}U \to I_\varphi$ such that*

$$(R\varphi) \circ c\delta_R|_{X(U)} = \varphi. \tag{9}$$

Proof. Indeed,

$$R\varphi = pr_{RU\varphi} \circ q_{RU} \circ \pi_{RU}|_{(cf_R)^{-1}U},$$

where $pr_{RU\varphi} : Z_{RU} \to I_\varphi$ is the projection of the Tychonoff product Z_{RU} onto its factor space I_φ, because (see (3) and (5))

$$\varphi = pr_{RU\varphi} \circ \Delta R_U = pr_{RU\varphi} \circ q_{RU} \circ \Delta(f, \Delta R_U)|_{X(U)} = pr_{RU\varphi} \circ q_{RU} \circ \pi_{RU} \circ \Delta_R|_{X(U)} =$$

$$(pr_{RU\varphi} \circ q_{RU} \circ \pi_{RU}|_{(cf_R)^{-1}U}) \circ c\delta_R|_{X(U)} = R\varphi \circ c\delta_R|_{X(U)}.$$

Uniqueness of $R\varphi$ follows from (9) and from the density of $c\delta_R X(U)$ in $(cf_R)^{-1}U$. □

Lemma 2.2 *If $U \in \theta, \psi \in C^*_{Uf}$ and*
(⋆⋆) *for any $y \in U$ and $\epsilon > 0$ there exist $V \in \mathcal{N}(y)$ and $\varphi \in R_V$ such that $V \subseteq U$ and $\|\psi|_{X(V)} - \varphi\| < \epsilon$, then there exists a unique continuous function $\xi : (cf_R)^{-1}U \to I_\psi$ with the property*

$$\xi \circ c\delta_R|_{X(U)} = \psi. \tag{10}$$

Proof. Fix $y \in U$ and for any $n \in \mathbf{N}$ fix $V(n,y) \in \mathcal{N}(y)$ and $\varphi_{ny} \in R_{V(n,y)}$ such that $V(n+1,y) \subseteq V(n,y) \subseteq U$ and $\|\psi|_{X(V(n,y))} - \varphi_{ny}\| < \frac{1}{n}$. Evidently, the functions $R\varphi_{ny}|_{(cf_R)^{-1}y}$ (see Lemma 2.1), $n \in \mathbf{N}$, form a Cauchy sequence with respect to the metric of uniform convergence on $C^*((cf_R)^{-1}y)$. Put $\xi_y = \lim_{n\to\infty} R\varphi_{ny}|_{(cf_R)^{-1}y}$. Clearly,

$$\psi|_{f^{-1}y} = \xi_y \circ c\delta_R|_{f^{-1}y}. \tag{11}$$

Now we can define ξ in the following way:

$$\xi|_{(cf_R)^{-1}y} = \xi_y,\ y \in U.$$

Then (10) follows from (11). Let us prove the continuity of ξ. Fix $t \in (cf_R)^{-1}U$ and $\epsilon > 0$. Let $y = cf_R(t)$. There exists $n \in \mathbb{N}$ such that

$$\|\psi_{X(V(m,y))} - \varphi_{my}\| < \frac{\epsilon}{4},\ m \geq n.$$

From (10) it follows that

$$\|\xi|_{(f_R)^{-1}V(m,y)} - R\varphi_{my}|_{(f_R)^{-1}V(m,y)}\| < \frac{\epsilon}{4},\ m \geq n. \tag{12}$$

We can also suppose that

$$|\xi(t) - R\varphi_{my}(t)| < \frac{\epsilon}{4},\ m \geq n, \tag{13}$$

because $\xi(t) = \lim_{m\to\infty} R\varphi_{my}(t)$.

Let $t' \in (cf_R)^{-1}V(n,y)$ and $y' = cf_R(t')$. Similarly as above we can find $\kappa(y')$ such that

$$\|\xi|_{(f_R)^{-1}V(m,y')} - R\varphi_{my'}|_{(f_R)^{-1}V(m,y')}\| < \frac{\epsilon}{4},\ m \geq \kappa(y'). \tag{14}$$

From (12) and (14) it follows for $W(m) = V(n,y) \cap V(m,y')$, $m \geq \kappa(y')$, that

$$\|R\varphi_{my'}|_{(f_R)^{-1}W(m)} - R\varphi_{ny}|_{(f_R)^{-1}W(m)}\| < \frac{2\epsilon}{4},\ m \geq \kappa(y'),$$

and

$$\|R\varphi_{my'}|_{(cf_R)^{-1}W(m)} - R\varphi_{ny}|_{(cf_R)^{-1}W(m)}\| \leq \frac{2\epsilon}{4},\ m \geq \kappa(y'),$$

because the set $(f_R)^{-1}W(m)$ is dense in $(cf_R)^{-1}W(m)$. In particular,

$$|R\varphi_{my'}(t') - R\varphi_{ny}(t')| \leq \frac{2\epsilon}{4},\ m \geq \kappa(y').$$

and hence

$$|\xi(t') - R\varphi_{ny}(t')| \leq \frac{2\epsilon}{4}. \tag{15}$$

Take a neighborhood G of t in $\text{cl}\Delta_R X$ such that $cf_R G \subseteq V(n,y)$ and

$$|R\varphi_{ny}(t') - R\varphi_{ny}(t)| < \frac{\epsilon}{4} \text{ for any } t' \in G. \tag{16}$$

Then, for any $t' \in G$, we have (see (15), (16) and (13)) the following inequality.

$$|\xi(t') - \xi(t)| \leq |\xi(t') - R\varphi_{ny}(t')| + |R\varphi_{ny}(t') - R\varphi_{ny}(t)| +$$

$$|R\varphi_{ny}(t) - \xi(t)| < \frac{2\epsilon}{4} + \frac{\epsilon}{4} + \frac{\epsilon}{4} = \epsilon.$$

The continuity of ξ is proved. The uniqueness of ξ follows from (10) and from the density of $c\delta_R X(U)$ in $(cf_R)^{-1}U$. □

Lemma 2.3 *If for subrings $R = \{R_U; U \in \theta\}$ and $\{S_U; U \in \theta\}$ of $C^*(f)$ we have that $S < R$, then there exist surjective morphisms $\lambda : f_R \to f_S$ and $c\lambda : cf_R \to cf_S$ such that $\lambda = c\lambda : f_R \to f_S$ (i.e. $c\lambda(x) = \lambda(x)$ for any $x \in \Delta_R X$) and*

$$\lambda \circ \delta_R = \delta_S, \; c\lambda \circ c\delta_R = c\delta_S,$$

i.e. $c\lambda$ is an f-canonical morphism from cf_R to cf_S and thus $cf_S < cf_R$.

Proof. Fix $U \in \theta$. By Lemma 2.2 for any $\psi \in S_U$ there exists a continuous function $\xi(\psi) : (cf_R)^{-1}U \to I_\psi$ such that

$$\xi(\psi) \circ c\delta_R|_{X(U)} = \psi. \tag{17}$$

Take the diagonal product
$\Delta_{RSU} = \Delta\{\xi(\psi) : \psi \in S_U\} : (cf_R)^{-1}U \to Z_{SU} = \prod\{I_\psi : \psi \in S_U\}$.
Then

$$pr_{SU\psi} \circ \Delta_{RSU} = \xi(\psi), \; \psi \in S_U, \tag{18}$$

where $pr_{SU\psi}$ is the projection of the Tychonoff product Z_{SU} onto its factor space I_ψ. Since (see (17) and (18))

$$pr_{SU\psi} \circ \Delta S_U = \psi = \xi(\psi) \circ c\delta_R|_{X(U)} = pr_{SU\psi} \circ \Delta_{RSU} \circ c\delta_R|_{X(U)}, \; \psi \in S_U,$$

we see that

$$\Delta S_U = \Delta_{RSU} \circ c\delta_R|_{X(U)}. \tag{19}$$

The diagonal product
$\Delta_{RS} = \Delta(cf_R, \{\Delta_{RSU} : U \in \theta\}) : \mathrm{cl}\Delta_R X \to P_S$ is a morphism from cf_R to p_S, i.e.

$$p_S \circ \Delta_{RS} \doteq cf_R. \tag{20}$$

From (20) and (19) it follows that

$$p_S \circ (\Delta_{RS} \circ \Delta_R) = cf_R \circ \Delta_R = f$$

and (see (5))

$$q_{SU} \circ \pi_{SU} \circ (\Delta_{RS} \circ \Delta_R)|_{X(U)} = q_{SU} \circ \pi_{SU} \circ \Delta_{RS}|_{(cf_R)^{-1}U} \circ \Delta_R|_{X(U)} =$$

$$\Delta_{RSU} \circ \Delta_R|_{X(U)} = \Delta_{RSU} \circ c\delta_R|_{X(U)} = \Delta S_U, \ U \in \theta.$$

Therefore by Section 1,(#),

$$\Delta_{RS} \circ \Delta_R = \Delta(f, \{\Delta S_U : U \in \theta\}) = \Delta_S. \tag{21}$$

Hence, $\Delta_S X = \Delta_{RS}(\Delta_R X)$ and $\Delta_{RS}\mathrm{cl}\Delta_R X \subseteq \mathrm{cl}(\Delta_{RS} \circ \Delta_R).X = \mathrm{cl}\Delta_S X$.

Put $\lambda = \Delta_{RS} : \Delta_R X \to \Delta_S X$ and $c\lambda = \Delta_{RS} : \mathrm{cl}\Delta_R X \to \mathrm{cl}\Delta_S X$.

From (20) it follows that λ and $c\lambda$ are morphisms from f_R to f_S and from cf_R to cf_S, respectively. We have already proved that the morphism λ is surjective and thus the morphism $c\lambda$ is dense. But the morphism cf_S is Hausdorff and the restriction of cf_S to $c\lambda(\mathrm{cl}\Delta_R X)$ is compact (by compactness of cf_R). Therefore the set $c\lambda(\mathrm{cl}\Delta_R X)$ is closed (and dense) in $\mathrm{cl}\Delta_S X$ and, consequently, coincides with it. The surjectivity of $c\lambda$ is also proved. From (21) we have $\lambda \circ \delta_R = \delta_S$ and $c\lambda \circ c\delta_R = c\delta_S$. □

Corollary 2.1 *If subrings R and S of $C^*(f)$ are equivalent, then the images f_R and f_S of f under the morphisms δ_R and δ_S generated by R resp. S are f-equivalent and the compactifications cf_R and cf_S of f_R and f_S generated by R resp. S are f-equivalent, too.*

Proof. By Lemma 2.3 there exist morphisms $\lambda : f_R \to f_S, c\lambda : cf_R \to cf_S$ and $\mu : f_S \to f_R$, $c\mu : cf_S \to cf_R$ such that $\lambda = c\lambda : f_R \to f_S, \mu = c\mu : f_S \to f_R, \lambda \circ \delta_R = \delta_S, c\lambda \circ c\delta_R = c\delta_S, \mu \circ \delta_S = \delta_R$ and $c\mu \circ c\delta_S = c\delta_R$.

Let $z \in \Delta_R X$. Then we have a point $x \in X$ with the property $\Delta_R x = z = \delta_R x$ and satisfying the following relations

$$(\mu \circ \lambda)(z) = (\mu \circ \lambda \circ \delta_R)(x) = (\mu \circ \delta_S)(x) = \delta_R x = z.$$

Analogously, $(\lambda \circ \mu)(s) = s$ for any $s \in \Delta_S X$. Hence, λ and μ are the f-equivalent homeomorphisms between f_R and f_S that are inverse to each other.

Let $z \in \mathrm{cl}\Delta_R X$. Then $(c\mu \circ c\lambda)(z) = z$, because the mapping cf_R is Hausdorff, $\Delta_R X$ is dense in $\mathrm{cl}\Delta_R X$ and $c\mu \circ c\lambda : \Delta_R X \to \Delta_R X$ is equal to $\mu \circ \lambda = \mathrm{id}_{\Delta_R X}$ (as we just proved). Hence $c\mu \circ c\lambda = \mathrm{id}_{\mathrm{cl}\Delta_R X}$. Similarly, $c\lambda \circ c\mu = \mathrm{id}_{\mathrm{cl}\Delta_S X}$. From this it follows that cf_R and cf_S are equivalent compactifications of f_R and f_S and $c\lambda$ and $c\mu$ are their corresponding canonical homeomorphisms that are inverse to each other. □

Now we shall determine when the relation $cf_S < cf_R$ implies the relation $S < R$ for subrings S and R of $C^*(f)$.

We shall say that a subring $R = \{R_U; U \in \theta\}$ of $C^*(f)$ *contains all constant functions* if R_U contains all constant functions on $X(U)$ for any $U \in \theta$.

Lemma 2.4 *Let $R = \{R_U; U \in \theta\}$ be a subring of $C^*(f)$ that contains all constant functions and suppose (using the notations introduced above) that $\psi \in C^*((cf_R)^{-1}U)$ for some $U \in \theta$. Then*

$(\star\star\star)$ *for any $y \in U$ and $\epsilon > 0$ there exist $V \in \mathcal{N}(y)$ and $\varphi \in R_V$ such that $V \subseteq U$ and (see Lemma 2.1)*

$$\|\psi|_{(cf_R)^{-1}V} - R\varphi\| < \epsilon \textit{ and } \|\psi \circ c\delta_R|_{X(V)} - \varphi\| < \epsilon. \tag{22}$$

Proof. Fix $y \in U$ and $\epsilon > 0$. The space $(cf_R)^{-1}y$ is Hausdorff and compact. For any $V \in \mathcal{N}(y)$ and $\varphi \in R_V$ put $R\varphi_{yV} = R\varphi|_{(cf_R)^{-1}y}$. Evidently the system $\overline{R}_y$ of all such functions $R\varphi_{y,V}$ contains all constant functions on $(cf_R)^{-1}y$. It is a subring of the ring $C^*((cf_R)^{-1}y)$.

Indeed, for $R\varphi(i)_{yV(i)}$, $i = 1, 2$, we can take $V = V(1) \cap V(2)$ and $\chi(i) = \varphi(i)|_{X(V)}$, $i = 1, 2$. Then $R\varphi(i)_{yV(i)} = R\chi(i)_{yV}$, $i = 1, 2$; $\chi(1) + \chi(2) \in R_V$ and $R\varphi(1)_{yV(1)} + R\varphi(2)_{yV(2)} = R\chi(1)_{yV} + R\chi(2)_{yV} = R(\chi(1) + \chi(2))_{yV}$. Analogously $R\varphi(1)_{yV(1)} \cdot R\varphi(2)_{yV(2)} = R(\chi(1) \cdot \chi(2))_{yV}$.

Furthermore the system $\overline{R}_y$ separates points of the compactum $(cf_R)^{-1}y$ (in the sense of [5, page 190]), because (see the proof of Lemma 2.1) $R\varphi_{yV} = pr_{RV\varphi} \circ q_{RV} \circ \pi_{RV}|_{(cf_R)^{-1}y}$ and the system of all mappings $q_{RV} \circ \pi_{RV}|_{p_R^{-1}y}$ separates points of the subset $p_R^{-1}y$ of the PTP P_R and thus the system of all functions $pr_{RV\varphi} \circ q_{RV} \circ \pi_{RV}|_{p_R^{-1}y}$, $y \in V, \varphi \in R_V$, separates points of $p_R^{-1}y$.

By the Stone-Weierstrass Theorem (see for example [5, page 191]) $\overline{R}_y$ is a dense subring of $C^*((cf_R)^{-1}y)$ where the latter ring carries the topology of uniform convergence.

Thus we can choose $R\chi_{yW}$ in such a way that $\|\psi|_{(cf_R)^{-1}y} - R\chi_{yW}\| < \frac{\epsilon}{2}$. From compactness of cf_R it follows that there exists $V \in \mathcal{N}(y)$ such that $V \subseteq W \cap U$ and $\|\psi|_{(cf_R)^{-1}V} - R\chi|_{(cf_R)^{-1}V}\| < \epsilon$. Then for $\varphi = \chi|_{X(V)}$ we have $R\varphi = R\chi|_{(cf_R)^{-1}V}$ and (22). □

Lemma 2.5 *Let $R = \{R_U : U \in \theta\}$ and $S = \{S_U : U \in \theta\}$ be subrings of $C^*(f)$, $cf_S < cf_R$ and R contain all constant functions. Then $S < R$.*

Proof. Fix $U \in \theta$, $\psi \in S_U, y \in U$ and $\epsilon > 0$. Let $\lambda : f_R \to f_S$ and $c\lambda : cf_R \to cf_S$ be f-canonical morphisms such that $\lambda = c\lambda : f_R \to f_S$.

Then (see Lemma 2.1) $S\psi \in C^*((cf_S)^{-1}U)$ and $\chi = S\psi \circ c\lambda|_{(cf_R)^{-1}U} \in C^*((cf_R)^{-1}U)$. By Lemma 2.4 there exist $V \in \mathcal{N}(y)$ and $\varphi \in R_V$ such that $V \subseteq U$ and

$\|\chi|_{(cf_R)^{-1}V} - R\varphi\| < \epsilon, \qquad \|\chi \circ c\delta_R|_{X(V)} - \varphi\| < \epsilon.$

But (because λ and $c\lambda$ are f-canonical and because of (9))

$$\chi \circ c\delta_R|_{X(V)} = S\psi \circ c\lambda|_{(cf_R)^{-1}U} \circ c\delta_R|_{X(V)} = S\psi \circ \lambda|_{f_R^{-1}U} \circ \delta_R|_{X(V)} = S\psi \circ \delta_S|_{X(V)} =$$

$$S\psi \circ c\delta_S|_{X(V)} = \psi|_{X(V)}. \ \square$$

Corollary 2.2 *Let R and S be subrings of $C^*(f)$ containing all constant functions and let the compactifications cf_R and cf_S be f-equivalent. Then R and S are also equivalent.*

For a subring R of $C^*(f)$ containing all constant functions let $[R]$ denote the set of all such subrings of $C^*(f)$ that are equivalent to R and let $[cf_R]$ denote the class of all Tychonoff compactifications of Tychonoff images of f that are f-equivalent to cf_R. Moreover let $\mathcal{R}$ be the set of all classes of equivalent subrings of $C^*(f)$ containing all constant functions and let $\mathcal{C}$ be the set of all classes of f-equivalent Tychonoff compactifications of Tychonoff images of f. Put

$[S] < [R]$ if $S < R$ for subrings S and R of $C^*(f)$ containing all constant functions;

$[c] < [d]$ for $[c], [d] \in \mathcal{C}$ where c and d are Tychonoff compactifications of Tychonoff images of f, $c \in [c], d \in [d]$ and $c < d$.

It is easily seen that these orders on $\mathcal{R}$ and $\mathcal{C}$ are well-defined.

Define the mapping $r : \mathcal{R} \to \mathcal{C}$ by putting $r([R]) = [cf_R]$ for any subring R of $C^*(f)$ containing all constant functions. From Corollaries 2.1 and 2.2 it follows that the mapping r is defined correctly and is injective. From Lemmas 2.3 and 2.5 it follows that r is an isomorphism between the ordered sets $\mathcal{R}$ and $r(\mathcal{R})$.

Let us prove that the mapping r is surjective.

Lemma 2.6 *For any Tychonoff compactification $cg : cT \to Y$ of a Tychonoff image $g : T \to Y$ of f under a morphism λ there exists a subring R of $C^*(f)$ containing all constant functions and such that cg and cf_R are f-equivalent.*

Proof. Let $D = C^*(cg) = \{D_U = C^*((cg)^{-1}U); U \in \theta\}$ be the ring of all real-valued bounded continuous functions on the mapping cg.

The system $\bigcup\{D_U : U \in \theta\}$ separates points of cg as well as points and closed sets of cg. Hence by the Main Lemma (from Section 1) the morphism $(cg)_D : cg \to p_D$ from cg to the long projection of the PTP $P_D = P(Y, \{Z_{DU} = \prod\{I_\psi = [\inf\psi, \sup\psi] : \psi \in D_U\}\}, \theta; \theta)$ is a topological embedding. From compactness of cg and the Hausdorff property of p_D it follows that the set $(cg)_D cT$ is closed in P_D. We shall identify cT and $(cg)_D cT$, cg and $p_D|_{(cg)_D cT}$ by means of $(cg)_D$. Then

$$f = g \circ \lambda = cg \circ \lambda = p_D \circ \lambda \text{ and } \psi = pr_{DU\psi} \circ q_{DU} \circ \pi_{DU} \circ (cg)_D|_{(cg)^{-1}U}$$

$$\equiv pr_{DU\psi} \circ q_{DU} \circ \pi_{DU}|_{(cg)^{-1}U}, \ \psi \in D_U, U \in \theta, \tag{23}$$

where $pr_{DU\psi}$ is the projection from the product Z_{DU} onto its factor space I_ψ. Clearly $R = \{R_U = \{\varphi(\psi) = \psi \circ \lambda|_{X(U)} : \psi \in D_U\}; U \in \theta\}$ is a subring of $C^*(f)$ containing all constant functions. Since $I_{\varphi(\psi)} \equiv I_\psi$, $\psi \in D_U$, we have identities

$$Z_{RU} \equiv Z_{DU}, \ U \in \theta; \ P_R \equiv P_D, p_R \equiv p_D; \pi_{RU} \equiv$$

$$\pi_{DU}; q_{RU} \equiv q_{DU}, \ U \in \theta; \ pr_{RU\varphi(\psi)} \equiv pr_{DU\psi}, \ \psi \in D_U, U \in \theta.$$

From (23) we establish the following relation

$\varphi(\psi) = \psi \circ \lambda|_{X(U)} = pr_{RU\varphi(\psi)} \circ (q_{RU} \circ \pi_{RU} \circ \lambda|_{X(U)})$ whenever $\psi \in D_U$ and $U \in \theta$, and thus $\Delta R_U = \Delta\{\varphi(\psi) : \psi \in D_U\} = (q_{RU} \circ \pi_{RU}) \circ \lambda|_{X(U)}$, $U \in \theta$.

Since $p_R \circ \lambda = p_D \circ \lambda = f$ (see (23)), by Section 1, (#), it follows that $\lambda = \delta_R = \Delta_R : X \to \Delta_R X$. Hence $\Delta_R X = \lambda X = T, \mathrm{cl}\Delta_R X = \mathrm{cl}T = cT$ and $cf_R = p_R|_{\mathrm{cl}\Delta_R X} = p_D|_{\mathrm{cl}T} = p_D|_{cT} \equiv cg$. □

Lemma 2.6 proves the surjectivity of $r : \mathcal{R} \to \mathcal{C}$. Thus we have verified the following statement.

Theorem 2.1 *Given any Tychonoff mapping f there exists an order isomorphism between the poset of all classes of equivalent subrings of the ring $C^*(f)$ that contain all constant functions and the poset of all classes of f-equivalent Tychonoff compactifications of Tychonoff images of f.*

Remark 2.1 The order isomorphism of Theorem 2.1 is obtained in the following way: *The image of any class p of equivalent subrings of $C^*(f)$ that contain all constant functions is the class $[cf_R]$ of f-equivalent compactifications of Tychonoff images of f for an arbitrary $R \in p$.*

A subring $R = \{R_U; U \in \theta\}$ of $C^*(f)$ will be called *separating* if the system $\bigcup\{R_U : U \in \theta\}$ separates points and closed sets of f. (Automatically, it separates points of f, too.) If the subring R of $C^*(f)$ is separating, then (by the Main Lemma) $\Delta_R : f \to p_R$ is a topological embedding and cf_R is a compactification of f. Vice versa if in Lemma 2.6 cg is a compactification of $f \equiv g$, then (with the notation of that lemma) the system $\bigcup\{R_U : U \in \theta\} = \bigcup\{\{\varphi(\psi) = \psi|_{X(U)} : \psi \in D_U\} : U \in \theta\}$ separates points and closed sets of f, because the system $\bigcup\{D_U : U \in \theta\}$ separates points and closed sets of cg. Therefore we can formulate the following corollary to Theorem 2.1.

Corollary 2.3 *For any Tychonoff mapping f there exists an order isomorphism between the poset of all classes of equivalent separating subrings of the ring $C^*(f)$ that contain all constant functions and the poset of all Tychonoff compactifications of f (up to the corresponding equivalence).*

Remark 2.2 The maximal Tychonoff compactification $\beta f : \beta_f X \to Y$ of a Tychonoff mapping $f : X \to Y$ was constructed in [17] as (using our notations) $cf_{C^*(f)} : \mathrm{cl}\Delta_{C^*(f)}X \to Y$.

Clearly $C^*(f)$ contains all constant functions and $R < C^*(f)$ for any subring R of $C^*(f)$. Hence Lemmas 2.3 and 2.6 imply the following assertion proved in [17] (which just expresses the property of maximality of βf among all Tychonoff compactifications of f): *For any Tychonoff compactification bf of f there exists the canonical morphism $\lambda_b : \beta f \to bf$.*

Having the results obtained above, we can prove the equivalence of conditions $(\beta 1) - (\beta 3)$ from Section 1 rather quickly.

$(\beta 1) \Rightarrow (\beta 3)$. Let a Tychonoff compact mapping $c : T \to Y$ and a morphism $\lambda : f \to c$ be given. Then $cg = c|_{\mathrm{cl}\lambda X}$ is a Tychonoff compactification of the image $g = c|_{\lambda X}$ of f (under the morphism $\mu = \lambda : f \to g$). By Lemmas 2.3 and 2.6 there exists an f-canonical morphism $\overline{\mu} : \beta f \to cg$ (extending μ). If i is the identical embedding of cg into c (i.e. it=t for any $t \in \mathrm{cl}\lambda X$), then the morphism $\overline{\lambda} = i \circ \overline{\mu} : \beta f \to c$ extends λ.

$(\beta 3) \Rightarrow (\beta 1)$. If a Tychonoff compactification bf of f has property $(\beta 3)$, then evidently $\beta f < bf$. But we have always $bf < \beta f$. Hence $bf \equiv \beta f$ up to the canonical homeomorphism.

($\beta 1$) $\Rightarrow$ ($\beta 2$). If $U \in \theta$ and $\varphi \in C^*(X(U))$, then (supposing that $X \subseteq \beta_f X \equiv \mathrm{cl}\Delta_R X \subseteq P_R$ where $R = C^*(f)$) the mapping $pr_{RU\varphi} \circ q_{RU} \circ \pi_{RU}|_{(cf_R)^{-1}U}$ is the required continuous extension of φ over $(bf)^{-1}U \equiv (cf_R)^{-1}U$.

($\beta 2$) $\Rightarrow$ ($\beta 1$). Let $R = C^*(f)$. From ($\beta 2$) it follows that for any $U \in \theta$ the diagonal product $\Delta_{RU} : f^{-1}U \to Z_{RU}$ can be continuously extended over $(bf)^{-1}U$. Let $\overline{\Delta}_{RU}$ be this extension. Then $\overline{\Delta} = \Delta(bf, \{\overline{\Delta}_{RU} : U \in \theta\}) : bf \to p_R \equiv p_{C^*(f)}$ is a morphism extending the morphism $\Delta_R \equiv \Delta_{C^*(f)} : f \to p_{C^*(f)}$. But $\beta f = p_{C^*(f)}|_{\mathrm{cl}\Delta_{C^*(f)}X}$. Thus $\beta f < bf$ and (as in ($\beta 3$) $\Rightarrow$ ($\beta 1$)) $bf \equiv \beta f$ up to the canonical homeomorphism.

A description of all Tychonoff compactifications of Tychonoff images of f (all Tychonoff compactifications of f) that is often more convenient than the one described in Theorem 2.1 and Corollary 2.3 can be given as follows:

Fix some equivalence class ρ of subrings of $C^*(f)$ that contain all constant functions. Introduce on the class ρ an order by putting for $R, S \in \rho$, $S <_\rho R$ if $S_U \subseteq R_U$ for any $U \in \theta$.

Lemma 2.7 *There exists a largest element M_ρ in ρ.*

Proof. Let $R = \{R_U; U \in \theta\} \in \rho$ and $D = C^*(cf_R) = \{D_U = C^*((cf_R)^{-1}U); U \in \theta\}$ be the ring of all real-valued bounded continuous functions on the compactification cf_R of the image f_R of f. Evidently

$$\overline{R} = \{\overline{R}_U = \{\varphi(\psi) = \psi \circ \delta_R|_{X(U)} : \psi \in D_U\}; U \in \theta\}$$

is a subring of $C^*(f)$ containing all constant functions. If $\varphi \in R_U$, then, by Lemma 2.1,

$$\varphi = R\varphi \circ c\delta_R|_{X(U)} = R\varphi \circ \delta_R|_{X(U)}, \text{ where } R\varphi \in D_U.$$

Therefore $\varphi = \varphi(R\varphi)$ and thus $\varphi \in \overline{R}_U$. From this it follows that $R_U \subseteq \overline{R}_U$, $U \in \theta$. As in the proof of Lemma 2.6 it can be shown that $cf_{\overline{R}}$ and cf_R are f-equivalent. Therefore, by Corollary 2.2, $\overline{R} \in \rho$ and $R <_\rho \overline{R}$.

If $S \in \rho$, then the compactifications cf_R and cf_S are f-equivalent, i.e. there exist homeomorphisms $c\lambda : cf_R \to cf_S$ and $\lambda : f_R \to f_S$ such that $\lambda = c\lambda : f_R \to f_S$ and $\lambda \circ \delta_R = \delta_S$.

If $\chi \in C^*((cf_S)^{-1}U)$, $U \in \theta$, then $\psi = \chi \circ c\lambda|_{(cf_R)^{-1}U} \in C^*((cf_R)^{-1}U)$ and

$$\chi \circ \delta_S|_{X(U)} = \chi \circ c\lambda \circ \delta_R|_{X(U)} = \chi \circ c\lambda|_{(cf_R)^{-1}U} \circ \delta_R|_{X(U)} = \psi \circ \delta_R|_{X(U)}.$$

From this relation it follows that $\overline{S}_U \subseteq \overline{R}_U$ and, symmetrically, $\overline{R}_U \subseteq \overline{S}_U$, i.e. $\overline{S}_U = \overline{R}_U, U \in \theta$. This means that $\overline{S} = \overline{R}$. Now it is clear that $M_\rho = \overline{R}$ for any $R \in \rho$. □

A subring of the ring $C^*(f)$ will be called *complete* if it is the largest element in some equivalence class of subrings of $C^*(f)$ that contain all constant functions.

With the help of Lemma 2.7, Theorem 2.1 and Corollary 2.3 we conclude the following.

Theorem 2.2 *For any Tychonoff mapping f there exists an order isomorphism between the poset of all complete subrings of the ring $C^*(f)$ and the poset of all Tychonoff compactifications of Tychonoff images of f (up to f-equivalence of these compactifications).*

Corollary 2.4 *For any Tychonoff mapping f there exists an order isomorphism between the poset of all complete separating subrings of the ring $C^*(f)$ and the poset of all Tychonoff compactifications of f (up to their equivalence).*

For the most important case of a Tychonoff mapping, namely the case of a continuous mapping between Tychonoff spaces, we have the following corollaries.

Corollary 2.5 In the class of Tychonoff spaces, *for any continuous mapping there exists an order isomorphism between the poset of all complete subrings of the ring $C^*(f)$ and the poset of all compactifications of continuous images of f (up to f-equivalence of these compactifications).*

Corollary 2.6 *For any continuous mapping f between Tychonoff spaces there exists an order isomorphism between the poset of all complete separating subrings of the ring $C^*(f)$ and the poset of all compactifications of f with Tychonoff domain (up to equivalence of these compactifications).*

Remark 2.3 In the case that Y is a one point space the space X is Tychonoff; Tychonoff compactifications of Tychonoff images of f (resp. Tychonoff compactifications of f) coincide with Hausdorff compactifications of continuous Tychonoff images of X (resp. with Hausdorff compactifications of X); $C^*(f) \equiv C^*(X)$; the equivalence of subrings R and S of $C^*(f) \equiv C^*(X)$ means (by the definition of this equivalence) that $\mathrm{cl}R = \mathrm{cl}S$ (where the closures are taken with respect to the topology of uniform convergence on $C^*(X)$) and therefore the completeness of a subring of $C^*(f) \equiv C^*(X)$ containing all constant functions means its closedness in $C^*(X)$ with respect to the topology of uniform convergence on $C^*(X)$. Since every space X can be continuously mapped onto the one-point space, we have the following well-known corollaries to Theorem 2.2 and Corollary 2.4 (see [8,22]).

Corollary 2.7 *For any Tychonoff space X there exists an order isomorphism between the poset of all (with respect to the topology of uniform convergence) closed subrings of the ring $C^*(X)$ that contain all constant functions and the poset of all Hausdorff compactifications of all continuous Tychonoff images of X (up to equivalence of these compactifications).*

Corollary 2.8 *For any Tychonoff space X there exists an order isomorphism between the poset of all (with respect to the topology of uniform convergence) closed subrings of the ring $C^*(X)$ that contain all constant functions and separate points and closed sets of X and the poset of all Hausdorff compactifications of X (up to their equivalence).*

Since complete subrings of the ring $C^*(f)$ are important for us, we next wish to characterize them among all subrings of $C^*(f)$.

For a subring $R = \{R_U; U \in \theta\}$ of the ring $C^*(f)$ and a function $\psi \in C^*(X(U))$, $U \in \theta$, we shall say that *ψ is uniformly approximable by R* if the condition $(\star\star)$ of Lemma 2.2 is fulfilled.

Theorem 2.3 *A subring $R = \{R_U; U \in \theta\}$ of the ring $C^*(f)$ that contains all constant functions is complete if and only if for each $U \in \theta$, any function $\psi \in C^*(X(U))$ that is uniformly approximable by R belongs to R_U.*

Proof. Let R be complete, $U \in \theta$ and a function $\psi \in C^*(X(U))$ be uniformly approximable by R. By Lemma 2.2 there exists a continuous function $\xi : (cf_R)^{-1}U \to I_\psi$ such that (10) is satisfied. Thus $\xi \in C^*((cf_R)^{-1}U)$ and (see the proof of Lemma 2.7) $\psi = \xi \circ c\delta_R|_{X(U)} = \xi \circ \delta_R|_{X(U)} \in \overline{R}_U$. But $\overline{R} = R$, because of the completeness of R (see the proof of Lemma 2.7). Therefore $\psi \in R_U$.

For the converse suppose that for each $U \in \theta$ any function $\psi \in C^*(X(U))$ that is uniformly approximable by R belongs to R_U. Then for any subring $S = \{S_U; U \in \theta\}$ of $C^*(f)$ such that $S < R$ we have (by the definition of the relation $S < R$) the relations $S_U \subseteq R_U$ whenever $U \in \theta$. Hence, the subring R is complete. □

As can be seen from Remark 2.2, the complete subring R of the ring $C^*(f)$ such that $cf_R \equiv \beta f$ is equal to $C^*(f)$.

For a locally compact mapping f we shall now describe the complete subring R of $C^*(f)$ for which $cf_R \equiv \alpha f$. Here αf is the analogue of the one-point (=Alexandroff) compactification of a locally compact space.

In [17] a Tychonoff mapping f was called *locally compact* if f is an open submapping of βf (i.e. X is open in $\beta_f X$). Let us recall that a Tychonoff space X is locally compact if and only if X is open in βX.

Remark 2.4 In his dissertation [11] Ishmakhametov noted that a Tychonoff mapping f is locally compact if and only if

(lc) for any point $x \in X$ there exist $U \in \mathcal{N}(fx)$ and an open neighborhood O of x in X such that the mapping $f : X(U) \cap \mathrm{cl}_X O \to U$ is compact.

Remark 2.5 In [12] the condition (lc) was taken as the definition of local compactness of any continuous mapping.

The following statement was proved in [17].

Theorem on One-Point-Compactifications. *For any Tychonoff locally compact mapping $f : X \to Y$ there exists its Tychonoff compactification $\alpha f : \alpha_f X \to Y$ such that the mapping $\alpha f : \alpha_f X \setminus X \to Y$ is a closed (i.e. the set $\alpha f(\alpha_f X \setminus X)$ is closed in Y) topological embedding and αf is the smallest Tychonoff compactification of f (i.e. $\alpha f < cf$ for any Tychonoff compactification cf of f).*

Let us note that $|(\alpha f)^{-1}y \setminus f^{-1}y| \leq 1$ for any $y \in Y$ and that $\alpha_f X \setminus X$ is closed in $\alpha_f X$. Evidently αf is unique up to canonical homeomorphisms.

For a Tychonoff locally compact mapping f put $NP(f) = \alpha f(\alpha_f X \setminus X)$. It is not difficult to prove that $NP(f)$ is the set of all points of nonperfectness of f (i.e. for any $y \in NP(f)$, either $f^{-1}y$ is not compact or y is not a point of closedness of f (i.e. there exists an open neighborhood O of $f^{-1}y$ such that $f^{-1}U \setminus O \neq \emptyset$ for any $U \in \mathcal{N}(y)$)).

A function $\varphi \in C^*(X(U)), U \in \theta$, will be called *stable at infinity* (briefly, an SI-function) if there exists a continuous function $\zeta : U \cap NP(f) \to \mathbf{R}$ with the property:

(∞) for any $y \in U \cap NP(f)$ and $\epsilon > 0$ there are $V \in \mathcal{N}(y)$ and $C \subseteq X(V)$ such that $V \subseteq U$, $f : C \to V$ is compact and $|\varphi(x) - \zeta(y)| < \epsilon$ for any $x \in X(V) \setminus C$.

It is easy to check that the set R^α_U of all SI-functions $\varphi \in C^*(X(U))$ is a subring of the ring $C^*(X(U))$ containing all constant functions whenever $U \in \theta$, and that if $V \subseteq U$ where $V, U \in \theta$ and $\varphi \in R^\alpha_U$, then $j_{UV}(\varphi) = \varphi|_{X(V)} \in R^\alpha_V$. Thus we can consider the subring $R^\alpha = \{R^\alpha_U : U \in \theta\}$ of the ring $C^*(f)$ containing all constant functions.

Theorem 2.4 *For any Tychonoff locally compact mapping $f : X \to Y$ the subring R^α of the ring $C^*(f)$ is complete and $cf_{R^\alpha} \equiv \alpha f$.*

Proof. Let $Gf = \alpha_f X \setminus X$. By the proof of Lemma 2.6 (in our situation λ is the identical embedding of X into $\alpha_f X$ and $f = \alpha f|_X$) we have $\alpha f \equiv cf_R$ where $R = \{R_U = \{\varphi(\psi) = \psi|_{X(U)} : \psi \in C^*((\alpha f)^{-1}U)\}; U \in \theta\}$.

Fix $U \in \theta$ and $\psi \in C^*((\alpha f)^{-1}U)$. Evidently there exists a homeomorphism $h : NP(f) \to Gf$ such that $\alpha f \circ h = \mathrm{id}_{NP(f)}$. Then $\zeta = \psi \circ h|_{U \cap NP(f)}$ is a continuous function on $U \cap NP(f)$. Take $\epsilon > 0$ and $y \in U \cap NP(f)$. For the point $x_0 = hy$ there exists an open neighborhood O such that $|\psi(x) - \psi(x_0)| < \epsilon$ whenever $x \in O$. Then also $|\psi(x) - \zeta(y)| = |\psi(x) - ((\psi \circ h)(y) = \psi(x_0))| < \epsilon$.

The set $K = (\alpha f)^{-1}y \setminus O$ is compact. By the Hausdorff property of αf it follows that there exist disjoint open neighborhoods $G \subseteq O$ of x_0 and H of K in $\alpha_f X$. By closedness of αf there is $V \in \mathcal{N}(y)$ such that $V \subseteq U \setminus \alpha f(Gf \setminus G)$ and $(\alpha f)^{-1}V \subseteq O \cup H$. Then the set $Q = G \cap (\alpha f)^{-1}V$ is an open neighborhood of the set $G \cap Gf \cap (\alpha f)^{-1}V = Gf \cap (\alpha f)^{-1}V$ and $Q \subseteq O$. Since

$$C = X(V) \setminus Q = (X(V) \setminus Q) \cup ((Gf \cap (\alpha f)^{-1}V) \setminus Q) = (\alpha f)^{-1}V \setminus Q,$$

the mapping $f : C \to V$ coincides with the mapping $\alpha f : C \to V$ and thus is compact. If $x \in X(V) \setminus C = Q \subseteq O$, then $|\psi(x) - \zeta(y)| < \epsilon$. We have proved that $\varphi(\psi) = \psi|_{X(U)}$ is an SI-function, i.e. $\varphi(\psi) \in R^\alpha_U$. Therefore $R_U \subseteq R^\alpha_U$ whenever $U \in \theta$.

Suppose now that $U \in \theta$ and $\varphi \in R^\alpha_U$. Then there exists a continuous function $\zeta : U \cap NP(f) \to \mathbf{R}$ having property (∞). Let $\psi(x)$ be equal to $\varphi(x)$ if $x \in X(U)$, and $(\zeta \circ \alpha f)(x)$ if $x \in Gf \cap (\alpha f)^{-1}U$. The function ψ is continuous in all points of the open subset $X(U)$ of $(\alpha f)^{-1}U$, because $\psi|_{X(U)} = \varphi$. Let $x_0 \in Gf \cap (\alpha f)^{-1}U$ and $y = \alpha f(x_0)$. Fix $\epsilon > 0$. Find such V and C as described in (∞). Then C is closed in $(\alpha f)^{-1}V$, because the mapping $\alpha f : (\alpha f)^{-1}V \to V$ is Hausdorff and the mapping $f : C \to V$ is compact. Hence, the set $O = (\alpha f)^{-1}V \setminus C$ is an open neighborhood of the point x_0 in $(\alpha f)^{-1}U$ and for any $x \in X(V) \cap O = X(V) \setminus C$ we have

$$|(\psi(x) = \varphi(x)) - (\psi(x_0) = (\zeta \circ \alpha f)(x_0) = \zeta(y))| < \epsilon.$$

Since ζ is a continuous function, there exists an open neighborhood G of x_0 in $Gf \cap (\alpha f)^{-1}U$ (and also in Gf) such that $|((\zeta \circ \alpha f)(x) = \psi(x)) - ((\zeta \circ \alpha f)(x_0) = \psi(x_0))| < \epsilon$ for any $x \in G$ and such that $\alpha f(G) \subseteq V$. Then $W = V \setminus \alpha f(Gf \setminus G) \in \mathcal{N}(y)$, the set $H = (\alpha f)^{-1}W \setminus C = G \cup (X(W) \setminus C) \subseteq G \cup (X(V) \setminus C)$ is an open neighborhood of x_0 in $(\alpha f)^{-1}U$ and $|\psi(x) - \psi(x_0)| < \epsilon$ for any $x \in H$. Hence $\psi \in C^*((\alpha f)^{-1}U)$ and $\varphi(\psi) = \psi|_{X(U)} = \varphi$. Therefore $R^\alpha_U \subseteq R_U$ and thus $R^\alpha_U = R_U$ whenever $U \in \theta$. We have proved that $R^\alpha = R$ and $\alpha f = cf_R = cf_{R^\alpha}$. It was also verified that $R^\alpha = R = \overline{R}$ (in the notation of Lemma 2.7). But this means (see the last sentence of the proof of Lemma 2.7) that the subring R^α is complete. □

3 Homeomorphic R-complete mappings

Let us recall two possible definitions of **R**-completeness ($\equiv$ **R**-compactness) of a space:

A Tychonoff space X is **R**-complete if any of the following (equivalent) conditions is fulfilled:

(1) for any point $x \in \beta X \setminus X$ there exists a countable functionally open (and locally finite) covering κ of X such that $x \notin \mathrm{cl}_{\beta X} O$ for any $O \in \kappa$;

(2) there exists a closed topological embedding of X into $\mathbf{R}^\tau$ for τ large enough.

The following generalization of the notion of **R**-completeness to mappings was introduced by the second author in [10].

A Tychonoff mapping is **R**-*complete* ($\equiv$ **R**-*compact*) if for any point $x \in \beta_f X \setminus X$ there exist $U \in \mathcal{N}(\beta f(x))$ and a countable functionally open (and locally finite in $f^{-1}U$) covering κ of $f^{-1}U$ such that $x \notin \mathrm{cl}_{(\beta f)^{-1}U} O$ for any $O \in \kappa$.

It is proved in [10] that for any Tychonoff mapping $f : X \to Y$ the following conditions are equivalent:

($1_{\mathbf{R}}$) f is **R**-complete;

($2_{\mathbf{R}}$) there exists a closed topological embedding of f into the long projection of a PTP all fibers of which are equal to the reals **R**;

($3_{\mathbf{R}}$) there exists a closed topological embedding of f into the long projection of a PTP all fibers of which are **R**-complete spaces.

Any Tychonoff space X has the Hewitt extension vX which is unique (up to canonical homeomorphisms) and is characterized by the following condition: vX is **R**-complete and X is C-embedded in vX.

For any Tychonoff spaces X and T and continuous mapping $\varphi : X \to T$ there exists a unique continuous extension $\varphi_v : vX \to vT$ of φ.

It is proved in [19] that

for any Tychonoff mapping $f : X \to Y$ there exists a unique (up to f-canonical homeomorphisms) **R**-*complete extension $vf : v_f X \to Y$ of f such that for any $U \in \theta$ and continuous function $\varphi : f^{-1}U \to \mathbf{R}$ there exists a (unique) continuous extension of φ over $(vf)^{-1}U$.*

Evidently vf is an analogue of vX and $v_f X \equiv vX$ if $|Y| = 1$.

It is also proved in [19] that $\eta f < vf$ for any **R**-complete extension ηf of f. Therefore vf is called the *maximal* **R**-*completion* ($\equiv$ *maximal* **R**-*compactification*) *of* f.

Hewitt showed [9] that the Hewitt extensions vX and vT of Tychonoff spaces are homeomorphic if and only if the rings $C(X)$ and $C(T)$ are isomorphic. We shall extend this assertion to the case of Tychonoff mappings.

Fix a Tychonoff mapping $f : X \to Y$. Similarly as in Section 2, let $C_{Uf} = C(f^{-1}U)$ for any $U \in \theta$; if $U, V \in \theta$ and $V \subseteq U$, then i_{VUf} denotes the identical embedding of $f^{-1}V$ into $f^{-1}U$ and $j_{UVf} = i^*_{VUf} : C_{Uf} \to C_{Vf}$. The system

$$C(f) = \{C_{Uf}, j_{UVf}; U \in \theta\}$$

forms a presheaf (and, in fact, a sheaf in the sense of [21, ch. 6, §7]) of rings on Y. It will be called *the ring of all continuous real-valued functions on* f.

We shall identify C_{Uf} and $C(\tau f^{-1}U)$ by means of $\tau^*_{f^{-1}U}$ for any $U \in \theta$. Then (see (8))

$$j_{UVf} = i^*_{VUf} \equiv (\tau i_{VUf})^* \text{ whenever } V, U \in \theta, V \subseteq U.$$

We shall also identify $C_{Uf} \equiv C(\tau f^{-1}U)$ with $C(v\tau f^{-1}U)$ by putting $\varphi \equiv \varphi|_{\tau f^{-1}U}$ for any $\varphi \in C(v\tau f^{-1}U)$, $U \in \theta$. Then

$$j_{UVf} = (\tau i_{VUf})^* \equiv ((\tau i_{VUf})_v)^* \text{ whenever } V, U \in \theta, V \subseteq U.$$

Let $\tau_{Uf} = \tau_{f^{-1}U}$, e_{Uf} be the identical embedding of $\tau f^{-1}U$ into $Z_{Uf} = v\tau f^{-1}U$ and $\Delta_{Uf} = e_{Uf} \circ \tau_{Uf}$. We can consider the PTP $P_f = P(Y, \{Z_{Uf}\}, \theta; \theta)$ and the diagonal product

$$\Delta_f = \Delta(f, \{\Delta_{Uf} : U \in \theta\}) : X \to P_f.$$

The systems $\{\tau_{Uf} : U \in \theta\}$ and $\{\Delta_{Uf} : U \in \theta\}$ separate points and closed sets of f, because f is Tychonoff. Therefore Δ_f is a topological embedding of f into the long projection p_f of P_f. Since all spaces Z_{Uf} are Tychonoff, the projection p_f is Tychonoff, too. For $U \in \theta$ let $\pi_{Uf} : P_f \to P_{Uf} = P(Y, Z_{Uf}, U)$, $q_{Uf} : U \times Z_{Uf} \to Z_{Uf}$ and $p_{Uf} : P_{Uf} \to Y$ be the short projections of the PTP P_f, the side projection and the projection of the EPTP P_{Uf}, respectively.

We identify X and $\Delta_f X$ by means of Δ_f. Then $f = p_f|_{X \equiv \Delta_f X}$ and $\overline{f} = p_f|_{\mathrm{cl}X \equiv \mathrm{cl}\Delta_f X}$ is a Tychonoff extension of f. From **R**-completenss of all fibers Z_{Uf} it follows (by the equivalence of conditions ($1_{\mathbf{R}}$) and ($3_{\mathbf{R}}$) formulated above) that the mapping $\overline{f}$ is **R**-complete. If $\varphi \in C_{Uf}$, $U \in \theta$, then (see (6)) $\varphi = \tau\varphi \circ \tau_{Uf}$ and there exists a continuous extension $\overline{\tau\varphi}$ of $\tau\varphi$ over Z_{Uf}. Consequently (see (5))

$$\varphi = \tau\varphi \circ \tau_{Uf} = \overline{\tau\varphi} \circ e_{Uf} \circ \tau_{Uf} =$$

$$\overline{\tau\varphi} \circ \Delta_{Uf} = \overline{\tau\varphi} \circ q_{Uf} \circ \pi_{Uf} \circ \Delta_f|_{f^{-1}U} \equiv \overline{\tau\varphi} \circ q_{Uf} \circ \pi_{Uf}|_{f^{-1}U}.$$

Then $\overline{\tau\varphi} \circ q_{Uf} \circ \pi_{Uf}|_{\overline{f}^{-1}U}$ is a continuous extension of φ over $\overline{f}^{-1}U$. We conclude that $\overline{f} \equiv vf$ (and $\mathrm{cl}\Delta_f X \equiv v_f X$).

Let us now consider another Tychonoff mapping $g : T \to Y$ such that the presheaves $\mathcal{C}(f)$ and $\mathcal{C}(g) = \{C_{Ug}, j_{UVg}; U \in \theta\}$ are *isomorphic*. Then we can identify $\mathcal{C}(f)$ and $\mathcal{C}(g)$ and put $\mathcal{C} = \mathcal{C}(f) \equiv \mathcal{C}(g)$, $C_U = C_{Uf} \equiv C_{Ug}$, $U \in \theta$, and $j_{UV} = j_{UVf} \equiv j_{UVg}$ whenever $U, V \in \theta$ and $V \subseteq U$.

Theorem 10.6 from [6] asserts that for $\mathbf{R}$-compact spaces R and S and for a ring homomorphism $t : C(R) \to C(S)$ with the property that $t(\mathbf{1}) = \mathbf{1}$ (where $\mathbf{1}$ denotes the constant function that is identically equal to 1) there is a unique continuous mapping $\varphi : S \to R$ such that $\varphi^* = t$.

Applying this result and Hewitt's theorem, we can identify Z_{Uf} and Z_{Ug} in such a manner that $(\tau i_{VUf})_v : Z_{Vf} \to Z_{Uf}$ and $(\tau i_{VUg})_v : Z_{Vg} \to Z_{Ug}$ will be identified, too, where $U, V \in \theta$ and $V \subseteq U$. Let $Z_U = Z_{Uf} \equiv Z_{Ug}$, $U \in \theta$, and $r_{VU} = (\tau i_{VUf})_v \equiv (\tau i_{VUg})_v$ whenever $U, V \in \theta$ and $V \subseteq U$.

Now we put $P = P_f \equiv P_g = P(Y, \{Z_U\}, \theta; \theta)$, $p = p_f \equiv p_g$, $\pi_U = \pi_{Uf} \equiv \pi_{Ug}$, $P_U = P_{Uf} \equiv P_{Ug}$, $p_U = p_{Uf} \equiv p_{Ug}$ and $q_U = q_{Uf} \equiv q_{Ug}$ for each $U \in \theta$.

We shall need the following relation for $U, V \in \theta, V \subseteq U$, (see (6))

$$\Delta_{Uf}|_{f^{-1}V} = e_{Uf} \circ \tau_{Uf}|_{f^{-1}V} = e_{Uf} \circ \tau_{Uf} \circ i_{VUf} =$$

$$e_{Uf} \circ \tau i_{VUf} \circ \tau_{Vf} = (\tau i_{VUf})_v \circ e_{Vf} \circ \tau_{Vf} = r_{VU} \circ \Delta_{Vf} = \tag{24}$$

$$r_{VU} \circ q_{Vf} \circ \pi_{Vf} \circ \Delta_f|_{f^{-1}V} \equiv r_{VU} \circ q_V \circ \pi_V|_{f^{-1}V}.$$

We wish to prove that $vf \equiv p|_{\mathrm{cl}X \equiv \mathrm{cl}\Delta_f X}$ and $vg \equiv p|_{\mathrm{cl}T \equiv \mathrm{cl}\Delta_g T}$ coincide. It is sufficient to prove that the sets $\mathrm{cl}X \equiv \mathrm{cl}\Delta_f X$ and $\mathrm{cl}T \equiv \mathrm{cl}\Delta_g T$ coincide.

Let $t \in \mathrm{cl}T \setminus \mathrm{cl}X$. Then there exists an open neighborhood O of t in P such that $O \cap \mathrm{cl}X = \emptyset$. We can suppose that there exist $U \in \theta$, $V(k)$ and $U(k) \in \theta$ with $V(k) \subseteq U(k)$ and $W(k)$ open in $Z_{U(k)}$ $(k = 1, \ldots, s)$ such that

$$O = p^{-1}U \cap \bigcap\{\pi^{-1}_{U(k)}(p^{-1}_{U(k)}V(k) \cap q^{-1}_{U(k)}W(k)) : k = 1, \ldots, s\}.$$

Put $G = U \cap \bigcap\{V(k) : k = 1, \ldots, s\}$. Then $G \subseteq U(k), k = 1, \ldots, s$,

$$O = p^{-1}G \cap \bigcap\{\pi^{-1}_{U(k)} q^{-1}_{U(k)} W(k) : k = 1, \ldots, s\}$$

and (see (24))

$$O \cap X \equiv \Delta_f^{-1} O = f^{-1}G \cap \bigcap\{(q_{U(k)} \circ \pi_{U(k)} \circ \Delta_f|_{f^{-1}U(k)})^{-1} W(k) : k = 1, \ldots, s\} =$$

$$\bigcap\{f^{-1}G \cap \Delta^{-1}_{U(k)f} W(k) : k = 1, \ldots, s\} =$$

$$\bigcap\{(r_{GU(k)} \circ q_G \circ \pi_G|_{f^{-1}G})^{-1} W(k) : k = 1, \ldots, s\} =$$

$$X \cap \pi_G^{-1} q_G^{-1}(\bigcap\{(r_{GU(k)})^{-1} W(k) : k = 1, \ldots, s\}).$$

Since $t \in g^{-1}G$ and (see (24))

$$(r_{GU(k)} \circ q_G \circ \pi_G)(t) = \Delta_{U(k)g}(t) =$$

$$(q_{U(k)g} \circ \pi_{U(k)g} \circ \Delta_g)(t) \equiv (q_{U(k)} \circ \pi_{U(k)})(t) \in W(k), k = 1, \ldots, s,$$

the point $(q_G \circ \pi_G)(t)$ belongs to the open set $H = \bigcap\{(r_{GU(k)})^{-1} W(k) : k = 1, \ldots, s\}$ of Z_G. The image $\Delta_{Gf}(f^{-1}G) = (e_{Gf} \circ \tau_{Gf})(f^{-1}G) = e_{Gf}(\tau f^{-1}G)$ is dense in $Z_G \equiv Z_{Gf} = v\tau f^{-1}G$. Therefore $H \cap \Delta_{Gf}(f^{-1}G) \neq \emptyset$. Hence,

$$O \cap X = X \cap \pi_G^{-1} q_G^{-1} H \equiv \Delta_f^{-1} \pi_G^{-1} q_G^{-1} H =$$

$$(\Delta_{Gf})^{-1} H = (\Delta_{Gf})^{-1}(H \cap \Delta_{Gf}(f^{-1}G)) \neq \emptyset.$$

This contradicts the choice of O. We have proved that $\mathrm{cl}T \subseteq \mathrm{cl}X$. Analogously $\mathrm{cl}X \subseteq \mathrm{cl}T$ and thus $\mathrm{cl}X = \mathrm{cl}T$. We conclude that we have verified the following analogue of Hewitt's theorem.

Theorem 3.1 *For two Tychonoff mappings f and g their maximal* **R***-completions vf and vg are homeomorphic if and only if the rings $C(f)$ and $C(g)$ are isomorphic (as presheaves).*

Corollary 3.1 *Two* **R***-complete mappings f and g are homeomorphic if and only if the rings $C(f)$ and $C(g)$ are isomorphic (as presheaves).*

Remark 3.1 Our proof of Theorem 3.1 is analogous to the proof in [2] of the following assertion (which is an analogue of the Gelfand-Kolmogoroff Theorem [7]):

Compact Tychonoff mappings f and g are homeomorphic if and only if the rings $C^*(f)$ and $C^*(g)$ (see Section 2) are isomorphic (as presheaves); in general, Tychonoff mappings f and g have homeomorphic maximal Tychonoff compactifications βf and βg if and only if the rings $C^*(f)$ and $C^*(g)$ are isomorphic (as presheaves).

Corollary 3.1 takes a particularly simple form in the basic case of continuous mappings between Tychonoff spaces. Let us first give a very simple characterization of **R**-completeness in this case.

A continuous mapping $\varphi : S \to T$ is said to be *closedly parallel to a space Z* (see [17]) if *φ has a closed topological embedding* into the projection $pr : T \times Z \to T$ of the topological product $T \times Z$ onto its factor space T. It is convenient to say that $pr : T \times Z \to T$ is *the mapping Z* (in $\mathbf{Top}_T$) and that *$\varphi : S \to T$ has a (closed) topological embedding into the mapping Z* if φ is (closedly) parallel to Z.

This terminology allows us to formulate the following well-known result in the usual way:

A continuous mapping between Tychnoff spaces is compact if and only if it has a closed topological embedding into the mapping I^τ (where I^τ is the Tychonoff cube of weight τ) for some τ.

Indeed, let $f : X \to Y$ be a compact mapping between Tychonoff spaces. Then there exists a topological embedding $e : X \to I^\tau$ for some τ. Evidently the diagonal product $\Delta = f\Delta e$ is a topological embedding of X into $Y \times I^\tau$ and, consequently, of f into the mapping I^τ. From compactness of f it follows that Δ is a closed topological embedding of X into $Y \times I^\tau$ and thus of f into the mapping I^τ.

Recall that a Tychonoff space is $\mathbf{R}$-complete if and only if it has a closed topological embedding into $\mathbf{R}^\tau$ for some τ.

Theorem 3.2 *A continuous mapping $f : X \to Y$ between Tychonoff spaces is $\mathbf{R}$-complete if and only if f has a closed topological embedding into the mapping $\mathbf{R}^\tau$ for some τ.*

Proof. Evidently $Y \times \mathbf{R}^\tau = P(Y, \{Z_\alpha = \mathbf{R}\}, \{O_\alpha = Y\}; A)$ where $|A| = \tau$. Thus if f has a closed topological embedding into the mapping $\mathbf{R}^\tau$, then f is $\mathbf{R}$-complete, because of the equivalence of conditions $(1_\mathbf{R})$ and $(2_\mathbf{R})$ (see the beginning of this section).

Now let f be $\mathbf{R}$-complete. Fix $x \in \beta_f X \setminus X$. It is proved in [10] that $U(x) \in \mathcal{N}(\beta f(x))$ and a continuous function $\varphi'_x : f^{-1}U(x) \to \mathbf{R}$ can be found such that φ'_x cannot be continuously extended to the point x. From this it follows that there exists a continuous function $\varphi_x : f^{-1}U(x) \to \overline{J}_x =]0,1[$ which cannot be continuously extended to the point x.

But by the characterization of βf given in [17] and cited in Section 1 there exists a continuous extension $\overline{\varphi}_x : (\beta f)^{-1}U(x) \to I_x = [0,1]$ of φ_x. Hence $\overline{\varphi}_x(x) \in I_x \setminus \overline{J}_x$.

Without loss of generality we can suppose that $\overline{\varphi}_x(x) = 1$. Take a functionally open set $V(x) \in \mathcal{N}(\beta f(x))$ and a continuous function $\psi_x : Y \to I_x = [0,1]$ such that $V(x) \subseteq U(x)$, $V(x) = \psi_x^{-1}]0,1]$ and $\psi_x(\beta f(x)) = 1$. Put $\lambda_x(t) = \overline{\varphi}_x(t) \cdot (\psi_x \circ \beta f)(t)$ if $t \in (\beta f)^{-1}U(x)$, and $\lambda_x(t) = 0$ if $t \in \beta_f X \setminus (\beta f)^{-1}U(x)$. It is easily seen that λ_x is continuous, $\lambda_x \beta_f X \subseteq I_x$, $\lambda_x(x) = 1$ and $\lambda_x(X) \subseteq [0,1[$. By the Tychonoff property of X there exist continuous functions $\lambda'_\gamma : X \to \overline{J}_\gamma =]0,1[, \gamma \in C$, such that $e' = \Delta\{\lambda'_\gamma : \gamma \in C\}$ is a topological embedding. Let $\lambda_\gamma : \beta_f X \to I_\gamma = [0,1]$ be a continuous extension of λ'_γ, $\gamma \in C$. Take the diagonal product $\overline{e} = \Delta\{\lambda_\alpha : \alpha \in A = (\beta_f X \setminus X) \cup C\} : \beta_f X \to I^A = \prod\{I_\alpha : \alpha \in A\}$.

Evidently $e = \overline{e}|X$ is a topological embedding, $eX \subseteq J^A = \prod\{J_\alpha : \alpha \in A\}$ where $J_\alpha = [0,1[\subseteq I_\alpha, \alpha \in A$, and $e(x) \in I^A \setminus J^A$ for any $x \in \beta_f X \setminus X$, i.e. $e(\beta_f X \setminus X) \subseteq I^A \setminus J^A$. Let $\overline{\Delta} = (\beta f)\Delta\overline{e} : \beta_f X \to Y \times I^A$. Then $\overline{\Delta}|_X = f\Delta e$ is a topological embedding of X into $Y \times I^A$, $\overline{\Delta}X \subseteq Y \times J^A$ and $\overline{\Delta}(\beta_f X \setminus X) \subseteq Y \times (I^A \setminus J^A)$. The mapping $\overline{\Delta}$ is compact, because βf is compact and because $\beta f = pr \circ \overline{\Delta}$ where $pr : Y \times I^A \to Y$ is the projection of the topological product $Y \times I^A$ onto its factor space Y. Therefore the mapping $\Delta = \overline{\Delta} : \overline{\Delta}^{-1}(Y \times J^A) \to Y \times J^A$ is compact, too. But $\overline{\Delta}^{-1}(Y \times J^A) = X$. Hence, $\Delta : X \to Y \times J^A$ is a topological embedding and the set ΔX is closed in $Y \times J^A$. Since $pr \circ \Delta = pr \circ \overline{\Delta}|_X = f$, the mapping Δ is a closed topological embedding of f into the mapping $J^A = \prod\{J_\alpha = [0,1[: \alpha \in A\}$. Hence, f is closedly topologically embeddable into $\prod\{{R_\alpha}^+ = [0,\infty[: \alpha \in A\}$ as well as into $\prod\{R_\alpha = \mathbf{R} : \alpha \in A\}$. □

Corollary 3.2 *Let continuous mappings $f : X \to Y$ and $g : Z \to Y$ have closed topological embeddings into some mapping $\mathbf{R}^\tau$ (i.e. f and g are closedly parallel to $\mathbf{R}^\tau$) and let the spaces X, Y and Z be Tychonoff. Then f and g are homeomorphic if and only if the rings $C(f)$ and $C(g)$ are isomorphic (as presheaves).*

Remark 3.2 It is not difficult to prove that *for Tychonoff spaces X, Y and a continuous mapping $f : X \to Y$, we have $vf = v^+f : (v^+f)^{-1}Y \to Y$ (and $(v^+f)^{-1}Y = v_fX$) where $v^+f : vX \to vY$ is the (unique) continuous extension of f (see [6, 8G].)*

Remark 3.3 Generalizing the notions of compactness and $\mathbf{R}$-completeness of mappings, in [1] the second author introduced the concepts of E-compactness and $\mathcal{E}$-compactness of mappings: For a space E (resp. for a class of spaces $\mathcal{E}$ that is closed under homeomorphisms) a continuous mapping $f : X \to Y$ is called *E-compact* (resp. *$\mathcal{E}$-compact*) if f is closedly embeddable into the long projection of a PTP $P(Y, \{Z_\alpha\}, \{O_\alpha\}; A)$ all fibers Z_α of which are homeomorphic to E (resp. belong to the class $\mathcal{E}$). The properties of E- and $\mathcal{E}$-compactness (as well as the related properties of E- and $\mathcal{E}$-regularity) were investigated in [1].

It was proved in [17] that compactness of Tychonoff mappings coincides with I-compactness where $I = [0, 1]$. In [10] for Tychonoff mappings the coincidence of $\mathbf{R}$-completeness and $\mathbf{R}$-compactness was asserted. In [3] and [19] Dieudonné complete mappings (i.e. $\mathcal{M}$-compact mappings where $\mathcal{M}$ consists of all metrizable spaces) were introduced and investigated. The property of $\mathcal{T}$-compactness (where $\mathcal{T}$ is the class of all Tychonoff spaces) turned out to be useful in [20]. In fact, it was proved in [20] that two Tychonoff $\mathcal{T}$-compact mappings $f : X \to Y$ and $g : Z \to Y$ are homeomorphic if and only if the presheaves $C_p(f)$ and $C_p(g)$ are (topologically) isomorphic. (Here $C_p(f)$ denotes $C(f)$ with all rings $C(f^{-1}U)$, $U \in \theta$, carrying the topology of pointwise convergence.) Let us note that every continuous mapping from a Tychonoff space into a T_1-space (in particular, every continuous mapping between Tychonoff spaces) is $\mathcal{T}$-complete [20].

References

[1] I.V. Bludova, *On $\mathcal{E}$-compactness of continuous mappings*, Candidate dissertation. MGPI, Moscow (1990) (in Russian).

[2] I.V. Bludova, V.I. Varankina, B.A. Pasynkov, *On a homeomorphism of continuous mappings*, Tartu Ülikooli Toimetised *940* (1992) 21–28.

[3] T.I. Buzulina, B.A. Pasynkov, *On Dieudonné complete mappings*, in: Geometry of immersed manifolds, Izdat. „Prometei" MGPI, Moscow (1989) 95–98 (in Russian).

[4] M.M. Clementino, E. Giuli and W. Tholen, *Compact objects and perfect morphisms*, preprint (preliminary version).

[5] R. Engelking, *General Topology*, PWN, Warszawa, 1977.

[6] L. Gillman, M. Jerison, *Rings of Continuous Functions*, Springer, New York. 1976.

[7] I.M. Gelfand, A.N. Kolmogoroff, *On rings of continuous functions on topological spaces*, Doklady Akad. Nauk SSSR *22* (1939) 11–15.

[8] I.M. Gelfand, A.D. Raikov, G.E. Shilov, *Commutative normed rings*, Uspekhi matemat. nauk *1* (1946) 48–146 (in Russian).

[9] E. Hewitt, *Rings of real-valued continuous functions, I*, Trans. Amer. Math. Soc. *64* (1948) 45–99.

[10] N.I. Il'ina, B.A. Pasynkov, *On* **R**-*complete mappings*, in: Geometry of immersed manifolds, Izdat. „Prometei" MGPI, Moscow (1989) 125–130 (in Russian).

[11] K. Ishmakhametov, *Compactifications of Tychonoff mappings*, Candidate dissertation. Frunze (1987) (in Russian).

[12] I.M. James, *Fibrewise Topology*, Cambridge Univ. Press, Cambridge, 1989.

[13] K. Morita, *Čech cohomology and covering dimension for topological spaces*, Fund. Math. *87* (1975) 31–52.

[14] J. Nagata, *On lattices of functions on topological spaces and of functions on uniform spaces*, Osaka Math. J. *1* (1949) 166–181.

[15] B.A. Pasynkov. *Partial topological products*, Doklady Akad. Nauk SSSR *154* (1964) 767–770.

[16] B.A. Pasynkov, *Partial topological products*, Trudy Moskov. Matem. Obshchestva *13* (1965) 136–245 (in Russian). English translation: Trans. Moscow Math. Soc. *13* (1965) 153–272.

[17] B.A. Pasynkov, *On extension to mappings of certain notions and assertions concerning spaces*, in: Mappings and functors, Izdat. MGU, Moscow (1984) 72–102 (in Russian).

[18] B.A. Pasynkov, *Proximities on mappings*, in: General Topology. Spaces and mappings, Izdat. MGU, Moscow (1989) 99–113 (in Russian).

[19] B.A. Pasynkov, *On completions of mappings*, in: Geometry of immersed manifolds, Izdat. „Prometei" MGPI, Moscow (1989) 131–136 (in Russian).

[20] B.A. Pasynkov, *On a theorem of Nagata*, Questions Answers Gen. Topology *12* (1994) 27–37.

[21] E.H. Spanier, *Algebraic Topology*, McGraw-Hill Book Company, New York, 1966.

[22] M.H. Stone, *Applications of the theory of Boolean rings to general topology*, Trans. Amer. Math. Soc. *41* (1937) 375–481.

[23] G.T. Whyburn. *Compactification of mappings*. Math. Ann. *166* (1966) 168–174.

Department of Mathematics, University of Berne, Sidlerstrasse 5, CH-3012 Berne, Switzerland
Chair of General Topology and Geometry, Mechanics and Mathematics Faculty, Moscow State University, Moscow 119899, Russia

Disconnectednesses: Two Examples

Harriet Lord
California State Polytechnic University at Pomona
Pomona, CA 91768 USA
hlord@csupomona.edu

Abstract. We define the T_1 property for objects in topological categories, and prove that in a topological category the full subcategory of T_1 objects is a disconnectedness. We then characterize the T_1 objects in the categories **Funsp** and **BiTop**. In addition, we investigate disconnectednesses and $\mathcal{A}$-regular morphisms in these categories.

Key words: connectedness, disconnectedness, topological category, bitopological space

Mathematics Subject Classification (1991): 18A20, 18A30,18B99, 54B30, 54C35

Introduction

The relationship between disconnectednesses and $\mathcal{A}$-regular morphisms was established by Cagliari and Cicchese in [3], where they showed that in **Top**, the category of topological spaces and continuous functions, a quotient reflective subcategory $\mathcal{A}$ is a disconnectedness if and only if the $\mathcal{A}$-regular morphisms are closed under composition. That this result holds in a number of other topological categories was shown in [11]. Hušek and Pumplün gave an example in [6] that shows that the result is not true in **Unif**, the category of uniform spaces and uniformly continuous functions.

A more complete discussion of the development of the theory of connectedness and disconnectedness can be found in [12].

In this paper we begin the investigation of disconnectednesses and $\mathcal{A}$-regular morphisms in the categories **Funsp** and **BiTop**.

In section 1, we present the definitions and basic properties of connectedness and disconnectedness. We then define T_1, and show that the full subcategory of T_1 objects is a disconnectedness.

In section 2, we introduce the category **Funsp**, a category of function spaces and function space preserving maps. We describe the T_1 objects in **Funsp**, and the $\mathbf{Funsp}_1$-regular morphisms, where $\mathbf{Funsp}_1$ is the full subcategory of T_1 objects in **Funsp**. Subcategories $\mathbf{Funsp}_{\mathcal{A}}$ of **Funsp** that are generated by subcategories $\mathcal{A}$ of **Top** are studied, and those subcategories $\mathbf{Funsp}_{\mathcal{A}}$ that are disconnectednesses in **Funsp** are characterized, as well as $\mathbf{Funsp}_{\mathcal{A}}$-regular morphisms.

In section 3, we study **BiTop**, the category of bitopological spaces and bicontinuous maps. We describe the T_1 objects, and go on to study various subcategories of **BiTop** generated by a subcategory $\mathcal{A}$ of **Top**. For those subcategories $\mathbf{BiTop}_{2\mathcal{A}}$ of **BiTop** whose objects are those bitopological spaces with both topological spaces in the subcategory $\mathcal{A}$ of **Top**, we characterize those that are disconnectednesses

Eraldo Giuli (ed.), Categorical Topology, 203–212.

in **BiTop** , and also characterize the $\mathbf{BiTop}_{2\mathcal{A}}$-regular morphisms. Other subcategories of **BiTop** that are also generated by subcategories of **Top** are investigated.

We will use the categorical terminology of [1] in this paper.

1 Preliminaries

Note: We assume that all subcategories are full and isomorphism closed, unless stated otherwise.

Definitions 1.1 Let $\mathcal{T}$ be a topological category over **Set**, and $\mathcal{A}$ and $\mathcal{B}$ subcategories of $\mathcal{T}$. We define two full subcategories $\mathcal{C}(\mathcal{A})$ and $\mathcal{D}(\mathcal{B})$ of $\mathcal{T}$ as follows.

An object $X \in \mathcal{C}(\mathcal{A})$ if and only if whenever f is a morphism from X to an object in $\mathcal{A}$, f must be constant. In this case, X is called $\mathcal{C}(\mathcal{A})$-connected. An object $Y \in \mathcal{D}(\mathcal{B})$ if and only if whenever f is a morphism from an object in $\mathcal{B}$ to Y, f must be constant.

$\mathcal{B}$ is called a *connectedness* in $\mathcal{T}$ if and only if $\mathcal{B} = \mathcal{C}(\mathcal{A})$ for some subcategory $\mathcal{A}$. $\mathcal{A}$ is called a *disconnectedness* in $\mathcal{T}$ if and only if $\mathcal{A} = \mathcal{D}(\mathcal{B})$ for some subcategory $\mathcal{B}$. (Note that $\mathcal{A}$ is a disconnectedness if and only if $\mathcal{A} = \mathcal{D}(\mathcal{C}(\mathcal{A}))$.)

Proposition 1.2 (See Satz 14.2.4 in [5].) *In a topological category T, every disconnectedness is quotient reflective.*

Definition 1.3 Let $\mathcal{T}$ be a topological category over **Set**. We say that an object X in $\mathcal{T}$ is $\mathbf{T_1}$ if and only if every two-element subobject of X is discrete.

The full subcategory of $\mathcal{T}$ whose objects are precisely the T_1 objects is denoted $\mathcal{T}_1$.

Theorem 1.4 *$\mathcal{T}_1$ is a disconnectedness in $\mathcal{T}$.*
Proof: We must show that $\mathcal{D}(\mathcal{C}(\mathcal{T}_1)) = \mathcal{T}_1$. Suppose that there exists an object Y such that Y is not T_1 and $Y \in \mathcal{D}(\mathcal{C}(\mathcal{T}_1))$. Since Y is not T_1, Y has a two-element subobject S that is not discrete. If we denote the two-element discrete object by D_2, then every morphism $f : S \to D_2$ must be constant. Consequently, $S \in \mathcal{C}(T_1)$. Since the inclusion $S \to Y$ is not constant, Y cannot be an object in $\mathcal{D}(\mathcal{C}(\mathcal{T}_1))$. •

2 Funsp

We begin by defining **Funsp**, the category of function spaces and function space preserving maps.

Definition 2.1 (See [8].) **Funsp** is the category whose object class consists of all pairs (X, H) such that X is a topological space and H is a linear subspace of $C(X)$, the space of real-valued continuous functions on X, that contains the constant functions. $f : (X, H) \to (Y, K)$ is a morphism in **Funsp** if and only if $f : X \to Y$ is a continuous function and $Kf = \{k \cdot f \mid k \in K\} \subset H$.

Note that our definition of **Funsp** differs from that in [7].

Proposition 2.2 **Funsp** *is a topological category over* **Set**.
Proof: The proof is based on the fact that **Top** is a topological category over **Set** with forgetful functor $V : \mathbf{Top} \to \mathbf{Set}$. Let $U : \mathbf{Funsp} \to \mathbf{Set}$ be the forgetful functor, where $U(X, H)$ is the underlying set of X. Then every U-structured source $(B \overset{f_\alpha}{\rightrightarrows} U(A_\alpha, K_\alpha))$ has a unique U-initial lift to $(X, H) \overset{\overline{f_\alpha}}{\rightrightarrows} (A_\alpha, K_\alpha)$, where $(X \overset{\overline{f_\alpha}}{\rightrightarrows} A_\alpha)$ is the V-initial lift of $(B \overset{f_\alpha}{\rightrightarrows} U(A_\alpha))$ and H is the linear span of $\cup\{K_\alpha f_\alpha\}$ in $C(X)$. The remainder of the proof is straightforward. •

Lemma 2.3 *Let* $f : X \to Y$ *be a continuous function.* $f : (X, H) \to (Y, constants)$ *is a morphism in* **Funsp** *for all linear subspaces* H *of* $C(X)$ *that contain the constant functions. In addition,* $f : (X, C(X)) \to (Y, K)$ *is a morphism in* **Funsp** *for all possible choices of* K.

Proposition 2.4 (S, m) *is a regular subobject of* X *in* **Top** *if and only if* $((S, Hm), m))$ *is a regular subobject of* (X, H) *in* **Funsp** *for all possible* H. *(Note that instead of writing* Hm*, we may write* $H|_S$*.)*
Proof: Suppose that (S, m) is a regular subobject of X in **Top**. Then there are two continuous functions f and g such that (S, m) is the equalizer of f and g, where $f, g : X \to Y$. $((S, Hm), m)$ is the equalizer of the morphisms $f, g : (X, H) \to (Y, constants)$ for any linear subspace H of $C(X)$ that contains the constant functions. Similarly, if $((S, H|_S), m)$ is the equalizer if two morphisms $f, g : (X, H) \to (Y, K)$ in **Funsp**, then (S, m) is the equalizer of $f, g : X \to Y$ in **Top**. •

The two-point discrete object in **Funsp** is $(D_2, C(D_2))$, where D_2 is the two-point discrete topological space. Thus, we have

Proposition 2.5 *The* T_1 *objects in* **Funsp** *are those objects* (X, H) *with the property that* H *separates the points of* X*; i.e., if* $x_1, x_2 \in X$*, with* $x_1 \neq x_2$*, then there exists* $h \in H$ *such that* $h(x_1) \neq h(x_2)$.

We denote the full subcategory of T_1 objects of **Funsp** by $\mathbf{Funsp_1}$. The following proposition is a consequence of Theorem 1.4.

Proposition 2.6 $\mathbf{Funsp_1}$ *is a disconnectedness in* **Funsp**.

In the following, we see that the **FH**-regular morphisms in **Top** can be obtained from the $\mathbf{Funsp_1}$-regular morphisms in **Funsp**. (**FH** is the category of Functionally Hausdorff Spaces and continuous functions. Recall that a topological space X is called Functionally Hausdorff if $C(X)$ separates the points of X.)

Proposition 2.7 (See [7] and [10].) *In the category* **Funsp**, *a morphism* $m : (S,K) \to (X,H)$ *is a* $\mathbf{Funsp}_1$*-regular morphism if and only if* $H - \text{aff}(m(S)) = m(S)$ *and* $K = Hm$, *where* $H-\text{aff}(m(S)) = \{x \in X \mid h(x) = 0$ *whenever* $h(m(s)) = 0$ *for all* $s \in S\}$.

Remark 2.8 In the category **FH**, $m : S \to X$ is a regular morphism if and only if m is an embedding and for all $x \notin S$, there exists $f : X \to \Re$ such that $f(x) \neq 0$ and $f(m(s)) = 0$ for all $s \in S$. (For details, see [9].)

If (S,m) is an **FH**-regular subobject of X, then $S = C(X) - \text{aff}(m(S)$.

It has been shown that the regular morphisms in **FH** are not closed under composition, while the regular morphisms in $\mathbf{Funsp}_1$ are closed under composition.

We now investigate subcategories of **Funsp** that are generated by subcategories of **Top**.

Definition 2.9 Let $\mathcal{A}$ be a subcategory of **Top**. $\mathbf{Funsp}_{\mathcal{A}}$ is the full subcategory of **Funsp** whose objects are pairs (X,H) with $X \in \mathcal{A}$.

Theorem 2.10 $\mathbf{Funsp}_{\mathcal{A}}$ *is a disconnectedness in* **Funsp** *if and only if* $\mathcal{A}$ *is a disconnectedness in* **Top**.

Proof:

For each $X \in \mathcal{C}(\mathcal{A})$, every continuous function $f : X \to A$ is constant whenever $A \in \mathcal{A}$. Thus $(X,H) \in \mathcal{C}(\mathbf{Funsp}_{\mathcal{A}})$ for all possible choices of H. If $X \notin \mathcal{C}(\mathcal{A})$, then there exists an $A \in \mathcal{A}$ and a continuous function $f : X \to A$ that is not constant. It follows that the morphism $f : (X,H) \to (A, \mathit{constants})$ is a non-constant morphism in **Funsp** for all possible choices of H, and so $\mathcal{C}(\mathbf{Funsp}_{\mathcal{A}}) = \mathbf{Funsp}_{\mathcal{C}(\mathcal{A})}$.

Suppose $\mathcal{A}$ is a disconnectedness in **Top**. Since $\mathcal{A}$ is a disconnectedness, we have that $\mathcal{D}(\mathcal{C}(\mathcal{A})) = \mathcal{A}$. Clearly, $\mathbf{Funsp}_{\mathcal{D}(\mathcal{C}(\mathcal{A}))} \subset \mathcal{D}(\mathcal{C}(\mathbf{Funsp}_{\mathcal{A}}))$ for any subcategory $\mathcal{A}$ of **Top**. Suppose that $Y \notin \mathcal{A}$. Then there exists $X \in \mathcal{C}(\mathcal{A})$ for which there is a non-constant continuous function $f : X \to Y$. It follows that $f : (X, C(X)) \to (Y,K)$ is a non-constant morphism in **Funsp** for every choice of K. Thus $(Y,K) \notin \mathcal{D}(\mathcal{C}(\mathbf{Funsp}_{\mathcal{A}}))$.

We have shown that if $\mathcal{A}$ is a disconnectedness in **Top**, then $\mathbf{Funsp}_{\mathcal{A}}$ is a disconnectedness in **Funsp**.

Now assume that $\mathcal{A}$ is not a disconnectedness in **Top**. Then there exists $Y \notin \mathcal{A}$ such that every continuous function to Y from every object $X \in \mathcal{C}(\mathcal{A})$ must be constant. Since we have shown that $\mathcal{C}(\mathbf{Funsp}_{\mathcal{A}}) = \mathbf{Funsp}_{\mathcal{C}(\mathcal{A})}$, it follows that every morphism in **Funsp** from (X,H) to (Y,K) must be constant for all possible choices of H and K. Thus $\mathbf{Funsp}_{\mathcal{A}}$ is not a disconnectedness in **Funsp**. •

Corollary 2.11 $\mathbf{Funsp}_{\mathcal{B}}$ *is a connectedness in* **Funsp** *if and only if* $\mathcal{B}$ *is a connectedness in* **Top**.

Proof: Suppose that $\mathcal{B}$ is a connectedness in **Top**. Then $\mathcal{B} = \mathcal{C}(\mathcal{A})$ for some subcategory $\mathcal{A}$ of **Top**. In the first paragraph of the above proof, we showed that $\mathcal{C}(\mathbf{Funsp}_{\mathcal{A}}) = \mathbf{Funsp}_{\mathcal{C}(\mathcal{A})}$. Thus we have that $\mathbf{Funsp}_{\mathcal{B}} = \mathbf{Funsp}_{\mathcal{C}(\mathcal{A})} = \mathcal{C}(\mathbf{Funsp}_{\mathcal{A}})$, and so $\mathbf{Funsp}_{\mathcal{B}}$ is a connectedness in **Funsp**.

The proof that $\mathcal{B}$ is a connectedness in **Top** whenever $\mathbf{Funsp}_{\mathcal{B}}$ is a connectedness in **Funsp** is accomplished by first showing that $\mathcal{D}(\mathbf{Funsp}_{\mathcal{B}}) = \mathbf{Funsp}_{\mathcal{D}(\mathcal{B})}$. This proof is similar to the proof that $\mathcal{C}(\mathbf{Funsp}_{\mathcal{A}}) = \mathbf{Funsp}_{\mathcal{C}(\mathcal{A})}$.

The remainder of the proof is similar to the proof that $\mathcal{A}$ is a disconnectedness in **Top** whenever $\mathbf{Funsp}_{\mathcal{A}}$ is a disconnectedness in **Funsp**. •

The following theorem shows that the $\mathbf{Funsp}_{\mathcal{A}}$-regular morphisms in $\mathbf{Funsp}_{\mathcal{A}}$ coincide with the $\mathcal{A}$-regular morphisms in **Top**.

Theorem 2.12 *The morphism $m : (S, H|_S) \to (X, H)$ is a $\mathbf{Funsp}_{\mathcal{A}}$-regular morphism in* **Funsp** *if and only if $m : S \to X$ is an $\mathcal{A}$-regular morphism in* **Top**.

Proof: Assume that $m : S \to X$ is an $\mathcal{A}$-regular morphism in **Top**. Then there exist morphisms $f, g : X \to Y$ such that (S, m) is an equalizer of f and g, and $Y \in \mathcal{A}$. For any linear subspace H of $C(X)$ that contains the constants, consider the morphism $m : (S, H|_S) \to (X, H)$. Then $((S, H|_S), m)$ is an equalizer of f and g, where $f, g : (X, H) \to (Y, \mathit{constants})$.

Now assume that $((S, H|_S), m)$ is an equalizer of $f, g : (X, H) \to (Y, K)$, with $(Y, K) \in \mathbf{Funsp}_{\mathcal{A}}$. Then (S, m) is an equalizer of $f, g : X \to Y$. •

We now look at the disconnectedness generated by $(D_2, \mathit{constants})$.

Theorem 2.13 *Let $\mathcal{A}$ be the full subcategory of* **Funsp** *whose objects are all pairs (X, H) isomorphic to $(D_2, \mathit{constants})$. $\mathcal{C}(\mathcal{A})$ is the full subcategory whose object class is $\{(X, H) \mid X$ is topologically connected$\}$. $\mathcal{D}(\mathcal{C}(\mathcal{A}))$ is the full subcategory whose object class is $\{(X, H) \mid X$ is totally disconnected$\}$.*

Proof: Note that $(D_2, C(D_2))$ is a mono-subobject of $(D_2, \mathit{constants})$, and therefore $(D_2, C(D_2)) \in \mathcal{D}(\mathcal{C}(\mathcal{A}))$. Suppose that X is topologically connected. Then every continuous function $f : X \to D_2$ must be constant. Therefore, every morphism $f : (X, H) \to (Y, K)$ must be constant for $(Y, K) \in \mathcal{A}$. Therefore, $(X, H) \in \mathcal{C}(\mathcal{A})$. If X is not topologically connected, then there exists a continuous function $f : X \to D_2$ that is not constant. Consequently, $f : (X, H) \to (D_2, \mathit{constants})$ is not constant, and so $(X, H) \notin \mathcal{C}(\mathcal{A})$. Thus, the object class of $\mathcal{C}(\mathcal{A})$ is $\{(X, H) \mid X$ is topologically connected$\}$.

If Y is totally disconnected, then every continuous function $f : X \to Y$ must be constant whenever X is topologically connected. If Y is not totally disconnected, then there exists a non-constant continuous function $f : X \to Y$. Thus the morphism $f : (X, H) \to (Y, \mathit{constants})$ is a non-constant morphism in **Funsp**. Therefore, $\{(X, H) \mid X$ is totally disconnected$\}$ is the object class of $\mathcal{D}(\mathcal{C}(\mathcal{A}))$. •

If $\mathcal{A}$ consists of those objects isomorphic to $(D_2, C(D_2))$, then characterizations of $\mathcal{C}(\mathcal{A})$ and $\mathcal{D}(\mathcal{C}(\mathcal{A}))$ are not as simple, as can be seen in the following proposition.

Proposition 2.14 *Let $\mathcal{A}$ be the full subcategory of* **Funsp** *whose objects are isomorphic to $(D_2, C(D_2))$.*

1. *$(D_2, constants) \in \mathcal{C}(\mathcal{A})$.*
2. *$(X, H) \in \mathcal{C}(\mathcal{A})$ whenever X is topologically connected.*
3. *$(X, constants) \in \mathcal{C}(\mathcal{A})$ whenever X is not topologically connected.*

3 BiTop

Definitions 3.1 **BiTop** is the category whose objects are all triples $(X, \mathcal{T}_1, \mathcal{T}_2)$, where X is a set, and $\mathcal{T}_1$ and $\mathcal{T}_2$ are topologies on X. $f : (X, \mathcal{T}_1, \mathcal{T}_2) \to (Y, \mathcal{S}_1, \mathcal{S}_2)$ is a morphism in **BiTop** if and only if $f : (X, \mathcal{T}_i) \to (Y, \mathcal{S}_i)$ is a continuous function, i.e. a morphism in **Top**, for $i = 1, 2$.

$(X, \mathcal{T}_1, \mathcal{T}_2)$ is called a *bitopological space* and the morphisms in **BiTop** are called *bicontinuous* functions.

Proposition 3.2 **BiTop** *is a topological category over* **Set**.

For information about bitopological spaces and bicontinuous functions, the reader is referred to [13], [14], [15], and [16].

The following is shown in [4].

Proposition 3.3 *In* **BiTop**, *$(X, \mathcal{T}_1, \mathcal{T}_2)$ is an indiscrete object if and only if $\mathcal{T}_i$ is the indiscrete topology for $i = 1, 2$. $(X, \mathcal{T}_1, \mathcal{T}_2)$ is a discrete object if and only if $\mathcal{T}_i$ is the discrete topology for $i = 1, 2$.*

Proposition 3.4 *The* $\mathbf{T}_1$ *objects in* **BiTop** *are the triples $(X, \mathcal{T}_1, \mathcal{T}_2)$, where $\mathcal{T}_i$ is a T_1 topology on X for $i = 1, 2$.*

The next theorem follows immediately from Theorem 1.4.

Theorem 3.5 $\mathbf{BiTop}_1$ *is a disconnectedness in* **BiTop**, *where* $\mathbf{BiTop}_1$ *is the full subcategory of* **BiTop** *whose objects are the* $\mathbf{T}_1$ *objects in* **BiTop**.

We now examine the full subcategory of **BiTop** whose objects are those triples that have the property that both topological spaces belong to a disconnectedness $\mathcal{A}$.

Definitions 3.6 Let $\mathcal{A}$ be a subcategory of **Top**. $\mathbf{BiTop}_{2\mathcal{A}}$ denotes the full subcategory of **BiTop** whose objects are all bitopological spaces $(X, \mathcal{T}_1, \mathcal{T}_2)$ such that $(X, \mathcal{T}_1) \in \mathcal{A}$ and $(X, \mathcal{T}_2) \in \mathcal{A}$.

$\mathbf{BiTop}_{1\mathcal{A}}$ denotes the full subcategory of **BiTop** whose objects are all bitopological spaces $(X, \mathcal{T}_1, \mathcal{T}_2)$ such that $(X, \mathcal{T}_1) \in \mathcal{A}$ or $(X, \mathcal{T}_2) \in \mathcal{A}$.

Theorem 3.7 $\mathcal{C}(\mathbf{BiTop}_{2\mathcal{A}}) = \mathbf{BiTop}_{1\mathcal{C}(\mathcal{A})}$. *Moreover,* $\mathbf{BiTop}_{2\mathcal{A}}$ *is a disconnectedness in* **BiTop** *if and only if* $\mathcal{A}$ *is a disconnectedness in* **Top**.

Proof: Clearly, $\mathbf{BiTop}_{1\mathcal{A}} \subset \mathcal{C}(\mathbf{BiTop}_{2\mathcal{A}})$. Let $(X, \mathcal{T}_1, \mathcal{T}_2) \in \mathcal{C}(\mathbf{BiTop}_{2\mathcal{A}})$. For every $(Y, \mathcal{S}_1, \mathcal{S}_2) \in \mathbf{BiTop}_{2\mathcal{A}}$, every morphism $f : (X, \mathcal{T}_1, \mathcal{T}_2) \to (Y, \mathcal{S}_1, \mathcal{S}_2)$ must be constant. Consequently, $(X, \mathcal{T}_1)$ or $(X, \mathcal{T}_2)$ must be in $\mathcal{C}(\mathcal{A})$, for if they were not, we could construct a non-constant morphism into $(Y \times Z, \mathcal{S} \times \mathcal{V}, \mathcal{S} \times \mathcal{V}) \in \mathbf{BiTop}_{2\mathcal{A}}$ for some $(Y, \mathcal{S}), (Z, \mathcal{V}) \in \mathcal{A}$. Thus we have that $\mathcal{C}(\mathbf{BiTop}_{2\mathcal{A}}) \subset \mathbf{BiTop}_{1\mathcal{A}}$, and so $\mathcal{C}(\mathbf{BiTop}_{2\mathcal{A}}) = \mathbf{BiTop}_{1\mathcal{A}}$

Now assume that $\mathcal{A}$ is a disconnectedness in **Top** and $(Y, \mathcal{S}_1, \mathcal{S}_2) \notin \mathbf{BiTop}_{2\mathcal{A}}$. Without loss of generality, assume that $(Y, \mathcal{S}_1) \notin \mathcal{A}$. Then there is a space $(X, \mathcal{T}) \in \mathcal{C}(\mathcal{A})$ such that there is a non-constant map $f : (X, \mathcal{T}) \to (Y, \mathcal{S}_1)$. Then $f : (X, \mathcal{T}, \mathcal{D}) \to (Y, \mathcal{S}_1, \mathcal{S}_2)$ is a non-constant morphism in **BiTop**, where $\mathcal{D}$ is the discrete topology on X. Thus $\mathbf{BiTop}_{2\mathcal{A}}$ is a disconnectedness.

Suppose now that $\mathbf{BiTop}_{2\mathcal{A}}$ is a disconnectedness in **BiTop** and $\mathcal{A}$ is not a disconnectedness in **Top**. Then there is a space $(Y, \mathcal{S}) \notin \mathcal{A}$ such that for all spaces $(X, \mathcal{T}) \in \mathcal{C}(\mathcal{A})$ and all morphisms $f : (X, \mathcal{T}) \to (Y, \mathcal{S})$, f must be constant. Now $(X, \mathcal{T}, \mathcal{T}) \in \mathcal{C}(\mathbf{BiTop}_{2\mathcal{A}})$. Therefore there is a non-constant morphism $g : (X, \mathcal{T}, \mathcal{T}) \to (Y, \mathcal{S}, \mathcal{S})$. $g : (X, \mathcal{T}) \to (Y, \mathcal{S})$ is non-constant, which contradicts $(Y, \mathcal{S}) \in \mathcal{D}(\mathcal{C}(\mathcal{A}))$. Therefore $\mathcal{A}$ must be a disconnectedness whenever $\mathbf{BiTop}_{2\mathcal{A}}$ is a disconnectedness. •

Theorem 3.8 *If* $\mathbf{BiTop}_{1\mathcal{A}}$ *is a disconnectedness in* **BiTop** *then* $\mathcal{A}$ *is a disconnectedness in* **Top**.

Proof: Suppose that $\mathbf{BiTop}_{1\mathcal{A}}$ is a disconnectedness in **BiTop** but $\mathcal{A}$ is not a disconnectedness in **Top**. Then there is a space $(Y, \mathcal{S}) \notin \mathcal{A}$ such that for all $(X, \mathcal{T}) \in \mathcal{C}(\mathcal{A})$, $f : (X, \mathcal{T}) \to (Y, \mathcal{S})$ must be constant for all morphisms f. Clearly, $(X, \mathcal{T}, \mathcal{T}) \in \mathcal{C}(\mathbf{BiTop}_{1\mathcal{A}})$, $(Y, \mathcal{S}, \mathcal{S}) \notin \mathbf{BiTop}_{1\mathcal{A}}$. Therefore, there is a non-constant morphism $g : (X, \mathcal{T}, \mathcal{T}) \to (Y, \mathcal{S}, \mathcal{S})$. $g : (X, \mathcal{T}) \to (Y, \mathcal{S})$ is a non-constant morphism in **Top**, and so $(Y, \mathcal{S}) \notin \mathcal{D}(\mathcal{C}(\mathcal{A}))$, and $\mathcal{A}$ must be a disconnectedness in **Top**. •

Proposition 3.9 $\mathbf{BiTop}_{2\mathcal{C}(\mathcal{A})} = \mathcal{C}(\mathbf{BiTop}_{1\mathcal{A}})$.

Proof: Clearly, $\mathbf{BiTop}_{2\mathcal{C}(\mathcal{A})} \subset \mathcal{C}(\mathbf{BiTop}_{1\mathcal{A}})$.

Suppose that $(X, \mathcal{T}_1, \mathcal{T}_2) \in \mathcal{C}(\mathbf{BiTop}_{1\mathcal{A}})$. If $(X, \mathcal{T}_1) \notin \mathcal{C}(\mathcal{A})$, then there exists $f : (X, \mathcal{T}_1) \to (Y, \mathcal{S})$ for some $(Y, \mathcal{S}) \in \mathcal{A}$. Then the bicontinuous function $f : (X, \mathcal{T}_1, \mathcal{T}_2) \to (Y, \mathcal{S}, \mathcal{I})$ is not constant, where $\mathcal{I}$ is the indiscrete topology on Y.

Since this implies that $(X, \mathcal{T}_1, \mathcal{T}_2) \notin \mathcal{C}(\mathbf{BiTop}_{1\mathcal{A}})$, we must have that $(X, \mathcal{T}_1) \in \mathcal{C}(\mathcal{A})$. Similarly, $(X, \mathcal{T}_2) \in \mathcal{C}(\mathcal{A})$. Consequently, $\mathbf{BiTop}_{2\mathcal{C}(\mathcal{A})} = \mathcal{C}(\mathbf{BiTop}_{1\mathcal{A}})$. •

We now look at the relationship between $\mathcal{A}$-regular morphisms in **Top** and $\mathbf{BiTop}_{2\mathcal{A}}$-regular morphisms in **BiTop** .

Proposition 3.10 *Suppose that $(S, \mathcal{T}_i|_S)$ is an $\mathcal{A}$-regular subobject of $(X, \mathcal{T}_i)$ for $i = 1, 2$. Then $(S, \mathcal{T}_1|_S, \mathcal{T}_2|_S)$ is a* $\mathbf{BiTop}_{2\mathcal{A}}$*-regular subobject of $(X, \mathcal{T}_1, \mathcal{T}_2)$ if $\mathcal{A}$ is closed under products of pairs.*

Proof: Suppose that $(S, \mathcal{T}_1|_S) = \mathrm{Eq}(f, g)$ and $(S, \mathcal{T}_2|_S) = \mathrm{Eq}(\alpha, \beta)$, where $f, g : (X, \mathcal{T}_1) \to (Y, \mathcal{S})$ and $\alpha, \beta : (X, \mathcal{T}_2) \to (Z, \mathcal{V})$. Then $(S, \mathcal{T}_1|_s, \mathcal{T}_2|_S) = \mathrm{Eq}(< f, \alpha >, < g, \beta >)$. •

Proposition 3.11 *If $(S, \mathcal{T}_1|_S, \mathcal{T}_2|_S)$ is a* $\mathbf{BiTop}_{2\mathcal{A}}$*-regular subobject of $(X, \mathcal{T}_1, \mathcal{T}_2)$, then $(S, \mathcal{T}_i|_S)$ is an $\mathcal{A}$-regular subobject of $(X, \mathcal{T}_i)$ for $i = 1, 2$.*

Proof: If $(S, \mathcal{T}_1|_S, \mathcal{T}_2|_S) = \mathrm{Eq}(f, g)$, where $f, g : (X, \mathcal{T}_1, \mathcal{T}_2) \to (Y, \mathcal{S}_1, \mathcal{S}_2)$, then $(S, \mathcal{T}_i|_S) = \mathrm{Eq}(f, g)$, where $f, g : (X, \mathcal{T}_i) \to (Y, \mathcal{S}_i)$. •

The following theorem is now immediate.

Theorem 3.12 *Let $\mathcal{A}$ be a subcategory of* **Top** *that is closed under products of pairs. $(S, \mathcal{T}_1|_S, \mathcal{T}_2|_S)$ is a* $\mathbf{BiTop}_{2\mathcal{A}}$*-regular subobject of $(X, \mathcal{T}_1, \mathcal{T}_2)$ if and only if $(S, \mathcal{T}_i|_S)$ is an $\mathcal{A}$-regular subobject of $(X, \mathcal{T}_i)$ for $i = 1, 2$.*

The definition of BiT_0 is due to Giuli and Salbany. (See [4].)

Definition 3.13 A bitopological space $(X, \mathcal{T}_1, \mathcal{T}_2)$ is called $\mathbf{BiT_0}$ if and only if every bicontinuous function $f : (2, \mathcal{I}, \mathcal{I}) \to (X, \mathcal{T}_1, \mathcal{T}_2)$ is a constant function, where $\mathcal{I}$ denotes the indiscrete topology. $\mathbf{BiTop}_0$ denotes the full subcategory of **BiTop** whose objects are BiT_0.

The following proposition follows from that fact that $\mathbf{BiTop}_0 = \mathcal{D}(\mathcal{B})$, where $\mathcal{B}$ is the full subcategory of **BiTop** whose only object is $(2, \mathcal{I}, \mathcal{I})$.

Proposition 3.14 $\mathbf{BiTop}_0$ *is a disconnectedness in* **BiTop** .

Definition 3.15 $\mathcal{T}_1 \vee \mathcal{T}_2$ is the weakest topology that contains both $\mathcal{T}_1$ and $\mathcal{T}_2$; i.e., it is the intersection of all topologies on X that contain both $\mathcal{T}_1$ and $\mathcal{T}_2$.

Proposition 3.16 (See [4].) *$(X, \mathcal{T}_1, \mathcal{T}_2)$ is BiT_0 if and only if $(X, \mathcal{T}_1 \vee \mathcal{T}_2)$ is a T_0 topological space if and only if $(X, \mathcal{T}_1, \mathcal{T}_2) \in$* $\mathbf{BiTop}_0$.

The definition of $\mathbf{BiTop}_{\vee\mathcal{A}}$ is motivated by the results of Giuli and Salbany for $\mathbf{BiTop}_0$.

Definition 3.17 Let $\mathcal{A}$ be a subcategory of **Top**. $\mathbf{BiTop}_{\vee\mathcal{A}}$ denotes the full subcategory of **BiTop** whose objects are all bitopological spaces $(X, \mathcal{T}_1, \mathcal{T}_2)$ such that $(X, \mathcal{T}_1 \vee \mathcal{T}_2) \in \mathcal{A}$.

The proof of the following is straightforward.

Lemma 3.18 *$f : (X, \mathcal{T}, \mathcal{T}) \to (Y, \mathcal{S}_1, \mathcal{S}_2)$ is bicontinuous if and only if $f : (X, \mathcal{T}) \to (Y, \mathcal{S}_1 \vee \mathcal{S}_2)$ is continuous.*

Theorem 3.19 *If $\mathcal{A}$ is a disconnectedness in* **Top** *then* $\mathbf{BiTop}_{\vee\mathcal{A}}$ *is a disconnectedness in* **BiTop** .

Proof: Assume that $\mathcal{A}$ is a disconnectedness in **Top**. Let $\mathcal{B}$ be the full subcategory of **BiTop** whose objects consist of all triples $(X, \mathcal{T}, \mathcal{T})$ such that $(X, \mathcal{T}) \in \mathcal{C}(\mathcal{A})$. We will show that $\mathbf{BiTop}_{\vee\mathcal{A}} = \mathcal{D}(\mathcal{B})$, which implies that $\mathbf{BiTop}_{\vee\mathcal{A}}$ is a disconnectedness.

Let $(Y, \mathcal{S}_1, \mathcal{S}_2) \in \mathbf{BiTop}_{\vee\mathcal{A}}$. For all $(X, \mathcal{T}) \in \mathcal{C}(\mathcal{A})$, every morphism $f : (X, \mathcal{T}) \to (Y, \mathcal{S}_1 \vee \mathcal{S}_2)$ is constant. Therefore every morphism from $(X, \mathcal{T}, \mathcal{T})$ to $(Y, \mathcal{S}_1, \mathcal{S}_2)$ is constant. Thus, $\mathbf{BiTop}_{\vee\mathcal{A}} \subset \mathcal{D}(\mathcal{B})$.

Now let $(Y, \mathcal{S}_1, \mathcal{S}_2) \in \mathcal{D}(\mathcal{B})$. If $(Y, \mathcal{S}_1 \vee \mathcal{S}_2) \notin \mathcal{A}$, then there exists $(X, \mathcal{T}) \in \mathcal{C}(\mathcal{A})$ such that there is a non-constant morphism $f : (X, \mathcal{T}) \to (Y, \mathcal{S}_1 \vee \mathcal{S}_2)$. Thus $f : (X, \mathcal{T}, \mathcal{T}) \to (Y, \mathcal{S}_1, \mathcal{S}_2)$ is non-constant, and so $(Y, \mathcal{S}_1, \mathcal{S}_2) \notin \mathcal{D}(\mathcal{B})$. Therefore, $(X, \mathcal{S}_1 \vee \mathcal{S}_2)$ must be in $\mathcal{A}$, and we have that $\mathcal{D}(\mathcal{B}) \subset \mathbf{BiTop}_{\vee\mathcal{A}}$. •

Theorem 3.20 *If $\mathcal{B}$ is a connectedness in* **Top** *then* $\mathbf{BiTop}_{\wedge\mathcal{B}}$ *is a connectedness in* **BiTop**, *where* $\mathbf{BiTop}_{\wedge\mathcal{B}}$ *is the full subcategory of* **BiTop** *whose objects are those triples $(X, \mathcal{T}_1, \mathcal{T}_2)$ with $(X, \mathcal{T}_1 \wedge \mathcal{T}_2) \in \mathcal{B}$.*

Proof: Let $\mathcal{B}$ be a connectedness in **Top**. We will show that $\mathcal{C}(\mathcal{K}) = \mathbf{BiTop}_{\wedge\mathcal{B}}$, where $\mathcal{K}$ is the full subcategory of **BiTop** whose objects are all triples $(X, \mathcal{T}, \mathcal{T})$ such that $(X, \mathcal{T}) \in \mathcal{D}(\mathcal{B})$.

Suppose $(Y, \mathcal{S}_1, \mathcal{S}_2) \in \mathbf{BiTop}_{\wedge\mathcal{B}}$. Then $(Y, \mathcal{S}_1 \wedge \mathcal{S}_2) \in \mathcal{B}$. Let $(X, \mathcal{T}) \in \mathcal{D}(\mathcal{B})$. If $f : (Y, \mathcal{S}_1, \mathcal{S}_2) \to (X, \mathcal{T}, \mathcal{T})$ is bicontinuous, then $f : (Y, \mathcal{S}_1 \wedge \mathcal{S}_2) \to (X, \mathcal{T})$ is continuous. Therefore f is constant, and so $(Y, \mathcal{S}_1, \mathcal{S}_2) \in \mathcal{C}(\mathcal{K})$.

Now assume $(Y, \mathcal{S}_1, \mathcal{S}_2) \in \mathcal{C}(\mathcal{K})$. If $f : (Y, \mathcal{S}_1 \wedge \mathcal{S}_2) \to (X, \mathcal{T})$ is continuous, then $f : (Y, \mathcal{S}_1, \mathcal{S}_2) \to (X, \mathcal{T}, \mathcal{T})$ is continuous. If $(X, \mathcal{T}) \in \mathcal{D}(\mathcal{B})$, then f must be constant, and so $(Y, \mathcal{S}_1 \wedge \mathcal{S}_2) \in \mathcal{B}$.

We have shown that if $\mathcal{B}$ is a connectedness in **Top** then $\mathbf{BiTop}_{\wedge\mathcal{B}}$ is a connectedness in **BiTop**. •

Remark 3.21 It is not known if the converses of Theorems 3.8, 3.19, and 3.20 are true.

References

1. J. Adámek, H. Herrlich, and G. Strecker, *Abstract and concrete categories*, John Wiley & Sons, Inc., New York, 1990.

2. A. V. Arhangel'skiĭ and R. Wiegandt, *Connectedness and disconnectedness in topology*, Topology and its Applications **5** (1975), 9–33.
3. Francesca Cagliari and M. Cicchese, *Disconnectednesses and closure operators*, Proceedings of the 13th. Winter School of Abstract Analysis, Section of Topology, Supplemento ai Rendiconti del Circolo Matematico di Palermo, Serie II numero 11, 1985, pp. 15–23.
4. Eraldo Giuli and Sergio Salbany, *$2T_0$ spaces and closure operators*, Seminarberichte aus dem Fachbereich Mathematik und Informatik, #29, 1988, pp. 11–40.
5. Horst Herrlich, *Topologische reflexionen und coreflexionen, lecture notes in mathematics*, Lecture Notes in Mathematics 78 (Berlin), Springer, 1968.
6. M. Hušek and D. Pumplün, *Disconnectedness*, Quaestiones Mathematicae **13** (1990), 449–459.
7. Harriet Lord, *Hull operators on a category of spaces of continuous functions on Hausdorff spaces*, Studia Mathematica **55** (1976), 225–237.
8. ———, *Factorizations, $\mathcal{M}$-separation, and extremal-epireflective subcategories*, Topology and it Applications **29** (1988), 241–253.
9. ———, *Functionally Hausdorff spaces*, Cahiers de Topologie et Géométrie Differéntielle Catégoriques **30** (1989), 247–256.
10. ———, *Factorization, diagonal separation and disconnectedness*, Topology and its Applications **47** (1992), 83–96.
11. ———, *Factorizations and disconnectednesses*, Recent Developments of General Topology and its Applications, International Conference in Memory of Felix Hausdorff(1868–1942) (Berlin) (W. Gähler, H. Herrlich, and G. Preuß, eds.), Akademie Verlag, 1992, pp. 197–202.
12. ———, *Connectednesses and disconnectednesses,*, Papers on General Topology and Applications: 9th Summer Conference at Slippery Rock University; Annals of the New York Academy of Science, Vol. 767 (New York), New York Academy of Science, 1995, pp. 115–139.
13. Gerhard Preuß, *Theory of topological structures, an approach to categorical topology*, D. Reidel, Dordrecht, Boston, Lancaster and Tokyo, 1988.
14. Sergio Salbany, *Bitopological spaces, compactifications and completions*, Ph.D. thesis, University of Cape Town, Cape Town, South Africa, 1970.
15. ———, *Reflective subcategories and closure operators*, Lecture Notes in Mathematics 540 (Berlin), Springer, 1975, pp. 548–565.
16. ———, *A bitopological view of topology and order,*, Categorical Topology: Proceedings of the International Conference at the University of Toledo,August 1-5,1983 (Berlin), Heldermann Verlag, 1984, pp. 481–504.

Objects with dense diagonals

Walter Tholen[*]

Department of Mathematics and Statistics, York University, Toronto, Canada

Introduction

The purpose of this note is tc show that, in a finitely complete category $\mathcal{X}$ with a proper $(\mathcal{E}, \mathcal{M})$-factorization system for morphisms and a closure operator c w.r.t. the class $\mathcal{M} \subseteq \text{Mono}(\mathcal{X})$ in the sense of [DG], the full subcategory $\nabla(c)$ of those objects $X \in \mathcal{X}$ for which the diagonal

$$\delta_X : X \to X \times X$$

is c-dense, satisfies all the stability properties that one expects a category of "connected" objects to have. In fact, subject to suitable conditions on the given data, we show that $\nabla(c)$ is closed in $\mathcal{X}$ under $\mathcal{E}$-images, c-dense extensions, direct products, and under chained sinks. The first three closure properties appear essentially in [DT], Section 7.8, but not the crucial fourth property, which exhibits $\nabla(c)$ as a component subcategory in the sense of [Ti]; see also [T] and [C].

Our terminology is essentially as in [DG].

1991 Mathematics Subject Classification: 18A32, 18B30, 54B30.
Keywords: closure operator, dense diagonal, chained sink, Lawvere-Tierney topology.

---

[*]Partial financial support from NSERC (Canada), the Australian Research Council and from a NATO Collaborative Research Grant are gratefully acknowledged. This work was completed while the author was a guest of Prof. G.M. Kelly at The University of Sydney.

Eraldo Giuli (ed.), Categorical Topology, 213–220.

1 Closedness under $\mathcal{E}$-images and c-dense extensions

We say that the class $\mathcal{E}$ is closed under finite products if with f and g also $f \times g$ belongs to $\mathcal{E}$. The closure operator c is said to be *finitely productive* if

$$c_{X \times Y}(m \times n) \cong c_X(m) \times c_Y(n)$$

for all $m \in \mathcal{M}/X$, $n \in \mathcal{M}/Y$; in this case the class $\mathcal{E}^c \cap \mathcal{M}$ of c-dense $\mathcal{M}$-morphisms is closed under finite products.

Proposition 1. *For a morphism $f : X \to Y$, $X \in \nabla(c)$ implies $Y \in \nabla(c)$ in each of the following two cases:*
(a) *$f \in \mathcal{E}$, and $\mathcal{E}$ is closed under finite products;*
(b) *$f \in \mathcal{E}^c \cap \mathcal{M}$, and c is finitely productive and idempotent.*

Proof. Consider the commutative diagram

$$\begin{array}{ccc} X & \xrightarrow{\ f\ } & Y \\ {\scriptstyle \delta_X}\downarrow & & \downarrow{\scriptstyle \delta_Y} \\ X \times X & \xrightarrow{\ f \times f\ } & Y \times Y \end{array}$$

In case (a) one has $f \times f \in \mathcal{E}$ and $\delta_X \in \mathcal{E}^c$, hence $\delta_Y \cdot f = (f \times f) \cdot \delta_X$ and then also δ_Y belongs to $\mathcal{E}^c$. In case (b), $f \times f \in \mathcal{E}^c$ and $\delta_X \in \mathcal{E}^c$, and since c is idempotent, also $(f \times f) \cdot \delta_X \in \mathcal{E}^c$, which implies $\delta_Y \in \mathcal{E}^c$ as before. □

Finite productivity of c usually follows from its idempotency. More precisely, referring to the elements of

$$\mathrm{pt}(X) = \mathcal{X}(1, X)$$

(with 1 the terminal object) as points of X, let us say that $\mathcal{X}$ *has enough points* if $1_X \cong \bigvee \mathrm{pt}(X)$ for every $X \in \mathcal{X}$. (Note that points are indeed $\mathcal{M}$-subobjects, due to our assumption $\mathcal{E} \subseteq \mathrm{Epi}(\mathcal{X})$.) We then have:

Lemma 1 (cf. [DT], 4.10). *If $\mathcal{X}$ has enough points, then an idempotent closure operator is finitely productive.*

Proof. For $X, Y \in \mathcal{X}$ one has $1_{X\times Y} \cong \bigvee \mathrm{pt}(X \times Y) \cong \bigvee\{s : p \cdot s = 1_X\}$, with $p : X \times Y \to X$ the projection, since every point $\langle x, y\rangle : 1 \to X \times Y$ satisfies $\langle x, y\rangle \leq s := \langle 1_X, y\rangle$. For subobjects $m \in \mathcal{M}/X$, $n \in \mathcal{M}/Y$ one derives from this successively

$$1_X \times n \cong \bigvee s \text{ and } m \times n \cong \bigvee s \cdot m, \qquad (1)$$

with s ranging over all sections s with $p \cdot s = 1_X$ and $s \leq 1 \times n$. For such s one has

$$s \cdot c_X(m) \leq c_{X\times Y}(s \cdot m) \leq c_{X\times Y}(m \times n),$$

hence $c_X(m) \times n \leq c_{X\times Y}(m \times n)$ when we apply (1) with $c(m)$ instead of m. Symmetrically, $m \times c_Y(n) \leq c_{X\times Y}(m \times n)$ follows, in fact

$$c_X(m) \times c_Y(n) \leq c_{X\times Y}(c_X(m) \times n)$$

when we trade m for $c_X(m)$ again. But since

$$c_{X\times Y}(c_X(m) \times n) \leq c_{X\times Y}(c_{X\times Y}(m \times n)),$$

with the idempotency of c we obtain

$$c_X(m) \times c_Y(n) \leq c_{X\times Y}(m \times n),$$

as desired; "$\geq$" holds trivially for every closure operator. □

Corollary 1. *If $\mathcal{X}$ has enough points and if c is idempotent, then $\nabla(c)$ is closed under c-dense extensions.*

Proof. Combine Proposition 1(b) and Lemma 1. □

2 Closedness under products

Closedness of $\nabla(c)$ under finite products in $\mathcal{X}$ is a triviality when c is finitely productive, since then $\mathcal{E}^c \cap \mathcal{M}$ is closed under finite products, and one has

$$\delta_{X\times Y} \cong \delta_X \times \delta_Y : X \times Y \to (X \times X) \times (Y \times Y).$$

With Lemma 1 we therefore obtain:

Proposition 2. *$\nabla(c)$ is closed under finite products in $\mathcal{X}$ whenever c is finitely productive, or if c is idempotent and $\mathcal{X}$ has enough points.* □

Turning to infinite products, we first observe that any existing product $X_I = \prod_{i\in I} X_i$ is an inverse limit of the finite products $X_F = \prod_{i\in F} X_i$ ($F \subseteq I$ finite), via the projections $p_F : X_I \to X_F$. One says that c has the *finite structure property for products* FSPP, (see [DT], [CGT]) if the closure operator preserves the inverse limit, that is: if

$$c_{X_I}(m) \cong \bigwedge_F p_F^{-1}(c_{X_F}(p_F(m))) \tag{2}$$

holds for all $m \in \mathcal{M}/X_I$. The product X_I is called non-trivial if all projections p_F belong to $\mathcal{E}$, and c is called *non-trivially productive* if

$$c_{X_I}(\prod_{i\in I} m_i) \cong \prod_{i\in I} c_{X_i}(m_i) \tag{3}$$

holds for all $m_i : M_i \to X_i$ in $\mathcal{M}$, $i \in I$, with $M_I = \prod_{i\in I} M_i$ a non-trivial product.

Lemma 2 (cf. [DT] 4.10). *An idempotent and finitely productive closure operator c is non-trivially productive, provided that there exists a closure operator $d \leq c$ with FSPP.*

Proof. With the composite of closure operators being defined "pointwise", from $d \leq c$ one has $c \leq dc \leq cc$, hence $c \cong dc$ since c is idempotent. For $m_I := \prod_{i\in I} m_i$ and $n_I = \prod_{i\in I} c_{X_i}(m_i)$ we must only show $n_I \leq c_{X_I}(m_I)$ since "$\geq$" holds trivially, and for that it suffices to show $n_I \leq d_{X_I}(c_{X_I}(m_I))$. In fact, since d satisfies FSPP, it is enough to verify

$$p_F(n_I) \leq d_{X_F}(p_F(c_{X_I}(m_I)))$$

for every finite $F \subseteq I$. After decomposing m_I as $m_I \cong m_F \times m_{I\setminus F}$, we obtain $c_{X_I}(m_I) \cong c_{X_F}(m_F) \times c_{X_{I\setminus F}}(m_{I\setminus F})$ from finite productivity of c, hence

$$p_F(c_{X_I}(m_I)) \cong c_{X_F}(m_F)$$

if the product M_I is non-trivial. Using finite productivity once again, we then have

$$p_F(n_I) \leq n_F \cong c_{X_F}(m_F) \leq d_{X_F}(p_F(c_{X_I}(m_I))),$$

as desired. □

Just as Lemma 1 implies Proposition 1, we now obtain:

Theorem 1. *Let c be idempotent and assume that there exists a closure operator $d \leq c$ with FSPP. Then $\nabla(c)$ is closed under non-trivial products in $\mathcal{X}$, provided that c is finitely productive, or that $\mathcal{X}$ has enough points.* □

Note that for a direct product $X_I = \prod_{i \in I} X_i$ to be non-trivial it is sufficient to have a point x_i in X_i for every $i \in I$.

3 Closedness under chained sinks

A morphism $f : X \to Y$ is *constant* (cf. [H]) if $\mathcal{X}(Z, f)$ is a constant function for all $Z \in \mathcal{X}$, i.e., if $f \cdot x = f \cdot y$ for all $x, y : Z \to X$ in $\mathcal{X}$. A sink $(g_i : X_i \to X)_{i \in I}$ in $\mathcal{X}$ is *chained* (cf. [Ti]) if a morphism $f : X \to Y$ is constant as soon as all $g_i \cdot f$ are constant. It is our aim to show that $X_i \in \nabla(c)$ for all $i \in I$ implies $X \in \nabla(c)$ for every chained sink $(g_i)_{i \in I}$, in which case we say that $\nabla(c)$ is *closed under chained sinks.*

We first deal with a type of closure operator that is rare in topology but frequent in commutative algebra and topos theory: an idempotent closure operator c with

$$f^{-1}(c_Y(n)) \cong c_X(f^{-1}(n))$$

for all $f : X \to Y$, $n \in \mathcal{M}/Y$ is called a *Lawvere-Tierney topology on $\mathcal{X}$ (w.r.t. $\mathcal{M}$).* Recall that every kernelpair of a morphism $f : X \to Y$ is an equivalence relation on X (see [J], for example); to say that an equivalence relation $(r_1, r_2 : R \to X)$ is *effective* means that it is the kernelpair of some morphism $f : X \to Y$.

Proposition 3. *Let $\mathcal{X}$ have coequalizers of equivalence relations, and let equivalence relations be effective in $\mathcal{X}$. Then, for a Lawvere-Tierney topology c on $\mathcal{X}$ w.r.t. $\mathcal{X}$, $\nabla(c)$ is closed under chained sinks.*

Proof. Barr [B] showed that for an LT-topology c of $\mathcal{X}$, with p_1, p_2 the product projections, $(p_1 c(\delta_X), p_2 c(\delta_X))$ is an equivalence relation of $\mathcal{X}$, hence it is the kernelpair of its coequalizer $q : X \to Q$, under our assumptions on $\mathcal{X}$. For a chained sink $(g_i : X_i \to X)_{i \in I}$ with δ_{X_i} c-dense for all $i \in I$, each morphism $g_i \times g_i$ in the commutative diagram

$$\begin{array}{ccc} X_i & \xrightarrow{g_i} & X \\ {\scriptstyle \delta_{X_i}}\downarrow & & \downarrow{\scriptstyle \delta_X} \\ X_i \times X_i & \xrightarrow{g_i \times g_i} & X \times X \end{array}$$

must factor through $c(\delta_X)$, as $g_i \times g_i = c(\delta_X) \cdot h_i$, say. For every $i \in I$ and any two morphisms $x, y : Z \to X_i$, we now have

$$\begin{aligned} q \cdot f_i \cdot x &= q \cdot f_i \cdot p_1^i \cdot \langle x, y \rangle \\ &= q \cdot p_1 \cdot (f_i \times f_i) \cdot \langle x, y \rangle \\ &= q \cdot p_1 \cdot c(\delta_X) \cdot h_i \cdot \langle x, y \rangle \\ &= q \cdot p_2 \cdot c(\delta_X) \cdot h_i \cdot \langle x, y \rangle \\ &= q \cdot f_i \cdot y, \end{aligned}$$

so that every $q \cdot f_i$ is constant. Hence q is constant; in particular, $q \cdot p_1 = q \cdot p_2$. Hence the pair (p_1, p_2) must factor through the kernelpair $(p_1 \cdot c(\delta_X),\ p_2 \cdot c(\delta_X))$ of q, which is possible only when $c(\delta_X)$ is iso. Consequently, δ_X is c-dense. □

By a *c-subobject classifier* we understand a morphism $t : 1 \to \Omega$ which classifies the c-closed $\mathcal{M}$-subobjects; hence for every $m : M \to X$ in $\mathcal{M}^c$ there is a unique morphism $\chi_m : M \to \Omega$ such that

$$\begin{array}{ccc} M & \xrightarrow{!_M} & 1 \\ {\scriptstyle m}\downarrow & & \downarrow{\scriptstyle t} \\ X & \xrightarrow{\chi_m} & \Omega \end{array} \tag{4}$$

is a pullback diagram.

Theorem 2. *Let c be an idempotent closure operator with a c-subobject classifier. Then $\nabla(c)$ is closed under those sinks $(g_i : X_i \to X)_{i \in I}$ for which $(g_i \times g_i)_{i \in I}$ is chained and X has a point.*

Proof. As in the proof of Proposition 3, we can establish a factorization $g_i \times g_i = c(\delta_X) \cdot h_i$ for every $i \in I$. Let $\chi : X \times X \to \Omega$ represent the c-closed subobject $m := c(\delta_X) : M \to X \times X$. Since

$$\chi \cdot (g_i \times g_i) = \chi \cdot m \cdot h_i = t \cdot !_M \cdot h_i$$

is constant for every $i \in I$, χ must be constant. Since X has a point, also M has a point x, and we obtain

$$\chi = \chi \cdot m \cdot x \cdot !_{X \times X} = t \cdot !_M \cdot x \cdot !_{X \times X} = t \cdot !_{X \times X}.$$

Now we can envoke the pullback property of (4) to produce a morphism $s : X \times X \to M$ with $m \cdot s = 1_{X \times X}$, so that $m = c(\delta_X)$ must be iso. This shows $X \in \nabla(c)$ whenever $X_i \in \nabla(c)$ for all $i \in I$. □

4 Applications

(1) In the category **Top** of topological spaces with its (surjective, embedding)-factorization system, the usual (Kuratowski-)closure gives a closure operator k. It is idempotent and satisfies FSPP, but k is not a Lawvere-Tierney topology. (There is in fact no LT-topology in **Top**, other than the trivial operator t with $t_X(M) = X$ for all $M \subseteq X$.) **Top** has enough points, and the Sierpinski space is a k-subobject classifier. Hence the category $\nabla(k)$ has the four closedness properties mentioned in the Introduction; it is the category of *irreducible spaces*, i.e., of those spaces X in which $X = F \cup G$ with closed sets F and G is possible only for $X = F$ or $X = G$.

(2) For $M \subseteq X \in$ **Top**, the quasi-component

$$q_X(M) = \cap\{A \subseteq X : A \text{ clopen}, M \subseteq A\}$$

defines an idempotent closure operator $q \geq k$, with $\nabla(q)$ containing exactly the *connected spaces*. An application of the categorical statements of this paper proves the four closure properties of the Introduction for connected spaces.

(3) A preradical r of R-modules is simply a subfunctor of the identity functor of R-**Mod**. For a submodule M of an R-module X, one defines the r-minimal closure of M in X by $c_X(M) = r(X) + M$. This gives an idempotent and finitely productive closure operator c w.r.t. the class of monomorphisms, and $\nabla(c)$ is easily recognized as the class of *r-torsion modules* (i.e., $r(X) = X$), which is therefore closed under images, dense extensions, and finite products, but not in general under arbitrary direct products. The closure operator c is an LT-topology if and only if r is hereditary ($r(M) = r(X) \cap M$ for $M \leq X$) and cohereditary

$(r(X/M) \cong (r(X) + M)/M)$. In this case Proposition 3 gives that r-torsion modules are closed under epic sinks (those $(g_i : X_i \to X)_{i \in I}$ with $X = \sum_{i \in I} g_i(X_i)$), which are precisely the chained sinks in R-**Mod**.

(4) After the appearance of a preprint of this paper, M.M. Clementino has shown under mild conditions on the data involved, that every left constant subcategory (cf. [C]) can be presented in the form $\nabla(c)$ for a suitable closure operator c. The proof will appear elsewhere.

References

[B] M. Barr, On categories with effective unions, in: Proc. Categorical Algebra and Its Applications, Louvain-la-Neuve 1987, Lecture Notes in Math. 1348 (Springer, Berlin 1988), pp. 19-35.

[C] M.M. Clementino, Constant morphisms and constant subcategories, preprint (University of Coimbra, 1994).

[DG] D.Dikranjan and E. Giuli, Closure Operators I, Topology Appl. 27 (1987) 129-143.

[DT] D. Dikranjan and W. Tholen, Categorical Structure of Closure Operators. With Applications to Topology, Algebra and Discrete Mathematics (Kluwer Academic Publishers, Dordrecht 1995).

[H] H. Herrlich, Topologische Reflexionen und Coreflexionen, Lecture Notes in Math. 78 (Springer, Berlin 1968).

[J] P.T. Johnstone, Topos Theory (Academic Press, New York 1977).

[T] W. Tholen, Factorizations, fibres and connectedness, in: Proc. Conf. Categorical Topology Toledo, Ohio 1983 (Heldermann Verlag, Berlin 1984), pp. 549-566.

[Ti] J.A. Tiller, Component subcategories, Quaestiones Math. 4 (1980) 19-40.

Ascoli–Arzelà–Theory based on continuous convergence in an (almost) non-Hausdorff setting

René Bartsh, Peter Dencker, Harry Poppe
University of Rostock, Dept. of Mathematics
Universitätsplatz 1 – 18055 Rostock
Germany

Mathematics Subject Classifications (1991). 54C35, 54D30, 54D50, 54E15.

Keywords. Relatively compact sets, continuous convergence, compactness on function spaces, compactness criteria of Ascoli-Arzelà-type, even continuity, equicontinuity, compact-open topology, uniform topology.

0 Introduction

In the sixties and at the beginning of the seventies by H. Poppe was constructed a unified approach to compactness criteria in topological and uniform functionspaces, especially in spaces of continuous functions. The main tools used for this approach were convergence spaces, generalized uniform spaces (based on coverings) and for the function spaces the convergence structure of continuous convergence. These constructions and results were summarized in the book [8]. From the recent papers and books where Ascoli-Arzelà theory is treated we want to mention: W. Gähler [1], J.W. Gray [2], H. Herrlich [3], R.A. McCoy and J. Ntantu [4], K. Morita [5], L.D. Nel [6], H. Render [11], [12].

In the present paper we want to outline the basic facts of a theory of compactness of Ascoli-Arzelà-type in spaces of continuous functions. By the use of relative compactness (see the definition below) instead of compactness the deduction will become somewhat smoother. We will explain that it is appropriate to work with continuous convergence. We also show where the Hausdorff axiom is needed and where we can avoid it. For simplicity our basic spaces X, Y will be topological spaces (and not for instance convergence spaces as in [8]). In a direct way we want to derive our main theorem which characterizes the relative compactness of subsets in $(C(X,Y), c-\lim)$, where $C(X,Y)$ denotes the set of all continuous functions from X to Y and $c-\lim$ is the convergence structure of continuous convergence.

We will not give the proofs of all our assertions. We start with the classical theorem of Ascoli-Arzeà and we will conclude the paper by deriving this result from our abstract criteria. Especially our paper will still consist of the following points:

1. The classical Ascoli-Arzelà-Theorem

2. Relatively compact subsets of a topological space

Eraldo Giuli (ed.), Categorical Topology, 221–240.

1 The classical Ascoli-Arzelà theorem

Let X be a compact metric space (Ascoli and Arzelà used $X = [0,1]$ with Euclidian topology); for $C(X) = C(X,\mathbb{R})$ we consider the uniform topology which can be induced by the Tschebysheff metric.

Theorem 1 *$H \subseteq C(X)$;*
($\forall$ sequences (f_n) from H $\exists$ subsequence (f_{n_k}) from (f_n) and $f \in C(X)$ such that $f_{n_k} \xrightarrow{\tau_u} f$)
$\Longleftrightarrow$

(α) H is (metrically) bounded

(β) H is equicontinuous.

Let us call such an H relatively (τ_u-) compact or relatively sequentially compact.

Definition 2 *H is called equicontinuous*
$:\Leftrightarrow \forall x \in X, \forall \varepsilon > 0 : \exists U \in \underline{U}(x) : |f(z) - f(x)| < \varepsilon \quad \forall (f,z) \in H \times U$
(here $\underline{U}(x)$ denotes the neighborhood filter of x).

Remark.

1. Of course this notion at once can be extended to the case that the range space is a (diagonal) uniform space.

2. For a proof of the theorem one especially uses the facts:

a) X is separable

b) A Tychonoff theorem:

(X_n) a countable family of topological spaces, $A_n \subseteq X_n \, \forall n$; then ΠA_n is relatively sequentially compact in ΠX_n with respect to the Tychonoff topology $\Leftrightarrow \forall n$, A_n is relatively sequentially compact in X_n.

2 Relatively compact subsets

Definition 3 *Let X be a topological space, $A \subseteq X$ is called relatively compact* $:\Leftrightarrow$ *each open cover of X has a finite subcover which covers A.*

Theorem 4 *Equivalent are:*

(1) *A is relatively compact*

(2) *Each ultrafilter on X containing A converges to some $x \in X$*

(3) *Each net from A has a subnet converging to some $y \in X$.*

Concerning standard operations: relative compactness is preserved by arbitrary products and by continuous maps. Moreover a subset of a relatively compact set is relatively compact too.

The comparison of relative compactness is simple:

Proposition 5 .

1. *$A \subseteq X$ compact $\Rightarrow$ A is relatively compact*

2. *A is relatively compact and closed $\Rightarrow$ A is compact.*

Comparing now relative compactness with the other similar notion we find:

Theorem 6 .

1. *Let X be a Hausdorff space and $A \subseteq X$; if the closure $\overline{A}$ is compact then A is relatively compact and $\overline{A}$ is as a subspace of X a T_3-space (here our notation is: T_1 and T_3 means regular).*

2. *If A is relatively compact and $\overline{A}$ is a T_3-space then $\overline{A}$ is compact.*

As a corollary for metric spaces we get:

Theorem 7 *Let (X, d) be a metric space, $A \subseteq X$, $A \neq \emptyset$; then are equivalent:*

(1) *A is relatively compact*

(2) *Each infinite subset of A has a cluster point* $x \in X$

(3) *Each sequence* (x_n) *from A has a subsequence converging to some* $y \in X$

(4) $\overline{A}$ *is compact.*

Corollary 8 *We consider Euclidian* $\mathbb{R}^n$. $A \subseteq \mathbb{R}^n$ *is relatively compact* $\Leftrightarrow$ *A is bounded.*

Remark. For proofs of the assertions of section 2 (and for further facts) see H. Poppe [10].

3 τ_p-relatively compact subsets $H \subseteq Y^X$

For topological spaces X, Y (or convergence spaces X, Y; for a definition of the notion of a convergence space see section 4) we denote by Y^X the set of all functions from X to Y, $C(X,Y)$ means all continuous functions and τ_p is the topology of pointwise convergence. By Tychonoff's theorem we get:

Theorem 9 $(X_i)_{i\in I}$ *a family of topological spaces (or convergence spaces),* $A \subseteq \Pi X_i$; *for* ΠX_i *we consider the Tychonoff topology; then holds: A is relatively compact* $\Longleftrightarrow \forall i \in I$, *the projections* $pr_i(A)$ *are relatively compact in* X_i.

Corollary 10 X, Y *topological spaces,* $H \subseteq Y^X$ *is relatively compact in* $(Y^X, \tau_p) \Leftrightarrow$ $\forall x \in X$, $H(x) := pr_x(H) = \{f(x) \mid f \in H\}$ *is relatively compact.*

Corollary 11 .

1. *Y Hausdorff,* $H \subseteq Y^X \tau_p$*-compact* $\Rightarrow$

   (1) *H is* τ_p*-closed*

   (2) $\forall x \in X$, $H(x)$ *is compact*

2. *Y an arbitrary topological space,* (1) *and* (2) *hold* $\Rightarrow$ *H is* τ_p*-compact.*

For a proof of theorem 9 see [8].

4 Continuous convergence

4.1 Convergence spaces

Definition 12 .

a) *X : set,* $F(X)$: *all filters on X, let "$\rightarrow$" be a binary relation on* $F(X) \times X$; $\psi \in F(X)$, $x \in X$;
   $\psi \rightarrow X$: "ψ *converges to x", $(X, \lim) = (X, \rightarrow)$ is called a convergence space (limit space):* $\Leftrightarrow$ *the following axioms hold:*

1. $\forall x \in X,\ [x] \to x$
2. $\phi,\ \psi \in F(X),\ \phi \leq \psi$ *and* $\phi \to x \Rightarrow \psi \to x$.

Remark. lim is called a convergence structure

b) $(X, \to)$ is called pseudotopological: $\Leftrightarrow \forall \psi \in F(X)$ holds: ($\forall \pi$, π ultrafilter on X such that $\psi \leq \pi \Rightarrow \pi \to x) \Rightarrow \psi \to x$

c) $(X, \to)$ is called pretopological: $\Leftrightarrow$ $\forall$ family $(\psi_i)_{i \in I} : \psi_i \to x \quad \forall i \in I \Rightarrow \bigcap_{i \in I} \psi_i \to x$.

Example. X topological space, "$\to$" topological convergence $\Rightarrow$ $(X, \to)$ is a convergence space.

4.2 Conjoining convergence structures or topologies for $C(X,Y)$

Definition 13 *X, Y sets; $\omega : Y^X \times X \to Y$, $\omega(f, x) := f(x)\ \forall (f, x) \in Y^X \times X$ is called the evaluation map.*

Definition 14 *X, Y topological spaces,* lim *a convergence structure for $C(X,Y)$ or $\tau = \tau - \lim$ a topology for $C(X,Y)$;* lim *or τ respectively are called conjoining for $C(X,Y)$:$\Leftrightarrow$ the evaluation map $\omega : C(X,Y) \times X \to Y$, $\omega(f,x) := f(x)$ is continuous.*

4.3 Continuous convergence

Definition 15
$(f_i)_{i \in I}$ net from Y^X, $f \in Y^X$;
$(f_i)_{i \in I}$ converges continuously to f, $f_i \xrightarrow{c} f$:$\Leftrightarrow \forall x \in X$, $\forall$ nets $(x_k)_{k \in K}$ in X, $x_k \to x \Rightarrow (f_i(x_k))_{(i,k) \in I \times K} \to f(x)$.

Characterization of continuous convergence by filters and neighborhoods.

Theorem 16 *X, Y topological spaces; $f \in Y^X$, $(f_i)_{i \in I}$ net from Y^X; let Φ be the filter on Y^X, generated by (f_i), $\Phi := [\{\{f_i \mid i \geq i_0\}, i_0 \in I\}]$. Then the following assertions are equivalent:*

(1) $f_i \xrightarrow{c} f$

(2) $\forall x \in X, \forall \psi$, *$\psi$ filter on X* $: \psi \to x \Rightarrow \omega(\Phi \times \psi) \to f(x)$

(3) $\forall x \in X$, $\forall V \in \underline{U}(f(x))\ \exists U \in \underline{U}(x)\ \exists i_0 = i_0\ (x, V) \in I$ *such that* $\forall i \geq i_0 \Rightarrow f_i(U) \subseteq V$.

Proof. "(1) $\Rightarrow$ (2)": let $x \in X$, ψ filter on X such that $\psi \to x$, $V \in \underline{U}(f(x))$; theorem of Bruns–Schmidt $\Rightarrow$ $\exists$ net $(x_k)_{k \in K}$ form X such that $x_k \to x$ and $[\{\{x_k \mid k \geq k_0\} \mid k_0 \in K\}] = \psi$

$$f_i(x_k) \to f(x) \Rightarrow \exists i_0 \in I\ , \quad \exists k_0 \in K\ , \quad \forall (i,k) \geq (i_0, k_0)\ f_i(x_k) \in V\ ;$$

$$\omega(\{f_i \mid i \geq i_0\} \times \{x_k \mid k \geq k_0\}) = \{f_i(x_k) \mid (i,k) \geq (i_0,k_0)\} \subseteq V\ .$$

Thus we have $\forall V \in \underline{U}(f(x))\ V \in \omega(\Phi \times \psi)$ and hence $\underline{U}(f(x)) \subseteq \omega(\Phi \times \psi)$.

"(2) $\Rightarrow$ (3)":

$$\begin{aligned}
&\text{let } x \in X,\ V \in \underline{U}(f(x)) \\
&\underline{U}(x) \to x \Rightarrow \omega(\Phi \times \underline{U}(x)) \to f(x) \Rightarrow \\
&\exists A \in \Phi \quad \exists U \in \underline{U}(x)\ \omega(A \times U) \subseteq V\ ; \\
&\exists i_0 \in I \{f_i \mid i \geq i_0\} \subseteq A \Rightarrow f_i(U) \subseteq V\ \forall i \geq i_0\ .
\end{aligned}$$

"(3) $\Rightarrow$ (1)":

$$\begin{aligned}
&\text{let } x \in X,\ \text{net from } X,\ x_k \to x\ , \quad V \in \underline{U}(f(x)) \\
&\exists U \in \underline{U}(x),\ \exists i_0 \in I : f_i(U) \subseteq V\ \forall i \geq i_0 \\
&\exists k_0 \in K : \forall k \geq k_0\ x_k \in U\ .
\end{aligned}$$

Thus we have $\forall\, (i,k) \geq (i_0, k_0)\ f_i(x_k) \in V$ and hence $f_i(x_k) \to f(x)$.

Definition 17 *X, Y topological spaces; Φ filter on Y^X, $f \in Y^X$, Φ converges continuously to f, $\Phi \xrightarrow{c} f$, $f \in c - \lim \Phi :\Leftrightarrow \forall x \in X$, $\forall$ filter ψ on $X : \psi \to x \Rightarrow \omega(\Phi \times \psi) \to f(x)$.*

Theorem 18 *X, Y topological spaces*

a) *If $(f_i)_{i \in I}$ a net from Y^X, then there exists a filter Φ on Y^X with the property: $f_i \xrightarrow{c} f \Leftrightarrow \Phi \xrightarrow{c} f$.*

b) *If Φ a filter on Y^X, then there exists a net $(f_i)_{i \in I}$ from Y^X with the property: $\Phi \xrightarrow{c} f \Leftrightarrow f_i \xrightarrow{c} f$.*

Proof.

a) We have shown this property of $\Phi := [\{\{f_i \mid i \geq i_0\} \mid i_0 \in I\}]$ in theorem 16.

b)

$$i := \{(g,F) \mid F \in \phi, g \in F\} \quad (g_1,F_1) \leq (g_2,F_2); \Leftrightarrow F_2 \subseteq F_1$$
$(I,\leq)$ is a directed set.

$$h : I \to Y^X \ , \quad h((g,F)) : g$$
$$f_i := h(i) \ , \quad i \in I$$
$(f_i)_{i\in I}$ is a net from Y^X .

Note that, if $i_0 = (g_0, F_0) \in I$, then $\{f_i \mid i \geq i_0\} = F_0$.
λ filter on X, $\lambda \to x \Rightarrow \exists$ net $(x_k)_{k\in K}$ from X, $x_k \to x$ such that

$$\lambda = [\{\{x_k \mid k \geq k_0\} \mid k_0 \in K\}]$$

let be λ a filter on X, $(x_k)_{k\in K}$ a net from X with the property $\lambda = [\{\{x_k \mid k > k_0\}, k_0 \in K\}]$

$$\begin{aligned}
&\omega(\Phi \times \lambda) \to f(x) \Leftrightarrow \\
&\underline{U}(f(x)) \subseteq \omega(\Phi \times \lambda) \Leftrightarrow \\
&\forall U \in \underline{U}(f(x)) \quad \exists F_0 \in \Phi \quad \exists k_0 \in K \quad \omega(F_0 \times \{x_k \mid k \geq k_0\}) \subseteq U \Leftrightarrow \\
&\forall U \in \underline{U}(f(x)) \ \exists i_0 = (g_0, F_0) \in I, \ \exists k_0 \in X : \\
&\qquad \{f_i(x_k) \mid i \geq i_0, \ k \geq k_0\} = \omega(F_0, \{x_k \mid k \geq k_0\}) \subseteq U \Leftrightarrow \\
&\forall U \in \underline{U}(f(x)) \ \exists (i_0, k_0) \in I \times K \quad \forall (i,k) \geq (i_0,k_0) : f_i(x_k) \in U \Leftrightarrow \\
&\qquad f_i(x_k) \to f(x) \ .
\end{aligned}$$

Now we obtain b) using the definitions 15 and 17.

Remark. The notion "continuous convergence" we also can define if X, Y are convergence spaces.

4.4 Some important properties of continuous convergence

We state the following theorem without proof.

Theorem 19 *Let X and Y be topological spaces*

1. $f \in Y^X$; $[f] \xrightarrow{c} f \Leftrightarrow f$ *is continuous*

2. ϕ, ψ *filter on* Y^X *(or on* $C(X,Y)$*),* $\phi \leq \psi$, $\phi \xrightarrow{c} f \Rightarrow \psi \xrightarrow{c} f$

3. $(C(X,Y), c - \lim)$ *is a pseudotopological, but in general not a pretopological convergence space*

4. $c-\lim$ *is conjoining for* $C(X,Y)$ *and* $c-\lim$ *is the smallest convergence structure with this property*

5. $\Phi \xrightarrow{c} f \Rightarrow \Phi \xrightarrow{\tau_p} f$ *in* Y^X

6. τ_k*: compact – open topology*

   a) $\Phi \xrightarrow{c} f \Rightarrow \Phi \xrightarrow{\tau_k} f$

   b) X *locally compact* $\Rightarrow \tau_k - \lim = c - \lim$ *on* $C(X,Y)$.
   Thus here $c-\lim$ *is a topological notion.*

7. Y *a uniform space,* τ_u*: uniform topology;*

$$\Phi \xrightarrow{\tau_u} f \Rightarrow \Phi \xrightarrow{c} f \quad \text{in } C(X,Y) \ ;$$

   if X *compact, then holds:* $\Phi \xrightarrow{c} f \Rightarrow \Phi \xrightarrow{\tau_u} f$ *in* $C(X,Y)$.

Corollary 20 .

1. *If* $C(X,Y) \subsetneqq M \subseteq Y^X$, *then* $(M, c-\lim)$ *is not a convergence space in our sense*

2. $\lim$ *conjoining for* $C(X,Y) \Leftrightarrow c-\lim \leq \lim$.

5 Even continuity of subsets $H \subset Y^X$

Definition 21 X, Y *topological spaces,* $H \subseteq Y^X$*;* H *is called evenly continuous:* $\Leftrightarrow \forall (x,y) \in X \times Y$, $\forall$ *filter* Φ *on* Y^X *such that* $H \in \Phi$, $\forall$ *filter* ψ *in* X *holds:*

$$\omega(\Phi \times [x]) \to y \quad \text{and } \psi \to x \Rightarrow \omega(\Phi \times \psi) \to y \ .$$

Theorem 22 *Equivalent are:*

(1) H *evenly continuous*

(2) $\forall (x,y,V) \in X \times Y \times \underline{U}(y) \quad \exists (U,W) \in \underline{U}(x) \times \underline{U}(y) : \forall f \in H,\ f(x) \in W \Rightarrow f(U) \subseteq V$ *[Condition of J.L. Kelley, A.P. Morse].*

Proof. "(1) $\Rightarrow$ (2)": If we suppose that assertion (2) does not hold, we find $x \in X$, $y \in Y$, $V \in \underline{U}(y)$ and a filter ψ on X with $\psi \to x$ such that:

$\forall (A,W) \in \psi \times \underline{U}(y)\ \exists f_{A,W} \in H$ with $f_{A,W}(x) \in W$ and $f_{A,W}(A) \not\subseteq V \Rightarrow$
$\forall (A,W) \in \psi \times \underline{U}(y)\ \exists x_{A,W} \in A : f_{A,W}(x_{A,W}) \notin V$;
$\forall (A_1,W_1),\ (A_2,W_2) \in \psi \times \underline{U}(y) : (A_1,W_1) \leq (A_2,W_2) :\Leftrightarrow A_2 \subseteq A_1$ and $W_2 \subseteq W_1$

$(\psi \times \underline{U}(y), \leq)$ is a directed set, $(f_{A,W})_{(A,W)\in(\psi\times\underline{U}(y),\leq)}$ is a net from Y^X let be $\underline{F}$ the filter on Y^X, generated by this net; $H \in \underline{F}$; we will show, that $\omega(\underline{F} \times [x]) \to y$; let $W_0 \in \underline{U}(y)$, $A_0 \in \psi$

$$\forall (A, W) \geq (A_0, W_0) \quad f_{A,W}(x) \in W \subseteq W_0$$
$$\omega(\{f_{A,W}, (A, W) \geq (A_0, W_0)\} \times \{x\}) \subseteq W_0 .$$

Thus we have $\underline{U}(y) \subseteq \omega(\underline{F} \times [x])$ and hence $\omega(\underline{F} \times [x]) \to y$.

H evenly continuous, $H \in \underline{F} \Rightarrow \omega(\underline{F} \times \psi) \to y \Rightarrow \exists F \in \underline{F}, \exists A_1 \in \psi$, $\omega(F \times A_1) \subseteq V$; let $W_1 \in \underline{U}(y)$.

Note that, if $(A, W) \geq (A_1, W_1)$, then $x_{A,W} \in A_1$; let $(\tilde{A}, \tilde{W}) \geq (A_1, W_1)$ and $f_{\tilde{A},\tilde{W}} \in F$.

We have got $f_{\tilde{A},\tilde{W}}(x_{\tilde{A},\tilde{W}}) \notin V$ and $\omega(F \times A_1) \subseteq V$.

Contradiction!

"(2) ⇒ (1)":

Let $x \in X$, $y \in Y$, $\underline{F}$ a filter on Y^X with $H \in \underline{F}$ and $\omega(\underline{F} \times [x]) \to y$ and let ψ be a filter on X such that $\psi \to x$.

Let $V \in \underline{U}(y)$

$$\exists W \in \underline{U}(y), \quad \exists A \in \psi : \forall f \in H : f(x) \in W \Rightarrow f(A) \subseteq V$$
$$\omega(\underline{F} \times [x]) \to y \Rightarrow W(\underline{F} \times [x]) \Rightarrow \exists F \in \underline{F} \quad \omega(\underline{F} \times \{x\}) \subseteq W ;$$

we have got $F \cap H \in \underline{F}$ and we show that $\omega((F \cap H) \times A) \subseteq V$:

$$g \in F \cap H \Rightarrow g(x) \in W, \quad g \in H \Rightarrow g(A) \subseteq V .$$

Thus we have got $\underline{U}(y) \subseteq \omega(\underline{F} \times \psi)$ and hence $\omega(\underline{F} \times \psi) \to y$.

SOME PROPERTIES OF EVEN CONTINUITY

Theorem 23 *X, Y topological spaces;*

1. *$H \subseteq C(X,Y)$, H finite $\Rightarrow$ H evenly continuous*
2. *H evenly continuous and $H_1 \subseteq H \Rightarrow H_1$ evenly continuous*
3. *$H \subseteq Y^X$ and H evenly continuous $\Rightarrow H \subseteq C(X,Y)$.*

Proof.

1. Let $(x, y) \in X \times Y$, let Φ be a filter on Y^X such that $H \in \Phi$ and $\omega(\Phi \times [x]) \to y$ and let ψ be an arbitrary filter on X with $\psi \to x$;
we will show, that $\omega(\Phi \times \psi) \to y$: let V be an arbitrary open set in $\underline{U}(y)$

$$\omega(\Phi \times [x]) \to y \Rightarrow \exists F \in \Phi : \omega(\underline{F} \times \{x\}) \subseteq V$$

$D := F \cap H \in \Phi$, $D \neq \emptyset$ and D finite, $D = \{f_1, \dots, f_k\}$

$$\omega(D \times \{x\}) \subseteq V \Rightarrow f_j(x) \in V \quad \forall j = 1, \dots, k$$

V open $\Rightarrow V \in \underline{U}(f_j(x))\ \forall j = 1, \dots, k$; $f_j \in C(X, Y)\ \forall j = 1, \dots, k$
$\exists$ open $U_j \in \underline{U}(x)$ $f_j(U_j) \subseteq V\ \forall j = 1, \dots, k$

$$\emptyset \neq U := U_1 \cap \dots \cap U_k \in \underline{U}(x)$$

$$f_j(U) \subseteq V \quad \forall j = 1, \dots, k \ ;$$

$$\begin{aligned} \psi \to x \ &\Rightarrow \ \underline{U}(x) \subseteq \psi \Rightarrow U \in \psi \\ &\Rightarrow \exists A \in \psi \quad A \subseteq U \ . \end{aligned}$$

Thus we have got $D \times A \in \Phi \times \psi$ such that $\omega(D \times A) \subseteq V$ and hence $\omega(\Phi \times \psi) \to y$.

2. If Φ is a filter on Y^X such that $H_1 \in \Phi$, then we have got $H \in \Phi$ too, because $H_1 \subseteq H$. Now we obtain 2. by using definition 21.

3. Let f_0 be an arbitrary function in H:
let $x \in X$, $y := f_0(x) \in Y$, let V be an arbitrary element of $\underline{U}(y)$ and $\psi := \underline{U}(x) \to x$.
If we use theorem 22, we find $U \in \psi$ and $W \in \underline{U}(y)$ with the property:
$f \in H$ and $f(x) \in W \Rightarrow f(U) \subseteq V$.
$f_0 \in H$ and $f_0(x) = y \in W \in \underline{U}(y) \Rightarrow f_0(U) \subseteq V$.
Thus we have got: $\forall x \in X\ \forall V \in \underline{U}(f_0(x))\ \exists U \in \underline{U}(x) : f_0(U) \subseteq V$ and hence f_0 is continuous.

6 Connection between even continuity and equicontinuity

As is well known we can define the notion of a uniform space by several ways. One important approach is based on coverings. And of course we can also weaken the axioms for these coverings, thus coming to the notion of a generalized uniform space (in the covering sense). For this see [7], [8] and [5]. Within this setting we can compare the (generalized) notions of even continuity and equicontinuity. In [7] (see also [8]) a general theorem on the connection of these two notions was proved. Using the notion of relative compactness we here want to generalize this result slightly.

Definition 24 *Let X be a set, Σ a family of coverings of X such that:*

1. $\forall \alpha, \beta \in \Sigma$, $\exists \gamma \in \Sigma : \gamma < \alpha$ *and* $\gamma < \beta$ *(γ is a common refinement of α, β)*

2. $\alpha \in \Sigma$, *β a covering of X*, $\alpha < \beta \Rightarrow \beta \in \Sigma$.

Then (X,Σ) is called a generalized uniform space.

(X,Σ) is called regular: $\Leftrightarrow$

1. $\forall x \in X$, $\forall A_1,\dots,A_n \in \bigcup\{\alpha \mid \alpha \in \Sigma\}$ *with* $x \in \bigcap A_i$ $\exists \beta \in \Sigma$ *such that* $\mathrm{St}(x,\beta) \subseteq \bigcap A_i$ *(weak regularity)*

2. $\forall \alpha \in \Sigma$ $\exists \beta \in \Sigma$ *with the property:* $\forall B \in \beta$ $\exists (\gamma_B, A_B) \in \Sigma \times \alpha$ *such that* $\mathrm{St}(B,\gamma_B) \subseteq A_B$.

Definition 25 *(X,Σ) a generalized uniform space; the uniform topology τ_Σ which is induced by (X,Σ) is defined by the subbase $\bigcup\{\alpha \mid \alpha \in \Sigma\}$.*

Definition 26 *X a topological space, (Y,Σ) a generalized uniform space;*

a) *$H \subseteq Y^X$ is called equicontinuous in $x \in X$* $:\Leftrightarrow \forall \alpha \in \Sigma$, $\exists U \in \underline{U}(x)$ *with the property:* $\forall f \in H$ $\exists A_f \in \alpha$ *such that* $f(U) \subseteq A_f$.

b) *H is equicontinuous on X* $:\Leftrightarrow \forall x \in X$, *$H$ is equicontinuous in x.*

Theorem 27 *X topological space, (Y,Σ) a generalized uniform space; let $H \subseteq Y^X$;*

1. *Let Σ be regular; we consider the following assertions:*

   (1) H is equicontinuous

   (2) $\forall (x,\alpha) \in X \times \Sigma$ $\exists U \in \underline{U}(x)$ such that: $\forall f \in H$, $f(U) \subseteq \mathrm{St}(f(x),\alpha)$

   (3) H is evenly continuous with respect to $X, (Y,\tau_\Sigma)$.
   Then holds: (1) $\Rightarrow$ (2) $\Rightarrow$ (3).

2. *H is evenly continuous, $x \in X$ and $H(x)$ is relatively compact $\Rightarrow$ H is equicontinuous in x. And clearly condition (2) of 1. also holds in x.*

Proof.

1. (1) $\Rightarrow$ (2) holds at once; (2) $\Rightarrow$ (3): we use the characterization 22, 2 of even continuity; let $(x,y) \in X \times Y$; if $V \in \underline{U}(y)$ we find $A_i \in \alpha_i \in \Sigma$, $i = 1,\dots,n$ such that $y \in A_1 \cap \dots \cap A_n \subseteq V$ and $\bigcap A_i \in \tau_\Sigma$; by the weak regularity of Σ we find $\alpha \in \Sigma$: $\mathrm{St}(y,\alpha) \subseteq \bigcap A_i$; by the second regularity condition for α there exists $\beta \in \Sigma : \forall B \in \beta$ $\exists(\gamma_B, A_B) \in \Sigma \times \alpha$ such that $\mathrm{St}(B,\gamma_B) \subseteq A_B$; since β is a covering of Y we find $B_0 \in \beta$ such that $y \in B_0$; for γ_{B_0} there exists $U \in \underline{U}(x) : \forall f \in H \Rightarrow f(U) \subseteq \mathrm{St}(f(x),\gamma_{B_0})$; B_0 is an open neighborhood of y; now let $f \in H$ and $f(x) \in B_0$; we then have: $f(U) \subseteq \mathrm{St}(f(x),\gamma_B) \subseteq \mathrm{St}(B_0,\gamma_{B_0}) \subseteq A_{B_0} \in \alpha$; now $y \in \mathrm{St}(B_0,\gamma_{B_0}) \Rightarrow y \in A_{B_0} \Rightarrow f(U) \subseteq A_{B_0} \subseteq \mathrm{St}(y,\alpha) \subseteq \bigcup A_i$, showing that H is evenly continuous.

2. Let $\alpha \in \Sigma$ and $x \in X$; we want to show that H is equicontinuous at x: since α is a covering of $Y : \forall y \in Y : \exists V_y \in \alpha$ such that $y \in V_y$; $V_y \in \tau_\Sigma \Rightarrow V_y \in \underline{U}(y)$; since H is evenly continuous we find: $\forall y \in Y$, for (x, y, V_y) $\exists (U^y, W_y) \in \underline{U}(x) \times \underline{U}(y)$, W_y open, such that: $\forall f \in H$, $f(x) \in W_y \Rightarrow f(U^y) \subseteq V_y$; $(W_y)_{y \in Y}$ is an open covering of Y and $H(x)$ is relatively compact, hence we find $y_1, \ldots, y_n \in Y$ such that $H(x) \subseteq W_{y_1} \cup \cdots \cup W_{y_n}$ by 3; setting $U := \bigcap_{i=1}^{n} U^{y_i}$ we have $U \in \underline{U}(x)$; now, $\forall f \in H$, $f(x) \in H(x) \Rightarrow \exists i_0 \in \{1, \ldots, n\}$ such that $f(x) \in W_{y_{i_0}}$; but $f(x) \in W_{y_{i_0}} \Rightarrow f(U) \subseteq f(U^{y_{i_0}}) \subseteq V_{y_{i_0}}$ and $V_{y_{i_0}} \in \alpha$ thus implying that H is equicontinuous in x by definition 26.

Corollary 28 *Let X be a topological space, (Y, Σ) a generalized uniform space, $H \subseteq Y^X$; if H is evenly continuous and $\overline{H(x)}$ is compact then H is equicontinuous at x.*

Proof. By 6, $\overline{H(x)}$ compact $\Rightarrow H(x)$ is relatively compact; hence the assertion follows from theorem 27, 2.

Remark. The assertion (1) $\Rightarrow$ (3) of theorem 27, 1 and the assertion of the corollary were proved in [7]; compare also [5].

Corollary 29 *X topological space, Y a (full) uniform space, $H \subseteq Y^X$; then hold:*

1. *H equicontinuous $\Rightarrow H$ evenly continuous*

2. *H evenly continuous, $x \in X$, $H(x)$ relatively compact $\Rightarrow H$ equicontinuous at x.*

Finally we consider a simple example:

$$(f_n)_{n \geq 1}, \quad f_n : [0,1] \longrightarrow \mathbb{R}; \quad f_1(x) := -x + 1; \quad n \geq 2:$$

$$f_n(x) = \begin{cases} 0, & x \in \left[\frac{1}{n}, 1\right] \\ -n^2 x + n, & x \in \left[0, \frac{1}{n}\right) . \end{cases}$$

Then in [9] is shown that $H = \{f_n\}$ is evenly continuous, but not equicontinuous (at $x = 0$); and of course $H(0) = \{1, 2, \ldots\}$ is not relatively compact.

7 The basic theorems

Theorem 30 *X, Y topological spaces; Φ filter on Y^X, Y T_3-space, $f \in Y^X$, $\Phi \xrightarrow{c} f \Rightarrow f \in C(X, Y)$.*

Proof. Let $(f_i)_{i\in I}$ be an arbitrary net from Y^X with $f_i \xrightarrow{c} f$. If we have shown, that $f \in C(X,Y)$, the theorem is proved, because of theorem 18 b).
Let $x \in X$, $V \in \underline{U}(f(x))$
we will show: $\exists U \in \underline{U}(x)\ f(U) \subseteq V$
Y T_3-space $\Rightarrow \exists W \in \underline{U}(f(x))$ with the property: W is a closed set and $W \subseteq V$
$f_i \xrightarrow{c} f$, theorem 16 $\Rightarrow \exists U \in \underline{U}(x)\ \exists i_0 \in I\ \forall i \geq i_0 \Rightarrow f_i(U) \subseteq W$
$f_i \xrightarrow{c} f \Rightarrow f_i \xrightarrow{\tau_p} f$
$z \in U \Rightarrow f_i(z) \in W\ \forall i \geq i_0$
$(f_i(z))_{\substack{i\in I\\ i\geq i_0}} \to f(z) \Rightarrow f(z) \in \overline{W} = W \subseteq V$. Thus we have got $f(U) \subseteq V$ and hence $f \in C(X,Y)$.

Theorem 31 *X, Y topological spaces; $H \subseteq Y^X$ evenly continuous, Φ filter on Y^X such that $H \in \Phi$ and $\Phi \xrightarrow{\tau_p} f \in Y^X$; then holds: $\Phi \xrightarrow{c} f$.*

Proof. Let $x \in X$, let ψ be an arbitrary filter on X with $\psi \to x$; we will show, that $\omega(\phi \times \psi) \to f(x)$:

$$\Phi \xrightarrow{\tau_p} f \Rightarrow pr_x\Phi = \omega(\Phi \times [x]) \to f(x)$$

H evenly continuous, $H \in \Phi$, $\psi \to x \Rightarrow \omega(\Phi \times \psi) \to f(x)$.

Theorem 32 *Let X be a topological space, Y a Hausdorff topological space and let $H \subseteq C(X,Y) \subseteq Y^X$. Let* lim *be a convergence structure for $C(X,Y)$ such that*

1. *H is in $C(X,Y)$ relatively compact, and*
2. lim *is conjoining for $C(X,Y)$.*

Then H is evenly continuous.

Proof. Let $x \in X$, $y \in Y$, let $\underline{F}$ be a filter on Y^X with $H \in \underline{F}$ and $\omega(\underline{F} \times [x]) \to y$; and let ψ be a filter on X with $\psi \to x$; we must show, that $\omega(\underline{F} \times \psi) \to y$ holds. If we have shown, that $\pi \to y$ for each ultrafilter π on Y with $\omega(\underline{F} \times \psi) \subseteq \pi$, then this is proved.

Let π be an arbitrary ultrafilter on Y:

$$\exists \text{ ultrafilter } \varrho \text{ on } Y^X \times X \text{ such that } \underline{F} \times \psi \subseteq \varrho \text{ and } \omega(\varrho) = \pi$$

$\varrho_1 := pr_{Y^X}\varrho$ is a ultrafilter on Y^X

$$\underline{F} \subseteq \varrho_1 \Rightarrow H \in \varrho_1 \Rightarrow C(X,Y) \in \varrho_1$$

$\widetilde{\varrho_1} := \varrho_1 \cap C(X,Y)$ is a ultrafilter on $C(X,Y)$:

$$\begin{aligned}\forall A \subseteq C(X,Y) : A \notin \widetilde{\varrho_1} &\Rightarrow A \notin \varrho_1 \Rightarrow Y^X \setminus A \in \varrho_1\\ &\Rightarrow C(X,Y) \setminus A = (Y^X \setminus A) \cap C(X,Y) \in \widetilde{\varrho_1}\\ \forall A \subseteq C(X,Y) &: A \in \widetilde{\varrho_1} \text{ or } C(X,Y) \setminus A \in \widetilde{\varrho_1}\end{aligned}$$

hence: $\widetilde{\varrho_1}$ is a ultrafilter on $C(X,Y)$;
$\varrho_2 := pr_X\varrho$ is a ultrafilter on X with $\varrho_2 \to x$, because of $\psi \subseteq \varrho_2$ and $\psi \to x$;

$$\omega(\underline{F} \times [x]) \subseteq \omega(\varrho_1 \times [x]) \Rightarrow \omega(\varrho_1 \times [x]) \to y \; ;$$

$C(X,Y) \in \varrho_1 \Rightarrow \widetilde{\varrho_1} \subseteq \varrho_1$ and $H \in \widetilde{\varrho_1}$
H is relatively compact in $(C(X,Y), \lim)$, $H \in \widetilde{\varrho_1}, \widetilde{\varrho_1}$ is ultrafilter on $C(X,Y) \Rightarrow$
$\exists f \in C(X,Y)$ such that $\widetilde{\varrho_1} \to f$;
let be λ an arbitrary filter on X with $\lambda \to x$
$\widetilde{\varrho_1} \to f \Rightarrow \tilde{\varrho} \times \lambda \to (f,x)$ on $(C(X,Y), \lim) \times X$
lim conjoining $\Rightarrow \omega(C(X,Y), \lim) \times X \to Y$ is continuous $\Rightarrow \omega(\widetilde{\varrho_1} \times \lambda) \to f(X) \Rightarrow$
$U(f(x)) \subseteq \omega(\widetilde{\varrho_1} \times \lambda) \subseteq \omega(\varrho_1 \times \lambda)$, because of $\widetilde{\varrho_1} \subseteq \varrho_1 \Rightarrow \omega(\varrho_1 \times \lambda) \to f(x)$;
we have got: $\forall$ filter λ on X with $\lambda \to x : \omega(\varrho_1 \times \lambda) \to f(x)$.
If we set $\lambda := [x]$, we get $\omega(\varrho_1 \times [x]) \to f(x)$ and we know that $\omega(\varrho_1 \times [x]) \to y$.
Y is Hausdorff $\Rightarrow f(x) = y$.
If we set $\lambda := \varrho_2$, we get $\omega(\varrho_1 \times \varrho_2) \to f(x) = y$.

$$\varrho_1 \times \varrho_2 \subseteq \varrho \Rightarrow \pi = \omega(\varrho) \to y \; .$$

Remarks.

1. More general theorem 32 holds if X is a convergence space, Y is a Hausdorff pseudotopological convergence space.

2. In our opinion within this approach in theorem 32 the Hausdorff property for Y cannot be dropped.

8 A (c–lim)-compactness criterion for $C(X, Y)$

Now we are able to formulate and prove our main theorem which characterizes the relatively compact subsets of $C(X,Y)$ with respect to the convergence structure $c - \lim$ of continuous convergence.

Theorem 33 *X, Y topological spaces, $H \subseteq C(X,Y)$, $H \neq \emptyset$; for H we consider the two conditions:*

(α) *$\forall x \in X$, $H(x)$ is relatively compact*

(β) *H is evenly continuous.*

Then hold:

1. *Let Y be Hausdorff; H relatively $c -$ lim-compact in $C(X,Y) \Rightarrow (\alpha), (\beta)$*

2. *Let Y be a T_3-space; (α), $(\beta) \Rightarrow H$ is in $(C(X,Y), c-\lim)$ relatively compact.*

Proof.

1. By 19.5 we know that continuous convergence is stronger that pointwise convergence, and hence the identity $id : (C(X,Y), c-\lim) \to (C(X,Y), \tau_p)$ is continuous; hence: H $c-\lim$-relatively compact $\Rightarrow H$ τ_p-relatively compact in $C(X,Y) \Rightarrow H$ is τ_p-relatively compact in $Y^X \Rightarrow (\alpha)$ by the Tychonoff theorem 10. Now, $c-\lim$ is conjoining by 19.4 and H is $c-\lim$-relatively compact in $C(X,Y)$; hence we get (β) by the basic theorem 32.
2. By the Tychonoff theorem, $(\alpha) \Rightarrow H$ is τ_p-relatively compact Y^X; let π be an ultrafilter in $C(X,Y)$ and $H \in \pi$; let $[\pi]$ be the extension of π to Y^X then: $[\pi]$ is ultrafilter in Y^X and $H \in [\pi]$; thus we find $g \in Y^X$, $[\pi] \xrightarrow{\tau_p} g$ implying by basic theorem 31 that $[\pi] \xrightarrow{c} g$, since H is evenly continuous; $[\pi] \xrightarrow{c} g$ and Y T_3-space $\Rightarrow g \in C(X,Y)$ by theorem 30; hence: $\pi \xrightarrow{c} g \Rightarrow H$ is in $(C(X,Y), c-\lim)$ relatively compact.

9 Immediate applications to classical τ_p- and τ_u-criteria

As is well known, a topological space X is called locally compact: $\Leftrightarrow \forall x \in X : \forall u \in \underline{U}(x) : \exists V \in \underline{U}(x) : V \subseteq U$ and V is compact. We know from section 4, that for topological spaces X, Y, where X is locally compact, holds $c-\lim = \tau_k-\lim$ on $C(X,Y)$. Thus from the $c-\lim$-criterion we get

Theorem 34 *Let X, Y be topological spaces: X locally compact; $H \subseteq C(X,Y)$.*

1) *Y Hausdorff, H τ_k-relatively compact*

$$\Longrightarrow \begin{array}{l} (\alpha)\ \forall x \in X : H(x) \text{ is relatively compact} \\ (\beta)\ H \text{ is evenly continuous}. \end{array}$$

2) *Y T_3-space and $(\alpha) \wedge (\beta) \Longrightarrow H$ is relatively compact in $(C(X,Y); \tau_k)$*

Corollary 35 *X a locally compact topological space, Y a Hausdorff uniform space; $H \subseteq C(X,Y)$; then holds:*
H is in $C(X,Y)$ τ_k-relatively compact
$\Longleftrightarrow$

(α) $\forall x \in X$, $H(x)$ is relatively compact

(β'') H is equicontinuous on X.

Proof. Y Hausdorff uniform space $\Rightarrow Y$ is a T_3-space; now using corollary 29 we can in theorem 34 replace condition (β) by condition (β'') of our corollary.

If X is a compact topological space and Y is an uniform space, then we have $c-\lim = \tau_u-\lim$. Again from the $c-\lim$-criterion follows

Theorem 36 .

1) *Y Hausdorff; H relatively compact in* $(C(X,Y),\tau_u) \Longrightarrow (\alpha) \wedge (\beta)$

2) $(\alpha) \wedge (\beta) \Longrightarrow$ *H relatively compact in* $(C(X,Y),\tau_u)$.

10 A τ_k-criterion for k-spaces

We remember, that a topological space X is called a <u>k-space</u>: $\Leftrightarrow$ ($\forall$ compact K : $(A \cap K)$ open in $K) \Longrightarrow A$ open in X.

Now let X and Y be topological spaces. We define

$$C_k(X,Y) := \{f \in Y^X \mid \forall \text{ compact } K \subseteq X : f|_K \text{ is continuous on } K\} .$$

We generally have $C(X,Y) \subseteq C_k(X,Y)$; and if X is a k-space, it is easy to see, that $C_k(X,Y) = C(X,Y)$.

With $H \subseteq Y^X$ and $A \subseteq X$, H is called evenly continuous on A :$\Leftrightarrow$ $H|_A; = \{f|_A \mid f \in H\}$ is evenly continuous in Y^A.

Using especially the maps $q_K : f \to f|_K$ for fixed compact $K \subseteq X$ and replacing condition (β) by (β'): "H is evenly continuous on each compact subset of X", we can transfer the τ_k-criterion of section 8 to $(C_k(X,Y),\tau_k)$.

Before doing this, we will prove two lemmas. The first lemma is a variation of theorem 30.

Lemma 37 *Let X and Y be topological spaces;* $H \subseteq Y^X$*; and let* Φ *be a filter on* Y^X*, such that* $H \in \Phi$.

a) *If* $\phi \xrightarrow{p} f \in Y^X$ *and H is evenly continuous on each compact subset of X, then* $\Phi \xrightarrow{\tau_k} f$. *If X is a k-space and Y is a* T_3*-space, we also have* $f \in C(X,Y)$.

b) *Let* $f \in Y^X$ *and let for each compact* $K \subseteq X$

$$q_K\Phi \xrightarrow{\tau_k} f|_K \text{ in } Y^K .$$

Then we have $\Phi \xrightarrow{\tau_k} f$ *in* Y^X.

Proof.

b) Let $K \subseteq X$ be compact and let G be an open subset of Y.
Let $f \in (K,G) = \{g \in Y^X \mid g(K) \subseteq G\}$.

We have $f(K) \subseteq G$ and hence $f|_K(K) \subseteq G$.

Since $q_K\Phi \xrightarrow{\tau_k} f|_K$ we find $A \in \Phi$ such that $q_K(A) \subseteq (K,G) \cap Y^X$, but then follows $A \subseteq (K,G)$ too.

a) By 31 we have $f|_K \in c-\lim q_K\Phi$ and hence $q_K\Phi \xrightarrow{\tau_k} f|_K$ for each compact $K \subseteq X$. Now the first part of a) follows from b).

If Y is a T_3-space, theorem 30 shows that $f \in C(K,Y)$ for each compact $K \subseteq X$, because $q_K\Phi \xrightarrow{c} f|_K$. This means $f \in C_k(X,Y)$. But X should be a k-space, hence we find $f \in C(X,Y) = C_k(X,Y)$.

Lemma 38 *Let X,Y be topological spaces and $D \subseteq Y^X$; let* lim *be a convergence structure on D satisfying the following properties:*

a) *If K is compact in X, then* lim *is also defined for $q_K(D) \subseteq Y^X$ and the map q_K from $(D, \lim)$ onto $(q_K(D), \lim)$ is continuous.*

b) *If K is compact in X, then* lim *and* $C-\lim$ *coincide on $q_K(D)$.*

Let $M \subseteq D \times X$ and let the projection $pr_X M$ be compact in X. Then the restriction of the evaluation map ω to M is continuous on M.

Proof. Let $\underline{F}$ be a filter on M such that $\underline{F} \to (f,x) \in M$. With $\Phi := pr_D\underline{F}$ and $\psi := pr_X\underline{F}$ ψ becomes a filter on D, ψ a filter on X with respectively $f \in \lim\Phi$, $pr_X M \in \psi$ and $\psi \to x$.

By assumption $K := pr_X M$ is compact. Let ψ_1 be the restriction of ψ to K.

(a)∧(b) $\Longrightarrow q_K(f) = f|_K \in \lim q_K\Phi = c-\lim q_K\Phi$ in $q_K(D)$. Since $\psi_1 \to x$ in K we have by definition of $c-\lim$:

$$\omega(q_K\Phi \times \psi_1) \to f|_K(x) = f(x) = \omega(f,x)\ . \qquad (*)$$

Clearly is $\omega(q_K\Phi \times \psi_1) \subseteq \omega(q_K\Phi \times \psi)$. $\qquad (**)$

Now let $A \in \Phi$ and $V \in \psi$ arbitrary elements of these filters. Since $\Phi \times \psi \subseteq \underline{F}$ and $K \in \psi$ we conclude $\omega(A \times (K \cap V)) \in \omega(\underline{F})$
and clearly $\omega(A \times (K \cap V)) \subseteq \omega(q_K(A) \times V)$
$\Longrightarrow \omega(q_K(A) \times V) \in \omega(\underline{F})$.
This holds for all $A \in \Phi$ and all $V \in \psi$,
so we see $\omega(q_K\Phi \times \psi) \subseteq \omega(\underline{F})$ $\qquad \binom{*}{**}$

$$(*) \wedge (**) \wedge \binom{*}{**} \Longrightarrow \omega(\underline{F}) \to \omega(f,x) \qquad \Longrightarrow \omega \text{ is continuous}\ .$$

Finally we can get our transferred criterion:

Theorem 39 *Let X be a Haudorff k-space, Y a topological space, and $H \subseteq C(X,Y)$.*

1) *Y Hausdorff; H is in $(C(X,Y), \tau_k)$ relatively compact*

$$\Longrightarrow \quad (\alpha)\ \ \forall x \in X : H(x) \text{ is relatively compact}$$
$$(\beta')\ \ H \text{ is evenly continuous on each compact } K \subseteq X\ .$$

2) *Y a T_3-space; then*

$$(\alpha) \wedge (\beta') \Longrightarrow H \text{ is relative compact in } (C(X,Y), \tau_k) .$$

Proof.

1) (α) As in the proof of the main theorem we apply the Tychonoff-theorem about relative compactness; which is possible because τ_k is stronger than the pointwise convergence.

(β) We want to use lemma 38.

Generally we have for topological spaces X and Y and a compact subset K of X

$$q_K(C_k(X,Y)) \subseteq C(X,Y) .$$

Furthermore $\tau_k - \lim$ and $c - \lim$ coincide on $q_K(C_k(X,Y))$, because a Hausdorff compact space is locally compact (see [8]).

We know, that here $C_k(X,Y) = C(X,Y)$ because X is a k-space.

It is easy to see, that the map

$$q_K : (C(X,Y), \tau_k) \rightarrow (q_K(C(X,Y)), \tau_k)$$

is continuous. Hence we can apply lemma 38 to

$$D = C(X,Y) \subseteq Y^X \text{ and } \lim = \tau_k - \lim .$$

We conclude: if K compact, then $\omega : (C(X,Y), \tau_k) \times K \rightarrow Y$ is continuous.

Since q_K is continuous, we find that $H|_K = q_K(H)$ is relatively t_k-compact in $C(K,Y)$ like H in $C(X,Y)$. Applying lemma 38 to $C(K,Y) \times K$ we see that τ_k is conjoining for $C(K,Y)$ and hence the assertion follows from theorem 32.

2) Let $\underline{F}$ be an ultrafilter on $C(X,Y)$ with $H \in \underline{F}$ and let $[\underline{F}]$ be the extension of $\underline{F}$ to Y^X; then $[\underline{F}]$ is an ultrafilter too.

By the Tychonoff-theorem and (α), H is relatively compact in (Y^X, τ_p); so we find $f \in Y^X : [\underline{F}] \xrightarrow{\tau_p} f$. By lemma 37 we have now $[\underline{F}] \xrightarrow{\tau_k} f$ and $f \in C(X,Y)$. For this reason $\underline{F}$ τ_k-converges to f in $C(X,Y)$ and H is now in fact relatively compact in $C(X,Y)$.

11 Coming back to the classical theorem

If we take X as a compact metric space, for Y a metric space; then for $C(X,Y)$ the uniform topology is metrizable and is generated by the Tschebycheff metric.

Hence: $H \subseteq C(X,Y)$ is relatively compact $\Leftrightarrow$ H is relatively sequentially compact.

Thus by the τ_k-criterion of section 8 we get

Theorem 40 *X compact metric space, Y metric space; $H \subseteq C(X,Y)$. Then: H is in $C(X,Y)$ relatively sequentially compact*

$\Leftrightarrow$ *(α) $\forall x \in X : H(x)$ relatively sequentially compact*

$\wedge$*(β) H is equicontinuous.*

By this theorem and by the following lemma we get back to the classical theorem of Ascoli-Arzeà.

Lemma 41 *Let X be a compact metric space, Y a metric space; $H \subseteq C(X,Y)$ and H equicontinuous. We consider the conditions*

(1) *$\forall x \in X : H(X)$ is relatively sequentially compact*

(2) *H is uniformly bounded*

(that means: bounded with respect to the Tschebyscheff–metric).

Then holds:

(a) (1) $\Rightarrow$ (2).

(b) *If for Y the Bolzano–Weierstraß–theorem holds, then* (2) $\Rightarrow$ (1).

Proof.

(a) By theorem 39 H is relatively sequentially τ_u-compact in $C(X,Y)$. Because $C(X,Y)$ is metrizable, from this follows H is relatively τ_u-compact in $C(X,Y)$. But then H is uniformly bounded.

(b) If H is uniformly bounded, then each $H(x)$ is bounded. But the theorem of Bolzano-Weierstraß shows now, that $H(x)$ is relatively sequentially compact.

We return to the

Theorem 42 (Ascoli and Arzelà) *Let X be a compact interval $[a,b]$ of reals and let Y be the metric space of the real or complex numbers; let $H \subseteq C(X,Y) = C([a,b])$. Then each sequence $(f_n)_{n\in\mathbb{N}}$ from H contains a subsequence uniformly converging to some function form $C(X,Y)$ if and only if H satisfies the conditions*

(α) H is uniformly bounded

and

(β) H is equicontinuous.

Proof. H is equicontinuous and uniformly bounded implies by Lemma 41 that $H(x)$ is relatively sequentially compact for each $x \in X$. By theorem 40 we find that H is relatively sequentially τ_u-compact in $C(X,Y)$.

If in reverse H is relatively sequentially τ_u-compact, then follows directly by theorem 40 that H is equicontinuous and $H(x)$ relatively sequentially compact for each $x \in X$, which shows by 41 that H is uniformly bounded.

References

1. W. Gähler, *Grundstrukturen der Analysis II*, Berlin 1978.
2. J.W. Gray, "Categorical Aspects of the Classical Ascoli Theorems", *Cahiers de Topologie et Geometrie Differentielle* vol. XXII-3, 1981, 337-342.
3. H. Herrlich, *Topologie II: Uniforme Räume*, Berlin 1988.
4. R.A. McCoy, I. Ntantu, "Topological properties of spaces of continuous functions,", *Lecture notes in mathematics* **1315**, Springer Verlag, Berlin Heidenlberg 1988.
5. K. Morita, Extensions of Mappings I in: K. Morita and j. Nagata (edit.) *Topics in general topology* Amsterdam 1989, 1-38.
6. L.D. Nel, "Introduction to categorical methods", Part Three, 1992, Carleton-Ottawa, *Mathematical Lecture Note* Series 12.
7. H. Poppe, "Ein Kompaktheitskriterium für Abbildungsräume mit einer verallgemeinerten uniformem Struktur", *Proceedings of the Second Prague Topological Symposium* 1966; Praha 1967, 284-289.
8. H. Poppe, *Compactness in General Function Spaces*, Berlin 1974.
9. H. Poppe, "Convergence of Evenly Continuous Nets in General Function Spaces, *Real Analysis Exchange* **18** (2), 1992/93, 1-6.
10. H. Poppe, "On locally defined topological notions", will appear in *Questions and Answers in Topology*.
11. H. Poppe, "Nonstandard topology on function spaces with applications to hyperspaces", *Trans. AMS* **336**, 1993, 101-119.
12. H. Poppe, Generalized uniform spaces and applications to function spaces, submitted.

A Subcategory of FIL

Ákos Császár

AMS Subj. Class. 54E17 (54B30)

Keywords. Filter space, screen, Chauchy space, M-screen, chain-complete screen.

0 Introduction

It is well-known that a *filter space* on a set X is a pair $(X, \mathfrak{S})$ where $\mathfrak{S}$ is a collection $\emptyset \neq \mathfrak{S} \subset \operatorname{Fil} X$ such that

$$\dot{x} \in \mathfrak{S} \quad \text{for } x \in X\ , \tag{1}$$

$$\mathfrak{s} \in \mathfrak{S}\ , \quad \mathfrak{s} \subset \mathfrak{s}' \in \operatorname{Fil} X \quad \text{imply } \mathfrak{s}' \in \mathfrak{S}\ . \tag{2}$$

Here $\operatorname{Fil} X$ is the set of all (proper or improper) filters in X and $\dot{x}$ is the ultrafilter fixed at x. A collection $\mathfrak{S}$ of this kind can be called a *screen* on X (see e.g. [2]).

If $(X, \mathfrak{S})$ and $(X', \mathfrak{S}')$ are filter spaces, the map $f : X \to X'$ is said to be $(\mathfrak{S}, \mathfrak{S}')$-*continuous* iff $\mathfrak{s} \in \mathfrak{S}$ implies $\operatorname{Fil}_Y f(\mathfrak{s}) \in \mathfrak{S}'$, where $f(\mathfrak{s}) = \{f(S) : S \in \mathfrak{s}\}$ and $\operatorname{Fil}_Y \mathfrak{t}$ is the filter in Y generated by the filter base $\mathfrak{t}$. The filter spaces and the continuous maps in the above sense constitute a topological category FIL.

If $\mathfrak{S}$ is a screen, a set $B \subset \mathfrak{S}$ is said to be a *base* for $\mathfrak{S}$ iff $\mathfrak{s} \in \mathfrak{S}$ implies the existence of $\mathfrak{s}_0 \in B$ such that $\mathfrak{s}_0 \subset \mathfrak{s}$.

A screen $\mathfrak{S}$ is a *Cauchy screen* (Cauchy structure) on X iff

$$\mathfrak{s}'\ , \quad \mathfrak{s}' \in \mathfrak{S}\ , \quad \mathfrak{s} \Delta \mathfrak{s}' \quad \text{imply } \mathfrak{s} \cap \mathfrak{s}' \in \mathfrak{S}\ , \tag{3}$$

where $a \Delta b$ means that $A \in a$, $B \in b$ imply $A \cap B \neq \emptyset$, $a \overline{\Delta} b$ iff there are $A \in a$, $B \in b$ with $A \cap B = \emptyset$.

A set $\mathfrak{C} \subset \operatorname{Fil} X$ is said to be a Δ-*system* iff $\mathfrak{s}, \mathfrak{s}' \in \mathfrak{C}$ implies $\mathfrak{s} \Delta \mathfrak{s}'$. A base $\mathfrak{B}$ for a screen $\mathfrak{S}$ is said to be a $\overline{\Delta}$-*base* iff $\mathfrak{s}, \mathfrak{s}' \in \mathfrak{B}$, $\mathfrak{s} \neq \mathfrak{s}'$ imply $\mathfrak{s} \overline{\Delta} \mathfrak{s}'$; $\mathfrak{B}$ is said to be *independent* iff $\mathfrak{s}', \mathfrak{s}' \in \mathfrak{B}$, $\mathfrak{s} \subset \mathfrak{s}'$ imply $\mathfrak{s} = \mathfrak{s}'$. A filter $\mathfrak{s}$ is said to be $\mathfrak{S}$-*minimal* for a screen $\mathfrak{S}$ iff $\mathfrak{s} \in \mathfrak{S}$ and $\mathfrak{s} \supset \mathfrak{s}' \in \mathfrak{S}$ implies $\mathfrak{s} = \mathfrak{s}'$ (see [3] for these definitions).

A screen $\mathfrak{S}$ *induces* a (Čech) *proximity* δ defined by

$$A \delta B \quad \text{iff there is } \mathfrak{s} \in \mathfrak{S} \quad \text{with } \{A\} \Delta \mathfrak{s} \Delta \{B\}\ . \tag{4}$$

The *closure* c induced by the above δ:

$$x \in c(A) \quad \text{iff } \{x\} \delta A \tag{5}$$

is said to be *induced* by $\mathfrak{S}$. We write $\delta = \delta(\mathfrak{S})$, $c = c(\delta) = c(\mathfrak{S})$. The closure $c = c(\mathfrak{S})$ is *symmetric* (i.e. $y \in c(\{x\})$ implies $x \in c(\{y\})$). Conversely, if δ

Eraldo Giuli (ed.), Categorical Topology, 241–244.

is a proximity, there is a coarsest (= largest) screen $\mathfrak{S}^0(\delta)$ inducing δ; if c is a symmetric closure, there is a coarsest screen $\mathfrak{S}^0(c)$ inducing c. If c is a T_1 closure (i.e. $c(\{x\}) = \{x\}$ for $x \in X$) then the c-neighbourhood filters $v_c(x)$ ($x \in X$) constitute a base for a screen $\mathfrak{S}^1_R(c)$ inducing c ($V \in v_c(x)$ iff $x \notin c(X - V)$) (see [4]). This is the case in particular if c is T_2 (i.e. $x \neq y$ implies $v_c(x)\overline{\Delta}v_c(y)$).

A screen $\mathfrak{S}$ in X is *Riesz* iff $v_c(x) \in \mathfrak{S}$ for $x \in X$ and $c = c(\mathfrak{S})$ (e.g. the screen $\mathfrak{S}^1_R(c)$ above is the finest Riesz screen inducing c).

If $\mathfrak{S}$ is a screen on X and $X_0 \subset X$ then $\mathfrak{S}|_{X_0} = \{\mathfrak{s}|_{X_0} : \mathfrak{s} \in \mathfrak{S}\}$ is the *restriction* of $\mathfrak{S}$ to X_0; here $\mathfrak{s}|_{X_0} = \{S \cap X_0 : S \in \mathfrak{s}\}$ is the *trace* of $\mathfrak{s}$ on X_0. The trace of a filter is a filter and the restriction of a screen is a screen (see [2]).

1 M-screens and CC-screens

The paper [1] contains the definition of a class of Cauchy screens; their description can be completed as follows:

Theorem 1 ([3], 1.1) *For a screen $\mathfrak{S}$ on X, the following are equivalent:*

(a) $\mathfrak{s}_0 \in \mathfrak{S}$ *implies* $\bigcap\{\mathfrak{s} \in \mathfrak{S} : \mathfrak{s} \subset \mathfrak{s}_0\} \in \mathfrak{S}$,

(b) *every* $\mathfrak{s}_0 \in \mathfrak{S}$ *contains a unique* $\mathfrak{S}$*-minimal filter,*

(c) $\mathfrak{S}$ *is Cauchy and every* $\mathfrak{s}_0 \in \mathfrak{S}$ *contains an* $\mathfrak{S}$*-minimal filter,*

(d) $\mathfrak{S}$ *admits a* $\overline{\Delta}$*-base,*

(e) *if* $\emptyset \neq \mathfrak{C} \subset \mathfrak{S}$ *is a* Δ*-system then* $\bigcap \mathfrak{C} \in \mathfrak{S}$.

(a) $\Leftrightarrow$ (b) $\Leftrightarrow$ (c) is contained in [1] (Corollaries 19 and 20 and Proposition 23), (a) $\Leftrightarrow$ (e) in [6], 7.1.

In [5] and [6], a screen fulfilling (a) to (e) is said to be *fully Cauchy* (cf. (c)). Our purpose is to examine two possible generalizations of this concept.

Let us say that a screen $\mathfrak{S}$ is an *M-screen* iff every $\mathfrak{s}_0 \in \mathfrak{S}$ contains an $\mathfrak{S}$-minimal filter (see (b) and (c)), and it is *chain-complete* (*CC*) iff the intersection of any proper chain contained in $\mathfrak{S}$ belongs to $\mathfrak{S}$ (see (e)). A set $\mathfrak{C} \subset \operatorname{Fil} X$ is said to be a *proper chain* iff $\mathfrak{C} \neq \emptyset$, $\exp X \notin \mathfrak{C}$, and $\mathfrak{s}$, $\mathfrak{s}' \in \mathfrak{C}$ implies either $\mathfrak{s} \subset \mathfrak{s}'$ or $\mathfrak{s}' \subset \mathfrak{s}$.

Theorem 2 ([3], 2.7) *A screen is M iff it admits an independent base.*

Compare (d) in Theorem 1.

Every CC screen is M ([3], 2.1). If δ is a proximity (c is a symmetric closure) then $\mathfrak{S}^0(\delta)$ ($\mathfrak{S}^0(c)$) is CC ([3], 2.5 and 2.3); if c is a T_1 closure then $\mathfrak{S}^1_R(c)$ is M but it need not be CC ([3], 2.8 and 2.9). A simpler example is the following one: let X be an infinite set, $\mathfrak{u}_n$ $\mathfrak{v}_n$ distinct free ultrafilters in X ($n \in \mathbf{N}$), and let a screen $\mathfrak{S}$ be determined by the base composed of $\dot{x}$ $(x \in X)$ and $\mathfrak{s}_n = \mathfrak{v}_n \cap \bigcap_1^n \mathfrak{u}_i$ $(n \in \mathrm{N})$. By Theorem 2, $\mathfrak{S}$ is an M-screen but, for $\mathfrak{t}_n = \bigcap_1^n \mathfrak{u}_i$, $\{\mathfrak{t}_n : n \in \mathrm{N}\} \subset \mathfrak{S}$ is a proper chain with $\bigcap_1^\infty \mathfrak{t}_n \notin \mathfrak{S}$.

However, a Cauchy screen is M iff it is CC iff it is fully Cauchy ([3], 2.2). Thus a Cauchy screen that is not fully Cauchy is not M. A simpler example for a non-M screen is given in 3, 2.12.

2 Embedding into M-spaces

The property of being an M-screen is not hereditary; on the contrary, *every* filter space $(X_0, \mathfrak{S}_0)$ is the subspace of a filter space $(X, \mathfrak{S})$ such that $\mathfrak{S}$ is M. In [3], 3.1 to 3.4, several constructions are given for $(X, \mathfrak{S})$ in the way that some properties of $\mathfrak{S}_0$ are conserved by $\mathfrak{S}$; in all of them, $c(X_0) = X$ for $c = c(\mathfrak{S})$ (i.e. X_0 is c-dense in X). Another kind of construction is presented by

Theorem 3 *If $(X_0, \mathfrak{S}_0)$ is an arbitrary filter space, there exists a filter space $(X, \mathfrak{S})$ such that $X_0 \subset X$, $\mathfrak{S}_0 = \mathfrak{S}|_{X_0}$, $\mathfrak{v}_c(p) = \dot{p}$ for $p \in X - X_0$ and $c = c(\mathfrak{S})$, and $\mathfrak{S}$ is M.*

Proof. Let $X \supset X_0$ be chosen on the manner that there is an injective map u from $\mathfrak{S}_0$ into the set of all free ultrafilters in $X - X_0$. Let a base for a screen $\mathfrak{S}$ on X be composed of the filters $\dot{p}$ $(p \in X - X_0)$, $\mathrm{Fil}_x \mathfrak{s}_0$ if $\mathfrak{s}_0 \in \mathfrak{S}_0$ is $\mathfrak{S}_0$-minimal and $(\mathrm{Fil}_X \mathfrak{s}_0) \cap (\mathrm{Fil}_X \mathrm{u}(\mathfrak{s}_0))$ if $\mathfrak{s}_0 \in \mathfrak{S}_0$ is not finer than any $\mathfrak{S}_0$-minimal filter. Clearly $\mathfrak{S}|_{X_0} = \mathfrak{S}_0$, the base is independent, and $\mathfrak{v}_c(p) = \dot{p}$ for $p \in X - X_0$. By Theorem 2, $\mathfrak{S}$ is an M-screen.

□

Corollary 1 *The screen $\mathfrak{S}$ constructed in the above proof is T_1 provided so is $\mathfrak{S}_0$. If $\mathfrak{S}_0$ is Riesz then $\mathfrak{S}$ is Riesz and $X_0 \in v_c(x)$ for $x \in X_0$ and $c = c(\mathfrak{S})$. If $\mathfrak{S}_0$ is Riesz and $c_0 = c(\mathfrak{S}_0)$ is a topology then c is a topology and x_0 is c-clopen.*

Proof. If $\mathfrak{S}$ is a screen on X, $\mathfrak{B}$ is a base for $\mathfrak{S}$, $c = c(\mathfrak{S})$, $x \in X$, then $v_c(x)$ is the intersection of all filters from $\mathfrak{B}$ fixed at x. Hence c is T_1 iff any $\mathfrak{s} \in \mathfrak{B}$ fixed at x is not fixed at any $y \neq x$. A filter $\mathfrak{v}_{c_0}(x)$ is $\mathfrak{S}_0$-minimal whenever it belongs to $\mathfrak{S}_0$, and $\mathfrak{v}_c(x) = \mathrm{Fil}_X \mathfrak{v}_{c_0}(x)$ in this case for $x \in X_0$.

□

Corollary 2 *If $c(\mathfrak{S}_0)$ is T_2, $(X, \mathfrak{S})$ in Theorem 3 can be chosen such that $c(\mathfrak{S})$ is T_2.*

Proof. We apply the same construction in the way that

$$X - X_0 = X' \cup \bigcup\{X(x) : x \in X_0\}$$

with disjoint members of the union, and $\mathfrak{u}(\mathfrak{s}_0) = \mathrm{Fil}_X \mathfrak{u}$ where $\mathfrak{u}$ is a free ultrafilter in $X(x)$ if $\mathfrak{s}_0$ is fixed at x and in X' if $\mathfrak{s}_0$ is free.

□

3 The category CCFIL

The property of being chain-complete is hereditary (see [3], 3.5). Moreover:

Theorem 4 ([3], 6.3) *The full subcategory* CCFIL *of* FIL *whose objects are the spaces* $(X, \mathfrak{S})$ *with a chain-complete* $\mathfrak{S}$ *is bireflective.*

Proof. For an arbitrary screen $\mathfrak{S}$ on X, let $\mathfrak{S}_{cc}$ denote the intersection of all CC screens containing $\mathfrak{S}$ (observe that $\mathrm{Fil}\,X$ is a CC screen). This is a CC screen ([3], 4.9) such that $\delta(\mathfrak{S}_{cc}) = \delta(\mathfrak{S})$ ([3], 4.10), and id_x is the reflector from $(X, \mathfrak{S})$ to the reflexion $(X, \mathfrak{S}_{cc})$ in CCFIL ([3], 6.2 and 6.3).

A transfinite construction for $\mathfrak{S}_{cc}$ can be obtained in the following way: let $\mathfrak{S}_0$ be equal to $\mathfrak{S}$, and suppose that a screen $\mathfrak{S}_\xi$ is defined for all ordinals ξ less than α. Put

$$\mathfrak{S}'_\alpha = \bigcup\{\mathfrak{S}_\xi : \xi < \alpha\}$$

and let $\mathfrak{S}_\alpha$ be composed of the intersections of the proper chains in $\mathfrak{S}'_\alpha$. Then $\mathfrak{S}_\gamma = \mathfrak{S}_{\gamma+1}$ for some γ and $\mathfrak{S}_{cc} = \mathfrak{S}_\gamma$ (see [3], 4.11).

If $\mathfrak{S}$ is Riesz then the same holds for $\mathfrak{S}_{cc}$ ([3], 4.13). Further interesting properties of chain-complete screens can be found in [3].

References

1. H.L. Bentley, H. Herrlich and E. Lowen-Colebunders, "Convergence", *Journ. Pure Appl. Algebra* **68**, (1990), 27-45.
2. Á. Császár, "Simultaneous extensions of screens", *Coll. Math. Soc. J. Bolyai* **55**, (1990), 285-300.
3. Á. Császár, "Chain-complete screens", *Acta Math. Hung.* (in print).
4. Á. Császár and J. Deák, "Simultaneous extensions of proximities, semi-uniformities, contiguities and merotopies I", *Math. Pannon* **1** N. 2, (1990), 67-90.
5. J. Deák, "Extending a family of Cauchy structures in a limit space I-II", *Studia Sci. Math. Hung.* (in print).

Epis in the Category of Pairwise-T_2 Spaces

J. Schröder *

University of the North
Department of Mathematics
Phuthaditjhaba 9866, South Africa

Abstract This note studies the subcategory of pairwise-T_2 spaces in the category of bitopological spaces. Epimorphisms are characterized by means of an additive closure operator $2h$. Non-cowellpoweredness is shown and properties of $2h$ are investigated.

AMS Subject classification: 54 E 55, 54 B 30, 54 A 05, 18 B 30.

Keywords: bitoplogical space, bipairwise pairwise T_2, separation axiom, epimorphism, closure operator, cowellpowered.

Introduction: Let **2Top** denote the category of bitopological spaces and bicontinuous maps, whose objects are triples $(X, \mathcal{X}_1, \mathcal{X}_2)$, where X is a set and $\mathcal{X}_1, \mathcal{X}_2$ are topologies an X. Morphisms in **2Top** are maps $f : X \to Y$ such that $f : (X, \mathcal{X}_1) \to (Y, \mathcal{Y}_1)$ and $f : (X, \mathcal{X}_2) \to (Y, \mathcal{Y}_2)$ are both continuous. **2Top** is a topological category. Initial and final structures (e.g. products, quotients, coproducts, subobjects, etc) are obtained by performing the corresponding constructions in **Top** for each component topology separately. Bitopological spaces gained importance in the study of non-symmetric topological structures (e.g. quasiuniformities) and in Categorical Topology because of subcategories with surprising properties (see [AR88]). The purpose of this paper is to provide information about another class of bitopological spaces, which could be used by categorical topologists. Pairwise-T_2 spaces were introduced by Salbany [Sal76] in 1976. The corresponding (quotient reflective) subcategory of **2Top** will be denoted by $\mathbf{pwT_2}$.

Definition 1 *A bitopological space $(X, \mathcal{X}_1, \mathcal{X}_2)$ is called pairwise -T_2 if for every pair of distinct points $x, y \in X$ there are disjoint $O \in \mathcal{X}_1, U \in \mathcal{X}_2$ such that $x \in O, y \in U$ or $y \in O, x \in U$.*

*Grants from the University of the North are acknowledged, despite incompetent management. Parts of this paper was written at the University of Cape Town

Eraldo Giuli (ed.), Categorical Topology, 245–248.

Remark 2

a) Please note that this is a rather weak separation axiom. It only implies that $\mathcal{X}_1, \mathcal{X}_2$ are T_0 topologies. For instance the upper- and lower-topology on the set of real numbers $\mathbf{R}$ define a pairwise-T_2 bitopological space.

b) To avoid unnecessary confusing indices, this paper adopts the following convention: Sets O, V are always elements of the first topology, U, W are open in the second topology. If the closure of an open set is taken, then it is always the closure in the other topology , e.g. $\overline{U}$ is the closure taken in the first topology of the open set U from the second topology.

Definition 3 *Let* $(X, \mathcal{X}_1, \mathcal{X}_2)$ *be a bitopological space. Define a closure operator* $2h : \mathcal{P}(X) \to \mathcal{P}(X)$ *by* $x \in 2h(D) \Leftrightarrow \forall_{O,U}\, x \in O\ capU \Rightarrow ((\overline{O} \cap U) \cup (O \cap \overline{U})) \cap D \neq \emptyset$. $2h^\infty$ *is the idempotent hull of* h.

Remark 4 $x \in 2h(D) \Leftrightarrow (\forall_{O,U}\, x \in O \cap U \Rightarrow \overline{O} \cap U \cap D \neq \emptyset) \vee (\forall_{O,U}\, x \in O \cap U \Rightarrow O \cap \overline{U} \cap D \neq \emptyset)$.

Proof: easy. □

Theorem 5 *Let* $f : (X, \mathcal{X}_1, \mathcal{X}_2) \to (Y, \mathcal{Y}_1, \mathcal{Y}_2)$ *be a morphism in* $\mathbf{pwT_2}$. *f is epimorphism if and only if* $2h^\infty(f[X]) = Y$.

Proof: " $\Leftarrow$ " assume we can find $\alpha, \beta : (Y, \mathcal{Y}_1, \mathcal{Y}_2) \to (Z, \mathcal{Z}_1, \mathcal{Z}_2)$ such that $\alpha \circ f = \beta \circ f$ and $\alpha \neq \beta$. Fetch $y \in Y$ fulfilling $\alpha(y) \neq \beta(y)$. We can find disjoint V, W separating $\alpha(y)$ and $\beta(y)$. For $O = \alpha^{-1}[V]$, $U = \beta^{-1}[W]$ we have $y \in O \cap U$ and $(\overline{O} \cap U \cup O \cap \overline{U}) \cap Eq(\alpha, \beta) = \emptyset$, because of $\overline{V} \cap W = V \cap \overline{W} = \emptyset$; where $Eq(\alpha, \beta)$ is the equalizer set of α and β. This shows $2h^\infty(Eq(\alpha, \beta)) = Eq(\alpha, \beta)$ and $Eq(\alpha, \beta) \neq Y$ is a contradiction.
" $\Rightarrow$ " assume $2h^\infty(f[X]) \neq Y$ and set $2h^\infty(f[X]) = C$. As usual form the pushout $Z\frac{Y \oplus Y}{C}$. To show: Z is pairwise-T_2. I shall discuss only the important case, where $x_0 = (x, 0), x_1 = (x, 1)$ lie above each other: 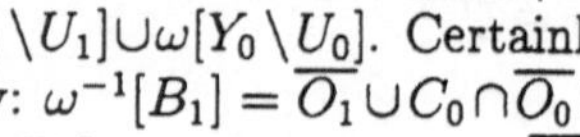 $= Z$
There is O, U such that $x \in O \cap U$ and $(\overline{O} \cap U \cup O \cap \overline{U}) \cap C = \emptyset$. Let $\omega : Y \oplus Y \to Z$ be the quotient map. In the sequel the subindex 0 (1) indicates a subset of the first (second) term in the coproduct $Y \oplus Y$. $V_1 := \omega[O_1] \cup \omega[Y_1 \setminus \overline{U_1}] \cup \omega[Y_0 \setminus \overline{U_0}]$ is an open neighbourhood of x_1, because $\omega^{-1}[V_1] = O_1 \cup (C_0 \cap O_0) \cup Y_1 \setminus \overline{U_1} \cup C_0 \cap (Y_0 \setminus \overline{U_0}) \cup Y_0 \setminus \overline{U_0} \cup C_1 \cap (Y_1 \setminus \overline{U_1}) = O_1 \cup Y_1 \setminus \overline{U_1} \cup Y_0 \setminus \overline{U_0}$. Note that $C_0 \cap O_0 \subseteq Y_0 \setminus \overline{U_0}$, because $C_0 \cap O_0 \cap \overline{U_0} = \emptyset$. Now look at $B_1 := \omega[\overline{O_1}] \cup \omega[Y_1 \setminus U_1] \cup \omega[Y_0 \setminus U_0]$. Certainly $V_1 \subseteq B_1$ and $x_0 \notin B_1$. B_1 is closed in the second topology: $\omega^{-1}[B_1] = \overline{O_1} \cup C_0 \cap \overline{O_0} \cup Y_1 \setminus U_1 \cup C_0 \cap (Y_0 \setminus U_0) \cup Y_0 \setminus U_0 \cup C_1 \cap (Y_1 \setminus U_1) = \overline{O_1} \cup Y_1 \setminus U_1 \cup Y_0 \setminus U_0$. Note that $C_0 \cap \overline{O_0} \subseteq Y_0 \setminus U_0$, because $C_0 \cap \overline{O_0} \cap U_0 = \emptyset$. This yields $x_1 \in V_1, x_0 \in Z \setminus B_1, V_1 \cap Z \setminus B_1 = \emptyset$. □

Remark 6 The above proof applies to the stronger bipairwise-T_2 axiom as well. This axiom is fulfilled if for every pair of distinct points x, y the first point possesses a neighbourhood in the first and the second point a neighbourhood in the second topology which are disjoint. The reason is the symmetric pushout construction. This phenomenon is not unique but also appears in the realm of T_0 and T_1 spaces. The b-closure used in T_1 spaces renders epimorphisms surjective.

Lemma 7 *Let $(X, \mathcal{X}_1, \mathcal{X}_2)$ be pairwise-T_2, $D \subseteq X$ infinite. Then $|2h(D)| \leq 2^{2^{|D|}}$.*

Proof: Decompose $2h(D)$ into two parts (which can overlap) by means of Lemma 4: $2h_1(D) = \{x | x \in O \cap U \Rightarrow D \cap \overline{O} \cap U \neq \emptyset\}$ and $2h_2(D) = \{y | y \in O \cap U \Rightarrow D \cap O \cap \overline{U} \neq \emptyset\}$. Then $h(D) = h_1(D) \cup h_2(D), D \subseteq h_1(D) \cap h_2(D)$. Take $a, b \in h_1(D), a \neq b$. There are without loss of generality disjoint $V, W; a \in V, b \in W$ separating a and b. We get $D \cap \overline{V} \cap X \neq \emptyset, D \cap \overline{X} \cap W \neq \emptyset$ and $\emptyset = D \cap \overline{V} \cap W = (D \cap \overline{V} \cap X) \cap (D \cap \overline{X} \cap W)$. Observe now that $\{D \cap \overline{O} \cap U | O \cap U \ni x\}$ is a filterbase on D. If $a \neq b$ then a and b generate different filterbases on D . Hence $|2h_1(D)| \leq 2^{2^{|D|}}$ and similarly $|2h_2(D)| \leq 2^{2^{|D|}}$, too. □

Theorem 8 $\mathbf{pwT_2}$ *is not cowellpowered.*

Proof: I make use of the spaces Y_β (which were used in the category **Ury**) to define a pairwise-T_2 space $(Y_\beta, \mathcal{Y}_\beta^1, \mathcal{Y}_\beta^2)$. $\mathcal{Y}_\beta^1$ is the topology taken from Y_β in [Sch83] without any change. $\mathcal{Y}_\beta^2$ is coarser than $\mathcal{Y}_\beta^1$: $U \in \mathcal{Y}_\beta^2 \Leftrightarrow \forall x \in U \exists O \in \mathcal{Y}_\beta^1 \ x \in O \wedge cl_{\mathcal{Y}_\beta^1}(O) \subseteq U$. In fact, this is the topology on Y_β generated by the idempotent hull of the θ-closure. In this specific case $(Y_\beta, \mathcal{Y}_\beta^1, \mathcal{Y}_\beta^2)$ is a pairwise-T_2 space Since $\overline{O} \supseteq cl_{\mathcal{Y}_\beta^1}(O)$ and for every $x \in O \cap U$ we can find V such that $x \in V$ and $cl_{\mathcal{Y}_\beta^1}(V) \subseteq U$ the results about θ-closure apply.□

Remark 9

a) $2h$ is additive, but not weakly hereditary. Take the example 1.3 (b) from [DG86] and as second topology the subspace topology with respect to the euclidean plane.

b) If $(X, \mathcal{X}_1, \mathcal{X}_2)$ is pairwise-T_2, then the diagonal $\Delta \subseteq X \times X$ is $2h$-closed. The reverse is problematic. This already follows from Remark 6, since one topological diagonal theorem cannot characterize two different categories. Here is the necessary specific counterexample: Take the euclidean plane $\mathbf{R} \times \mathbf{R}$ and a dense subset $H \subseteq \mathbf{R} \times \mathbf{R}$ such that both projections restricted to H are injective. You can easily obtain such a set through rotating the rational points $\mathbf{Q} \times \mathbf{Q}$ by a suitable angle around the origin. (There are uncountably many angles, but only countably many points in $\mathbf{Q} \times \mathbf{Q}$.) Let $(a, b) \subseteq \mathbf{R}$ be any interval. Then the first topology on H is generated by sets of the form $H \cap ((a, b) \times \mathbf{R})$, whereas the second by sets $H \cap (\mathbf{R} \times (a, b))$. Both topologies are T_2 in their own right. This implies that the diagonal in $H \times H$ is $2h$-closed. On the other hand, each non-empty open set of one topology intersects every non-empty open set of the other.

c) Starting with a pairwise-T_2 space X, what kind of space is $(X, 2h)$ (i.e. the space with $2h$-closed sets as a base for the closed sets)? Using ideas from [DG84;3.3,3.4], $(X, 2h)$ cannot always be T_2, otherwise $\mathbf{pwT_2}$ would be cowellpowered. For the same reason $2h$ generally does not commute with the product (otherwise the diagonal would be closed and $(X, 2h)$ is T_2). It is easy to show that $(X, 2h)$ is always T_1. $(Y_\beta, 2h)$ (see the proof of Th. 8) witnesses for a space which is not T_2.

References

[Sal76] Salbany, S: Bitopological spaces, compactifications and completions, Math Monographs Univ of Cape Town 1 (1976)

[Sch83] Schröder, J: The category of Urysohn spaces is not cowellpowered, Top Appl 16 (1983) 237-241

[DG84] Dikranjan, D & E Giuli: Epimorphisms and co-(well-po weredness) of epireflective subcategories of Top, Rend Circolo Mat Palermo 6 (19 84) 121-136

[DG86] Dikranjan, D & E Giuli: Closure operators induced by topological epireflections, Coll Math Soc J Bolyai 41 (1986) 233-246

[AR88] Adamek, J & J Rosicky: Intersections of reflective s ubcategories, Proc Am Math Soc 103,3 (1988) 710-712

Clone segments in Top and in Unif

Věra Trnková: TRNKOVA@karlin.MFF.CUNI.CZ

Abstract. Continuous and uniformly continuous maps of finite powers of metric spaces are investigated, e.g. for every $0 \leq m \leq n \leq \infty$, a metric space X is constructed such that the category of all continuous maps of the spaces $X^0 = \{\emptyset\}$, $X^1 = X$, $X^2 = X \times X, \ldots, X^k$ and the category of all their uniformly continuous maps are:

equal	exactly when $k \leq$ m and
isomorphic	exactly when $k \leq n$.

1992 Mathematics Subject Classification: 08A40, 54C05.

Key Words: Clones, products, continuous maps, uniformly continuous maps.

1. Preliminaries

Let us recall that an *algebraic theory* (in the sense of Lawvere) with a generic object a is a category k with finite products such that the set of all objects of k is precisely the set $\{a^i \mid i = 0, 1, 2, \ldots\}$ of all finite powers of a, $a^i \neq a^j$ for $i \neq j$. Its *n-segment* k_n is its full subcategory generated by $\{a^i \mid i = 0, 1, \ldots, n\}$. A *model* of k is any finite products preserving functor $F : k \to$ Set, Set being the category of all sets and all maps. A *clone* is a pair (k, F), where k is an algebraic theory and F its *faithful* model. An *n-segment* of a clone (k, F) is a pair (k_n, F_n) where k_n is an n-segment of k and F_n is the domain restriction of F to it.

If (k, F), (k', F') are clones or clone segments, we say that they are isomorphic (or that (k, F) can be fully embedded into (k', F')) if the categories k, k' are isomorphic (or k can be fully embedded into k'), isomorphism is denoted by $\simeq$; the equality

$$(k, F) = (k', F')$$

means that there exists an isofunctor Φ of k onto k' such that $F' \circ \Phi = F$.

If $\mathcal{K}$ is a concrete category (over Set) with finite products preserved by its forgetful functor, then every its object a determines a clone in the evident way, let us denote it by Clo(a) (or Clo($a, \mathcal{K}$) if necessary).

If X is a metric (or uniform) space, let us denote by Clo(X, Top) its clone of continuous maps and by Clo(X, Unif) its clone of uniformly continuous maps (and by $\text{Clo}_n(X, \text{Top})$, $\text{Clo}_n(X, \text{Unif})$ their n-segments).

Financial support of the Grant Agency of the Czech Republic under the grant no 201/93/0950 and of the Grant Agency of the Charles University under the grant GAUK 349 is gratefully acknowledged.

Eraldo Giuli (ed.), Categorical Topology, 249–268.

Clones are extensively investigated in universal algebra, see e.g. [8]. The present investigations were inspired by the monograph [10], namely by the problem to find topological spaces X, Y such that $\mathrm{Clo}_1(X) \simeq \mathrm{Clo}_1(Y)$ but $\mathrm{Clo}(X) \not\simeq \mathrm{Clo}(Y)$. This was solved in [11]. The method of [11] admits modifications which give further results. For instance, in [12], for every natural number n, a metric space X is constructed such that $\mathrm{Clo}_n(X, \mathrm{Top}) = \mathrm{Clo}_n(X, \mathrm{Unif})$ but $\mathrm{Clo}_{n+1}(X, \mathrm{Top}) \not\simeq \mathrm{Clo}_{n+1}(X, \mathrm{Unif})$. In the present paper, we develop the methods of [9], [11] and [12] and we enrich them by some further reasoning to get stronger, more general and more complex results. The contents of the paper is described in the next paragraph.

2. The Main Results

2.1. Let us investigate two metrics ϱ_1, ϱ_2 on a set P, let $X_i = (P, \varrho_i)$, $i = 1, 2$. Let us put

$$\begin{aligned} c \; &= \sup\{n \mid \mathrm{Clo}_n(X_1, \mathrm{Top}) = \mathrm{Clo}_n(X_2, \mathrm{Top})\}\,, \\ u \; &= \sup\{n \mid \mathrm{Clo}_n(X_1, \mathrm{Unif}) = \mathrm{Clo}_n(X_2, \mathrm{Unif})\}\,, \\ s_i &= \sup\{n \mid \mathrm{Clo}_n(X_i, \mathrm{Top}) = \mathrm{Clo}_n(X_i, \mathrm{Unif})\}\,, \quad i = 1, 2\,. \end{aligned}$$

In 2.5 below, we present a simple proof of the necessary conditions for c, u, s_1, s_2, namely

$$(*) \qquad \begin{cases} c \neq u & \Rightarrow \min(s_1, s_2) \leq \min(u, c) \text{ and} \\ s_1 \neq s_2 & \Rightarrow \min(u, c) \leq \min(s_1, s_2)\,. \end{cases}$$

The necessary conditions $(*)$ are also sufficient in the following strong sense.

Theorem 1. *Let c, u, s_1, s_2 are elements of $\{0, 1, \ldots, \infty\}$ which satisfy the conditions $(*)$. Then there exist metric spaces $X_1 = (P, \varrho_1)$, $X_2 = (P, \varrho_2)$ such that all the statements below are fulfilled:*

if $n \leq c$, then	$\mathrm{Clo}_n(X_1, \mathrm{Top}) = \mathrm{Clo}_n(X_2, \mathrm{Top})$ *but*
if $n > c$, then	*neither* $\mathrm{Clo}_n(X_1, \mathrm{Top})$ *is fully embeddable into* $\mathrm{Clo}(X_2, \mathrm{Top})$, *nor* $\mathrm{Clo}_n(X_2, \mathrm{Top})$ *into* $\mathrm{Clo}(X_1, \mathrm{Top})$;
if $n \leq u$, then	$\mathrm{Clo}_n(X_1, \mathrm{Unif}) = \mathrm{Clo}_n(X_2, \mathrm{Unif})$ *but*
if $n > u$, then	*neither* $\mathrm{Clo}_n(X_1, \mathrm{Unif})$ *is fully embeddable into* $\mathrm{Clo}(X_2, \mathrm{Unif})$, *nor* $\mathrm{Clo}_n(X_2, \mathrm{Unif})$ *into* $\mathrm{Clo}(X_1, \mathrm{Unif})$;
if $n \leq s_i$, then	$\mathrm{Clo}_n(X_i, \mathrm{Top}) = \mathrm{Clo}_n(X_i, \mathrm{Unif})$ *but*
if $n > s_i$, then	*neither* $\mathrm{Clo}_n(X_i, \mathrm{Top})$ *is fully embeddable into* $\mathrm{Clo}(X_i, \mathrm{Unif})$, *nor* $\mathrm{Clo}_n(X_i, \mathrm{Unif})$ *into* $\mathrm{Clo}(X_i, \mathrm{Top})$, *for $i = 1, 2$.*

2.2. If $\mathrm{Clo}_n(X, \mathrm{Unif}) = \mathrm{Clo}_n(X, \mathrm{Top})$ and $\mathrm{Clo}_{n+1}(X, \mathrm{Unif}) \neq \mathrm{Clo}_{n+1}(X, \mathrm{Top})$, is then necessarily $\mathrm{Clo}_{n+1}(X, \mathrm{Unif}) \not\simeq \mathrm{Clo}_{n+1}(X, \mathrm{Top})$? NO, as it is stated in

Theorem 2. *For every $m, n \in \{0, 1, \ldots, \infty\}$, $m \leq n$, there exists a metric space X such that simultaneously*

$$m = \sup\{k \mid \mathrm{Clo}_k(X, \mathrm{Top}) = \mathrm{Clo}_k(X, \mathrm{Unif})\} \text{ and}$$
$$n = \sup\{k \mid \mathrm{Clo}_k(X, \mathrm{Top}) \simeq \mathrm{Clo}_k(X, \mathrm{Unif})\}.$$

Remark. In fact, we prove stronger statement about n:

if $k \leq n$, then $\mathrm{Clo}_k(X, \mathrm{Top}) \simeq \mathrm{Clo}_k(X, \mathrm{Unif})$ but
if $k > n$, then neither $\mathrm{Clo}_k(X, \mathrm{Top})$ is fully embeddable into $\mathrm{Clo}(X, \mathrm{Unif})$, nor $\mathrm{Clo}_k(X, \mathrm{Unif})$ into $\mathrm{Clo}(X, \mathrm{Top})$.

The possible connection between non-isomorphism and mutual non-(full-embeddability) is not investigated in this paper. This problem seems to be not solvable by the methods of this paper, see paragraph 4. On the other hand, having a pair of metric spaces X_1, X_2 on a set, they determine 8 elements of $\{0, 1, \ldots, \infty\}$, [namely c, u, s_1, s_2 defined in 2.1 and $\tilde{c}$, $\tilde{u}$, $\tilde{s}_1$, $\tilde{s}_2$ defined analogously but equality $=$ replaced by isomorphism $\simeq$] and the author believes that the necessary and sufficient conditions of their interrelations *could* be solved by the present methods. However this more complex problem has not been attacked.

2.3. Let us describe briefly the contents of the paper. The proof of the necessity of the conditions $(*)$ is quite easy and it is presented in 2.4–5 below. On the other hand, the proofs of Theorem 1 and Theorem 2 are rather involved. These theorems are in fact applications of a general method for constructing metric spaces with some prescribed properties. The general method developed here is applied in the proofs of Theorem 1 and Theorem 2 presented in paragraph 6. The general method consists of a construction of metrics on initial Σ-algebras, where Σ is a finitary signature of (mono-sorted) universal algebras, whenever Σ has "enough" zero-ary operational symbols. The key of this method is the Main Theorem, formulated in paragraph 3 (but its topological proof is postponed into the last paragraph 7) and rather technical Main Lemma, deduced from it. By means of Main Lemma, we can work with continuous (or uniformly continuous) maps $X^m \to X$ of the constructed space X as with suitable Σ-terms in m variables. In paragraph 4, we introduce a notion of a rigid point in a concrete category and prove that the isomorphism and full embeddability coincide in clones and clone segments with all constants, the generic objects of which have at least three distinct rigid points. Since our constructed metric spaces have always at least $2^{\aleph_0}$ rigid points, it gives the results about "non-(full-embeddability)" in Theorem 1 and Theorem 2. In paragraph 5, we introduce the notion of a cell in a clone with constants, given in the form evidently invariant with respect to clone isomorphism. Its "internal characterization" for the spaces constructed by means of the Main Theorem allows partial recognition of the data Σ and κ, by means of which the space was originally constructed, from its clone. This is used in paragraph 6 in the proofs of Theorem 1 and Theorem 2, where the spaces with the required properties are *always* obtained from the Main Theorem by a specific choice of the "parameters" Σ and κ.

A simpler version of the general method presented here appears already in [9] (some special cases already in [ ]). In [9] and [11], clones and clone segments are investigated only in Top. On the other hand, the deeper applications of the method in [9] and in [11] give results also about elementary equivalences of clones and clone segments. This could be investigate here, in the more complex setting of clones in Top and in Unif, too. Thus, the general method described here admits also other applications than only Theorem 1 and Theorem 2, e.g. about elementary equivalences or, as the author believes, about the eight numbers c, u, s_1, s_2, $\tilde{c}$, $\tilde{u}$, $\tilde{s}_1$, $\tilde{s}_2$ mentioned in 2.2 and possibly some others.

2.4. Observation. Let $\mathcal{K}$ be a concrete category with finite products preserved by its forgetful functor, $a \in \operatorname{obj}\mathcal{K}$ (and let us suppose that $a^i \neq a^j$ whenever $i \neq j$). Clearly, $\mathrm{Clo}_n(a)$ is fully determined by the sets $\mathcal{K}(a^i, a)$, $i = 0, 1, \ldots, n$. Hence if $(\mathcal{K}_1, U_1)$ and $(\mathcal{K}_2, U_2)$ are concrete categories with finite products preserved by their forgetful functors U_1 and U_2 and if $a_1 \in \operatorname{obj}\mathcal{K}_1$, $a_2 \in \operatorname{obj}\mathcal{K}_2$ are objects such that

$$U_1(a_1) = U_2(a_2) = P \text{ and}$$
$$\mathrm{Clo}_n(a_1, \mathcal{K}_1) = \mathrm{Clo}_n(a_2, \mathcal{K}_2) \text{ but } \mathrm{Clo}_{n+1}(a_1, \mathcal{K}_1) \neq \mathrm{Clo}_{n+1}(a_2, \mathcal{K}_2),$$

then there exists a map $f : P^{n+1} \to P$ which carries a $\mathcal{K}_1$-morphism $a_1^{n+1} \to a_1$ but not a $\mathcal{K}_2$-morphism $a_2^{n+1} \to a_2$ or vice versa.

2.5. Let $X_1 = (P, \mathcal{U}_1)$, $X_2 = (P, \mathcal{U}_2)$ be two uniform spaces on a set P. Define c, u, s_1, s_2 as in 2.1. We show in the three lemmas below that the conditions (*) must be satisfied.

Lemma. *Let $c < u$. Then* $\min(s_1, s_2) \leq c$.

Proof. By 2.4, there exists a map $f : P^{c+1} \to P$ which is continuous as $X_1^{c+1} \to X_1$ but not continuous as $X_2^{c+1} \to X_2$ (or vice versa; we may suppose the first case). Since $c + 1 \leq u$, $f : X_1^{c+1} \to X_1$ cannot be uniformly continuous so that $s_1 \leq c$.

Lemma. *Let $u < c$. Then* $\min(s_1, s_2) \leq u$.

Proof. By 2.4, there exists a map $f : P^{u+1} \to P$ which is uniformly continuous as a map $X_1^{u+1} \to X_1$ but it is not uniformly continuous as a map $X_2^{u+1} \to X_2$ (or vice versa; we may suppose the first case). Hence $f : X_1^{u+1} \to X_1$ is continuous and, since $u + 1 \leq c$, f must be continuous also as a map $X_2^{u+1} \to X_2$, so that $s_2 \leq u$.

Lemma. *Let $s_1 < s_2$. Then* $\min(c, u) \leq s_1$.

Proof. By 2.4, there exists a continuous map $f : X_1^{s_1+1} \to X_1$ which is not uniformly continuous. Since $s_1 + 1 \leq s_2$, then $f : X_2^{s_1+1} \to X_2$ is either uniformly continuous, and then $u \leq s_1$, or it is not continuous, and then $c \leq s_1$.

3. The Main Theorem and the Main Lemma

3.1. Let X be a topological space, let $B \subseteq X$. We recall (see [11]) that the space X is called *B- semirigid* if every continuous selfmap $f : X \to X$ is either constant or the identity or it maps the whole X into B. Clearly, if X is B-semirigid and $X \setminus B \neq \emptyset$, then X must be connected.

Generalizing Herrlich's results in [3], [4] and observation in [10], the following proposition is proved in [11].

Proposition. *Let X be a topological space, $B \subseteq X$ and* $\operatorname{card}(X \setminus B) \geq 3$. *Let either α be a natural number or α be an arbitrary cardinal number and X be a Hausdorff space. If X is B-semirigid, then every continuous map $f : X^\alpha \to X$ is constant or it is a projection or it maps the whole X^α into B.*

We use this proposition below several times.

3.2. Let $n \geq 1$ be a natural number, let P be a set and $B \subseteq P$. Let us denote by

$$\begin{aligned} P^n[i,B] &= \{(x_0,\dots,x_{n-1}) \in P^n \mid x_i \in B\} \text{ for } i \in n\,, \\ P^n[i,j] &= \{(x_0,\dots,x_{n-1}) \in P^n \mid x_i = x_j\} \text{ for } i,j \in n,\ i \neq j \text{ and}\,, \\ P^n[i,c] &= \{(x_0,\dots,x_{n-1}) \in P^n \mid x_i = c\} \text{ for } i \in n,\ c \in P\,. \end{aligned}$$

We shall call the sets $P^n[i,B]$, $P^n[i,j]$, $P^n[i,c]$ and their subsets *small subsets of* P^n.

Clearly, if $n = 1 = \{0\}$, then $P^1[0,B] = B$, there are no sets $P^1[i,j]$ and $P^1[0,c] = \{c\}$.

Let $X = (P,\mathcal{U})$ be a uniform space, $B \subseteq P$ and let $f : P^n \to P$ be a map. Depending on the uniformity $\mathcal{U}$ and the set B, we define that

f is a suitable map of type 1 *(or type* 2 *or type* 3*)*

if it is one-to-one, its inverse $f(P^n) \to P^n$ is a uniformly continuous map of $(f(P^n),\mathcal{U}/f(P^n))$ onto X^n, the domain-restriction of f to any small subset of X^n (i.e. to any $P^n[i,B]$, $P^n[i,j]$, $P^n[i,c]$ for $i,j \in n$, $i \neq j$, $c \in P$) is uniformly continuous, but

type 1: f itself is not continuous;

type 2: f itself is continuous but not uniformly continuous;

type 3: f is uniformly continuous.

Remark. Let $X = (P,\mathcal{U})$ be a uniform space, let $m \geq n$ be natural numbers, $\psi : n \to m$ an injective map; let $g : P^m \to P^n$ be the map given by

$$g(z_0,\dots,z_{m-1}) = (z_{\psi(0)},\dots,z_{\psi(n-1)})\,.$$

If $f : X^n \to X$ is not (uniformly) continuous, then $f \circ g : X^m \to X$ is also not (uniformly) continuous.

3.3. In the rest of this paragraph, we use some basic notions of universal algebra. They can be found in any monograph about universal algebra, see e.g. [2] or [8]. We describe our notation briefly, now.

Let $\Sigma = \bigcup_{n=0}^{\infty} \Sigma_n$ be a finitary signature of (mono-sorted) universal algebras, i.e. Σ_n is a set of n-ary operational symbols. If $\sigma \in \Sigma_n$, we write $\operatorname{ar}\sigma = n$. Let

$\mathbf{P} = (P, \{p_\sigma \mid \sigma \in \Sigma\})$ be an initial Σ-algebra, i.e. an absolutely free Σ-algebra over the empty set of generators. As usual, if $\sigma \in \Sigma_0$, then the zero-ary operation p_σ is just an element of P. Denote

$$G_0 = \{p_\sigma \mid \sigma \in \Sigma_0\}, \quad B = P \setminus G_0 .$$

Then for every $\sigma \in \Sigma_n$ with $n \geq 1$, $p_\sigma : P^n \to P$ is a one-to-one map. Let us denote $B_\sigma = p_\sigma(P^n)$. Denote $\Gamma = \bigcup_{n=1}^{\infty} \Sigma_n = \Sigma \setminus \Sigma_0$. Then, as it is well-known (see e.g. [2]),

$$\begin{array}{rl} B = & \bigcup_{\sigma\in\Gamma} B_\sigma \quad \text{and } B_{\sigma_1} \cap B_{\sigma_2} = \emptyset \text{ whenever } \sigma_1, \sigma_2 \in \Gamma,\ \sigma_1 \neq \sigma_2 \\ P = & \bigcup_{k=0}^{\infty} G_k, \quad B_\sigma = \bigcup_{k=1}^{\infty} B_{\sigma,k} \text{ where} \\ & G_0 \text{ is as above,} \\ & G_{k+1} = G_k \cup \bigcup_{\sigma\in\Gamma} B_{\sigma,k+1} \text{ and} \\ & \text{for every } \sigma \in \Gamma \text{ with } \operatorname{ar}\sigma = n,\ p_\sigma \text{ maps } G_0^n \text{ onto } B_{\sigma,1} \text{ and} \\ & G_k^n \setminus G_{k-1}^n \text{ onto } B_{\sigma,k+1} \text{ for } k = 1, 2, \dots . \end{array}$$

3.4. Now, we are ready to formulate our

Main Theorem. *Let $\Sigma = \bigcup_{n=0}^{\infty} \Sigma_n$ be a finitary signature of universal algebras, $\Gamma = \Sigma \setminus \Sigma_0$. Let $(P, \{p_\sigma \mid \sigma \in \Sigma\})$ be an initial Σ-algebra, $G_0 = \{p_\sigma \mid \sigma \in \Sigma_0\}$, $B = P \setminus G_0$. If*

$$\operatorname{card} \Sigma_0 \geq 2^{\aleph_0} \cdot \operatorname{card} \Gamma ,$$

then for every map

$$\kappa : \Gamma \to \{1, 2, 3\}$$

there exists a metric ϱ_κ on the set P such that the metric space $X = (P, \varrho_\kappa)$

(a) is B-semirigid,

(b) if $\sigma_1, \sigma_2 \in \Gamma$, $\sigma_1 \neq \sigma_2$, then $\varrho_\kappa(B_{\sigma_1}, B_{\sigma_2}) = 1$ (where B_σ is as in 3.3*) and*

(c) for every $\sigma \in \Gamma$, the operation $p_\sigma : X^{\operatorname{ar}\sigma} \to X$ is a suitable map of type $\kappa(\sigma)$.

We postpone the proof of the Main Theorem into the paragraph 7. In the rest of this paragraph, we formulate the Main Lemma (see 3.6 below) and show that it is implied by the Main Theorem (3.7–3.10 below).

3.5. Let $\Sigma = \bigcup_{n=0}^{\infty} \Sigma_n$ be a finitary signature of universal algebras, $\Gamma = \Sigma \setminus \Sigma_0$, let $\mathbf{P} = (P, \{p_\sigma \mid \sigma \in \Sigma\})$ be an initial Σ-algebra, let G_0, B, G_k, B_σ, $B_{\sigma,k}$ be as in 3.3.

For every natural number m, the set $M^{(m)}$ of all maps $P^m \to P$ admits naturally a structure $\{\sigma^{(m)} \mid \sigma \in \Sigma\}$ of a Σ-algebra as follows:

for $\sigma \in \Sigma_0$, $\sigma^{(m)}$ is the constant map $P^m \to P$ with the value p_σ;
for $\sigma \in \Sigma_n$, $n \geq 1$, $\sigma^{(m)} : (P^m \to P)^n \to (P^m \to P)$ is defined by

$$\sigma^{(m)}(f_0, \dots, f_{n-1}) = p_\sigma \circ (f_0 \dot{\times} \dots \dot{\times} f_{n-1})$$

where the map $g = f_0 \dot{\times} \ldots \dot{\times} f_{n-1} : P^m \to P^n$ is defined by

$$g(z) = (f_0(z), \ldots, f_{n-1}(z)).$$

Hence the Σ-algebra $\mathbf{M} = (M^{(m)}, \{\sigma^{(m)} \mid \sigma \in \Sigma\})$ is just the direct power $\mathbf{P}^{P^m}$ of the initial algebra $\mathbf{P}$. Let

$$(H^{(m)}, \{\sigma^{(m)} \mid \sigma \in \Sigma\})$$

be the subalgebra of $\mathbf{M}$ (we denote the restrictions of the operations $\sigma^{(m)}$ to the subalgebra $H^{(m)}$ by $\sigma^{(m)}$ again) generated by the projections $\pi_i^{(m)} : P^m \to P$, $i \in m$ (i.e. $\pi_i^{(m)}(z_0, \ldots, z_{m-1}) = z_i$). Then $H^{(m)} = \bigcup_{k=0}^{\infty} H_k^{(m)}$ where

$$\begin{aligned} H_0^{(m)} &= \{\pi_0^{(m)}, \ldots, \pi_{m-1}^{(m)}\} \cup \{\sigma^{(m)} \mid \sigma \in \Sigma_0\}, \\ H_{k+1}^{(m)} &= H_k^{(m)} \cup \bigcup_{\sigma \in \Gamma} \{\sigma^{(m)}(f_0, \ldots, f_{\operatorname{ar}\sigma-1}) \mid f_i \in H_k^{(m)} \text{ for all } i \in \operatorname{ar}\sigma\}. \end{aligned}$$

For any $f : P^m \to P$, we denote by

$r(f)$ the least integer k for which $f(P^m) \cap G_k \neq \emptyset$.

It is easy to see that for $f \in H^{(m)}$,

$f \in H_0^{(m)}$ exactly when $r(f) = 0$

and that any $f = p_\sigma \circ (f_0 \dot{\times} \ldots \dot{\times} f_{n-1})$ with $\sigma \in \Sigma_n$, $n \geq 1$ and $f_0, \ldots, f_{n-1} \in H^{(m)}$, satisfies $r(f) > r(f_i)$ for all $i = 0, \ldots, n-1$. An inductive argument based on the "rank" $r(f)$ leads to the following (well-known) claim.

Statement. *$(H^{(m)}, \{\sigma^{(m)} \mid \sigma \in \Sigma\})$ is an absolutely free Σ-algebra over the set of generators $\{\pi_0^{(m)}, \ldots, \pi_{m-1}^{(m)}\}$.*

3.6. The standard construction of an absolutely free Σ-algebra over m variables $x_0, \ldots, x_{m-1}$ uses Σ-terms, introduced by the following inductive definition (see e.g. [2]):

every variable x_i, $i = 0, \ldots, m-1$, is a Σ-term and
every $\sigma \in \Sigma_0$ is a Σ-term;
if $\sigma \in \Sigma_n$, $n \geq 1$ and $t_0, \ldots, t_{n-1}$ are Σ-terms, then $\sigma(t_0, \ldots, t_{n-1})$ is a Σ-term;

(and there are no other Σ-terms than those obtained by finitely many applications of the above rules).

All the Σ-terms with the operations $\bar{\sigma}$, $\sigma \in \Sigma$, defined by

$$\begin{aligned} &\bar{\sigma} = \sigma \text{ if } \sigma \in \Sigma_0, \\ &\bar{\sigma}(t_0, \ldots, t_{n-1}) = \text{ the term } \sigma(t_0, \ldots, t_{n-1}) \text{ for } \sigma \in \Sigma_n, n \geq 1 \end{aligned}$$

are well-known to form the absolutely free Σ-algebra over $x_0, \ldots, x_{m-1}$. Since free algebras are unique up to isomorphisms, we get that there is a one-to-one correspondence λ of the set of all Σ-terms in variables $x_0, \ldots, x_{m-1}$ onto the set $H^{(m)}$ of maps $P^m \to P$ such that

$$\lambda(x_i) = \pi_i^{(m)} \text{ for all } i \in m$$
$$\lambda(\sigma) = \sigma^{(m)} \text{ for all } \sigma \in \Sigma_0$$
$$\lambda(\sigma(t_0, \ldots, t_{n-1})) = p_\sigma \circ (f_0 \dot{\times} \ldots \dot{\times} f_{n-1})$$

whenever $\sigma \in \Sigma_n$, $n \geq 1$ and $\lambda(t_i) = f_i$ for $i = 0, \ldots, n-1$.

Main Lemma. *Let $\Sigma = \bigcup_{n=0}^{\infty} \Sigma_n$ be a finitary signature, let $\mathbf{P} = (P, \{p_\sigma \mid \sigma \in \Sigma\})$, Γ, G_0, B, G_k, B_σ, $B_{\sigma,k}$ be as in 3.3. Let $\operatorname{card} \Sigma_0 \geq 2^{\aleph_0} \cdot \operatorname{card} \Gamma$. Let $\kappa : \Gamma \to \{1, 2, 3\}$, ϱ_κ, $X = (P, \varrho_\kappa)$ be as in the Main Theorem. Then for every natural number m, the set $C(X^m, X)$ (or $U(X^m, X)$) of all continuous (or uniformly continuous) maps of X^m into X is precisely the set of all maps f in $H^{(m)}$ for which the corresponding term $t = \lambda^{-1}(f)$ contains no subterm*

> *$\sigma(x_{\psi(0)}, \ldots, x_{\psi(n-1)})$ with $\sigma \in \Sigma_n$, $n \geq 1$ and $\kappa(\sigma) = 1$ (or $\kappa(\sigma) \in \{1, 2\}$) where $\psi : n \to m$ is a one-to-one map (i.e. $x_{\psi(0)}, \ldots, x_{\psi(n-1)}$ are* **distinct variables***).*

Remark. For $\kappa : \Gamma \to \{1, 3\}$, the statement about $C(X^m, X)$ has been proved in [9].

3.7. The proof of the Main Lemma is presented in 3.8–3.10 below. The proof for $C(X^m, X)$ is analogous to that for $U(X^m, X)$. To be able to speak about both the cases simultaneously, let us call the terms $\sigma(x_{\psi(0)}, \ldots, x_{\psi(n-1)})$ with $\sigma \in \Sigma_n$, $n \geq 1$ and $\psi : n \to m$ one-to-one

forbidden terms

if $\kappa(\sigma) = 1$ whenever we investigate $C(X^m, X)$ and
if $\kappa(\sigma) \in \{1, 2\}$ whenever we investigate $U(X^m, X)$.

3.8. First, we show that every map $f \in H^{(m)}$ corresponding to a Σ-term t containing no forbidden subterm $\sigma(x_{\psi(0)}, \ldots, x_{\psi(n-1)})$ is (uniformly) continuous. We show it by the induction on the depth $d(t)$ of t (we recall that $d(x_i) = 0$, $d(\sigma) = 0$ for $\sigma \in \Sigma_0$, $d(t) = 1 + \max\{d(t_0), \ldots, d(t_{n-1})\}$ whenever $t = \sigma(t_0, \ldots, t_{n-1})$): If $d(t) = 0$, then either $t = x_i$ or $t = \sigma \in \Sigma_0$; then f is either $\pi_i^{(m)}$ or the constant map $\sigma^{(m)}$ so that it is (uniformly) continuous. Let $t = \sigma(t_0, \ldots, t_{n-1})$, $\operatorname{ar} \sigma = n \geq 1$. Since t does not contain a forbidden subterm, none of the terms $t_0, \ldots, t_{n-1}$ contains a forbidden subterm, so that the maps $f_0, \ldots, f_{n-1}$ corresponding to them are (uniformly) continuous hence $g = f_0 \dot{\times} \ldots \dot{\times} f_{n-1} : X^m \to X^n$ is (uniformly) continuous. Since t itself is not a forbidden term, either $\kappa(\sigma)$ is in $\{2, 3\}$ (or $\kappa(\sigma) = 3$) so that $p_\sigma : X^n \to X$ is continuous (or uniformly continuous) hence $f = p_\sigma \circ g$ is continuous (or uniformly continuous) or $t_0, \ldots, t_{n-1}$ are not distinct variables. In the later case g maps X^m into a small subset of P^n (see 3.2 for the definitions of *small* subsets of P^n) and p_σ restricted to any small subset of X^n is uniformly continuous regardless the value of $\kappa(\sigma)$ (see the definition of a suitable map of any type in 3.2) so that $f = p_\sigma \circ g$ is (uniformly) continuous again.

3.9. Now, we show that if a Σ-term t contains a forbidden subterm, then the corresponding map f is **not** (uniformly) continuous. We prove it by induction in the depth $d(t)$ of t again.

a) If $d(t) = 1$, that means that $t = \sigma(x_{\psi(0)}, \dots, x_{\psi(n-1)})$ where $x_{\psi(0)}, \dots, x_{\psi(n-1)}$ are distinct variables, $\sigma \in \Sigma_n$ with $n \geq 1$ and $\kappa(\sigma) = 1$ (or $\kappa(\sigma) \in \{1, 2\}$), then $f = p_\sigma \circ (\pi^{(m)}_{\psi(0)} \dot{\times} \dots \dot{\times} \pi^{(m)}_{\psi(n-1)})$, hence $f = p_\sigma \circ g$ where g is given by the formula $g(z_0, \dots, z_{m-1}) = (z_{\psi(0)}, \dots, z_{\psi(n-1)})$, $\psi : n \to m$ is a one-to-one map. Then $f = p_\sigma \circ g$ is not (uniformly) continuous whenever p_σ is not (uniformly) continuous, by Remark 3.2.

b) Let $d(t) > 1$. Then $t = \sigma(t_0, \dots, t_{n-1})$ and one of the terms, say t_j, contains a forbidden subterm, so that, by the induction hypothesis, the corresponding map f_j is not (uniformly) continuous. Hence the map $g = f_0 \dot{\times} \dots \dot{\times} f_{n-1}$ of X^m into X^n is not (uniformly) continuous. However p_σ^{-1} is always (regardless the value of $\kappa(\sigma)$) a uniformly continuous map of $p_\sigma(X^n)$ onto X^n, see the definition of suitable maps in 3.2. If f is (uniformly) continuous, so is the composite $g = p_\sigma^{-1} \circ f$ — a contradiction.

3.10. Now, we show that every (uniformly) continuous map $f : X^m \to X$ corresponds to a unique Σ-term without a forbidden subterm. We recall that $r(f)$ is defined as the least integer k for which $f(P^m) \cap G_k \neq \emptyset$, see 3.5. We show, by induction in k, that

$f \in H_k^{(m)}$ and the corresponding Σ-term contains
no forbidden subterm

for every (uniformly) continuous f with $r(f) \leq k$.

For $k = 0$, we have $f(P^m) \cap G_0 \neq \emptyset$. Since $X = (P, \varrho_\kappa)$ is B-semirigid, the (uniformly) continuous map f is either a projection or a constant whose value lies in G_0, see 3.1. Thus $f \in H_0^{(m)}$ and the corresponding Σ-term is x_j (whenever $f = \pi_j^{(m)}$) or $\sigma \in \Sigma_0$ (whenever f is the constant map with the value $p_\sigma \in G_0$). Next assume that

$g \in H_k^{(m)}$ and the corresponding Σ-term does
not contain a forbidden subterm

for any (uniformly) continuous $g : X^m \to X$ with $r(g) \leq k$ and let $f : X^m \to X$ be (uniformly) continuous with $r(f) = k + 1$. Then $f(P^m) \cap G_0 = \emptyset$, hence $f(P^m) \subseteq B$, by 3.1 again. Since X is B-semirigid, it is connected so that $f(P^m)$ is also connected. Since the distance $\varrho_\kappa(B_{\sigma_1}, B_{\sigma_2}) = 1$ whenever $\sigma_1, \sigma_2 \in \Gamma$, $\sigma_1 \neq \sigma_2$ (see 3.4), necessarily there exists a unique $\sigma \in \Gamma$ such that $f(P^m) \subseteq B_\sigma$. Denote $n = \operatorname{ar} \sigma$. We put $f_j = \pi_j^{(n)} \circ p_\sigma^{-1} \circ f$ so that $f = p_\sigma \circ (f_0 \dot{\times} \dots \dot{\times} f_{n-1})$. Since $r(f) = k + 1$, there exists $c \in P^m$ such that $f(c) \in G_{k+1}$. But then $(f_0(c), \dots, f_{n-1}(c)) = p_\sigma^{-1}(f(c))$ is in $(G_k)^n$ — this follows from the definition of G_{k+1}, see 3.3. Hence $r(f_j) \leq k$ for all $j \in n$. Hence, by the induction hypothesis,

f_j is in $H_k^{(m)}$ and the corresponding term $t_j = \lambda^{-1}(f_j)$
contains no forbidden subterm.

Then f is in $H_{k+1}^{(m)}$ and $t = \lambda^{-1}(f) = \sigma(t_0, \dots, t_{n-1})$. Since f is (uniformly) continuous, the term t itself is not a forbidden subterm, by 3.9, whenever $k = 0$ (if $k > 0$, t does not contain a forbidden subterm because no t_i, $i = 0, \dots, n-1$, contains a forbidden subterm).

4. Rigid points in concrete categories, isomorphism and full embedding

In the paragraph 4, we introduce rigid points and show that if a clone (or clone-segment) (k, F) with all constants, the generic object a of which has at least three distinct rigid points, can be fully embedded into a clone (k', F'), then the full embedding (=full faithful functor) is already isomorphism of (k, F) onto (k', F') or onto its segment.

4.1. A concrete category $(\mathcal{K}, U)$ (over Set)

has all constants

if for every $a, b \in \operatorname{obj} \mathcal{K}$, every constant map $U(a) \to U(b)$ carries a $\mathcal{K}$-morphism $a \to b$.

Let $(\mathcal{K}, U)$ be a concrete category with all constants, $x \in U(a)$. We say that x is *a rigid point* of a (in $(\mathcal{K}, U)$) if for every $f \in \mathcal{K}(a, a)$ with $x \in \operatorname{Im} U(f)$ (where $\operatorname{Im} h$ denotes the image $h(X)$ for arbitrary map $h : X \to Y$) either $f = 1_a$ or $U(f)$ is the constant map with the value x.

If $(\mathcal{K}_1, U_1)$, $(\mathcal{K}_2, U_2)$ are concrete categories with all constants and $\Phi : \mathcal{K}_1 \to \mathcal{K}_2$ is a full embedding (i.e. a full and faithful functor), then it is easy to see (see e.g. [7] for semigroup formulation) that the functors U_1 and $U_2 \circ \Phi$ are naturally equivalent, denote by $\tau = \{\tau_a \mid a \in \operatorname{obj} \mathcal{K}_1\} : U_1 \to U_2 \circ \Phi$ the natural equivalence. Then the formula

$$U_2(\Phi(f)) = \tau_b \circ U_1(f) \circ \tau_a^{-1} \text{ for every } f \in \mathcal{K}_1(a, b)$$

implies easily all the statements below:

$$\begin{aligned} U_1(f) : U_1(a) \to U_1(b) \quad & \text{is one} - \text{to} - \text{one (or nonconstant or surjective)} \\ & \text{iff } U_2(\Phi(f)) \text{ has this property;} \\ \text{if } f, g \in \mathcal{K}_1(a, b), \quad & \text{then } \operatorname{Im} U_1(f) \cap \operatorname{Im} U_1(g) \neq \emptyset \text{ (or } \operatorname{Im} U_1(f) \subseteq \operatorname{Im} U_1(g)) \\ & \text{iff } \operatorname{Im} U_2(\Phi(f)) \cap \operatorname{Im} U_2(\Phi(g)) \neq \emptyset \\ & \text{(or } \operatorname{Im} U_2(\Phi(f)) \subseteq \operatorname{Im} U_2(\Phi(g))) \\ x \quad & \text{is a rigid point of } a \text{ in } (\mathcal{K}_1, U_1) \\ & \text{iff } \tau_a(x) \text{ is a rigid point of } \Phi(a) \text{ in } (\mathcal{K}_2, U_2). \end{aligned}$$

4.2. Let $(\mathcal{K}, U)$ be a category with finite concrete products (i.e. $\mathcal{K}$ has and U preserves finite products) and all constants.

Lemma A. *If* $b \in \operatorname{obj} \mathcal{K}$ *is isomorphic to* $a_1 \times a_2$ *with* $\operatorname{card} U(a_i) > 1$ *for* $i = 1, 2$, *then b has no rigid point.*

Proof. We may suppose that $b = a_1 \times a_2$, let $\pi_i : b \to a_i$ be the product projections. Let $x_i \in U(a_i)$ be arbitrary, $x = (x_1, x_2) \in U(b)$. We show that x is not a rigid point of b. In fact, we have $x \in U(g)$ for $g \in \mathcal{K}(b,b)$ where $[U(g)](y_1,y_2) = (y_1,x_2)$ is non-constant and non-identical.

Lemma B. *If $a \in \operatorname{obj}\mathcal{K}$ has at least 3 distinct rigid points, then for every natural number $n \geq 1$, every morphism $f \in \mathcal{K}(a^n,a)$ is either a product-projection or $U(f)$ is constant or $\operatorname{Im} U(f)$ contains no rigid point of a.*

Proof. Although not short, the proof is *quite* analogous to the proof of the Proposition in [11], quoted here in 3.1; therefore we omit it.

Corollary. If $a \in \operatorname{obj}\mathcal{K}$ has at least 3 distinct rigid points, then, for every natural number $n \geq 1$, the set $\mathcal{K}(a^n,a)$ contains precisely n distinct morphisms f with $U(f)$ surjective (namely the product projections).

4.3. Let (k_1,F_1) and (k_2,F_2) be clones or clone segments with all constants, let a_1 and a_2 denote their generic objects.

Proposition. *Let a_1 has at least three distinct rigid points. Let*

$$\Phi : k_1 \to k_2$$

be a full embedding. Then

$\Phi(a_1) = a_2$, $\Phi(a_1^n) = a_2^n$ *for all* a_1^n *in* k_1 *and*
for every n (with a_1^n in k_1) there is a permutation $\psi : n \to n$ such that
$\Phi({}_1\pi_i^{(n)}) = {}_2\pi_{\psi(i)}^{(n)}$,

where ${}_j\pi_i^{(n)} : a_j^n \to a_j$ are product projections, $j = 1,2$.

Proof. We have $\Phi(a_1^0) = a_2^0$ because a_i^0 is a unique object of k_i with the underlying set of cardinality 1. By 4.2, no a_2^n with $n > 1$ has a rigid point, hence $\Phi(a_1) = a_2$. Moreover, $\Phi(a_1^n) = a_2^n$ for $n > 1$, by Lemma B in 4.2. This lemma also implies the statement about ${}_1\pi_i^{(n)}$ and ${}_2\pi_i^{(n)}$.

Corollary. If k_1 admits a full embedding into k_2, then either (k_1,F_1) and (k_2,F_2) are isomorphic clones or (k_1,F_1) is isomorphic to a segment of the clone (or clone segment) (k_2,F_2).

5. Clone cells

5.1. Let (k,F) be a clone or a clone segment with a generic object a, let $n \geq 1$ be a natural number. A morphism $f \in k(a^n,a)$ is called

an *n-cell*

if it is

(a) *one-to-one* in the sense that $F(f)$ is a one-to-one map and $f \neq 1_a$ whenever $n = 1$;

(b) *indecomposable* in the sense that if $f = f_2 \circ f_1$ and $f_2 \in k(a,a)$ then

$$f_2 = 1_a \quad \text{whenever } n > 1,$$
$$f_2 = 1_a \text{ or } f_1 = 1_a \quad \text{whenever } n = 1;$$

(c) *maximal* in the sense that if $g \in k(a^n, a)$, g is not a projection and $\operatorname{Im} F(f) \cap \operatorname{Im} F(g) \neq \emptyset$ then $\operatorname{Im} F(g) \subseteq \operatorname{Im} F(f)$

(where Im denotes the image of the map).

Let (k_1, F_1) and (k_2, F_2) be clones or clone segments with all constants, a_i a generic object of (k_i, F_i), $i = 1, 2$, and let $\Phi : k_1 \to k_2$ be an isomorphism of k_1 onto k_2 or onto its segment with $\Phi(a_1^n) = a_2^n$. Then (see 4.1)

for every $n \geq 1$, $f \in k_1(a_1^n, a_1)$ is an n-cell iff $\Phi(f) \in k_2(a_2^n, a_2)$ is an n-cell.

5.2 The notion of a cell is a key notion for the proofs of Theorem 1 and Theorem 2. We have to give its "internal characterization" first. In the rest of paragraph 5, let Σ, $\Gamma = \Sigma \setminus \Sigma_0$, $\mathbf{P} = (P, \{p_\sigma \mid \sigma \in \Sigma\})$, G_0, G_k, B, B_σ, $B_{\sigma,k}$ be as in 3.3, let $\operatorname{card} \Sigma_0 \geq 2^{\aleph_0} \cdot \operatorname{card} \Gamma$ and let $\kappa : \Gamma \to \{1, 2, 3\}$, ϱ_κ, $X = (P, \rho_\kappa)$ be as in Main Theorem 3.4 and Main Lemma 3.6. Then every $p_\sigma \in G_0$ is a rigid point of X both in Top and Unif, so that X has at least three distinct rigid points and 4.2–4.4 can be applied.

5.3. Lemma. *Let (k, F) be a clone or clone segment of X in* Top *(or in* Unif*). Let $f = p_\sigma \circ h \in k(X^m, X)$, $\operatorname{ar} \sigma = n > 1$. Let $h(X^m)$ be a small subset of X^n (in the sense of 3.2). Then f is not maximal (in the sense of 4.4(c)).*

Proof. Let $h = h_0 \dot\times \ldots \dot\times h_{n-1}$. We choose $z \in G_0$ and put $g = p_\sigma \circ (g_0 \dot\times \ldots \dot\times g_{n-1})$ where $g_i = \pi_i^{(m)}$ and g_s is the constant map $X^m \to X$ with the value $h_s(z, \ldots, z)$ for all $s \in n$, $s \neq i$, whenever $h(X^m) \subseteq P^n[i, B]$ or $h(X^m) \subseteq P^n[i, j]$ or $h(X^m) \subseteq P^n[i, c]$ (where $P^n[i, B]$, $P^n[i, j]$, $P^n[i, c]$ are as in 3.2). Since $n > 1$, $\operatorname{Im}(g_0 \dot\times \ldots \dot\times g_{n-1})$ is a small subset of P^n so that p_σ is uniformly continuous on it (regardless $\kappa(\sigma)$), hence $g \in k(X^m, X)$. One can verify easily that $g(X^m) \cap f(X^m) \neq \emptyset$ but $g(X^m)$ is not a subset of $f(X^m)$.

5.4. Characterization lemma. *Let (k, F) be a clone or a clone segment of X in* Top *(or in* Unif*). Then $f \in k(X^n, X)$ is an n-cell iff*

$$f = p_{\sigma_1} \circ p_{\sigma_2} \circ \ldots \circ p_{\sigma_s} \circ p_\sigma \circ (\pi_{\psi(0)}^{(n)} \dot\times \ldots \dot\times \pi_{\psi(n-1)}^{(n)})$$

where $\psi : n \to n$ is a permutation, σ is in Σ_n and $\kappa(\sigma) \in \{2, 3\}$ (or $\kappa(\sigma) = 3$) and all $\sigma_1, \ldots, \sigma_s$ are in Σ_1 and $\kappa(\sigma_i) = 1$ (or $\kappa(\sigma_i) \in \{1, 2\}$) for $i = 1, \ldots, s$.

Remark. In the lemma, s could be zero so that the part $p_{\sigma_1} \circ \ldots \circ p_{\sigma_s}$ could be missing in the expression for f. The statement about κ says that p_σ is a morphism of k but none of p_{σ_i}, $i = 1, \ldots, s$, is a morphism of k. However $p_{\sigma_1} \circ \ldots \circ p_{\sigma_s} \circ p_\sigma$ (where the composition is in Set) is a morphism of k because every p_{σ_i} is uniformly continuous on small sets (see 3.2) and $\operatorname{Im}(p_{\sigma_{i+1}} \circ \ldots \circ p_{\sigma_s} \circ p_\sigma)$ is always a small set.

Proof. If $f \in k(X^n, X)$ has the form described in the Lemma, then it is easy to verify (a), (b), (c) in 5.1, hence f is an n-cell. (For the proof of (b), i.e. the indecomposability of f, we recall that none of $p_{\sigma_1}, \ldots, p_{\sigma_s}$ is a morphism of k and, for arbitrary $\tau, \tau' \in \Gamma$, $\tau \neq \tau'$, $\operatorname{Im} p_\tau$ is disjoint with $\operatorname{Im} p_{\tau'}$, see 3.3). We prove the converse. Thus, let $f \in k(X^n, X)$ be an n-cell. We proceed by the induction in $r(f)$ (see 3.5). The case $r(f) = 0$ is impossible. In fact, if $r(f) = 0$, then f is either a projection or a constant — this contradicts (a) in 5.1. Thus, let us suppose $r(f) > 0$, i.e. $f = p_\tau \circ h$ where $\operatorname{ar} \tau = m$, $h = h_0 \dot{\times} \ldots \dot{\times} h_{m-1}$. We discuss the following possibilities:

α) $m = 1$: hence $h = h_0$.

α_1) if $p_\tau \in k(X, X)$, then necessarily $h = 1_X$, by (b) in 5.1, hence $f = p_\tau$ and $\kappa(\tau) \in \{2,3\}$ (or $\kappa(\tau) = 3$), so that f has the required form.

α_2) if p_τ is not in $k(X, X)$ (i.e. $\kappa(\tau) = 1$ in the case of Top or $\kappa(\tau) \in \{1,2\}$ in the case of Unif), we use the induction hypothesis on h (the verification that h is an n-cell again is quite easy).

β) $m > 1$: if $\operatorname{Im} h$ is a small subset of X^m (in the sense of 3.2), then f is not maximal, see 5.3 — a contradiction. Since X is B-semirigid (see Main Theorem 3.4), every $h_j : X^n \to X$ is either a constant or a projection or $\operatorname{Im} h_j \subseteq B$ so that $\operatorname{Im} h$, being not a small subset of X^m, is necessarily the whole X^m. This is possible only when $m \leq n$ and there is a one-to-one map $\psi : m \to n$ such that $h_j = \pi^{(n)}_{\psi(j)}$ for all $j \in m$. If ψ is not surjective, then h is not one-to-one, hence $f = p_\tau \circ h$ is not one-to-one — a contradiction. Hence $\psi : n \to n$ is a permutation. Since $f \in k(X^n, X)$, necessarily $\kappa(\tau) \in \{2,3\}$ (or $\kappa(\tau) = 3$) so that $f = p_\tau \circ (\pi^{(n)}_{\psi(0)} \dot{\times} \ldots \dot{\times} \pi^{(n)}_{\psi(n-1)})$ has the required form.

5.5. Remark. Let $f, f' \in k(X^n, X)$ be n-cells. We say that they are equivalent if $\operatorname{Im} f = \operatorname{Im} f'$.

Let (k_i, F_i) be a clone or a clone segment of a space X_i in Top or in Unif, X_i as in the Main Theorem 3.4, $i = 1, 2$. Let $\Phi : (k_1, F_1) \to (k_2, F_2)$ be a full embedding. If $f, f' \in k_1(X_1^n, X_1)$ are equivalent n-cells, then $\Phi(f), \Phi(f') \in k_2(X_2^n, X_2)$ are also equivalent n-cells, by 4.1–4.3 and 5.1. We shall work with equivalent classes of n-cells. The distinct number of them will serve as a tool for the proof of the non-(full-embeddability). We shall choose the representants with the permutation $\psi : n \to n$ being the identity. Hence the distinct number of maps of the form

$$p_{\sigma_1} \circ \ldots \circ p_{\sigma_s} \circ p_\sigma$$

[where $\sigma \in \Sigma_n$ and $\kappa(\sigma) \in \{2,3\}$ (or $\kappa(\sigma) = 3$) and all $p_{\sigma_1}, \ldots, p_{\sigma_s}$ in Σ_1 with $\kappa(\sigma_i) = 1$ (or $\kappa(\sigma_i) \in \{1,2\}$) for all $i = 1, \ldots, s$] in $k_1(X_1^n, X_1)$ and in $k_2(X_2^n, X_2)$ will demonstrate the non-full-embeddability of $\operatorname{Clo}_m(X_1, \mathcal{K}_1)$ into $\operatorname{Clo}_m(X_2, \mathcal{K}_2)$ whenever $m \geq n$ [where $\mathcal{K}_i$ is either Top or Unif, and (k_i, F_i) is the clone or clone segment of X_i in $\mathcal{K}_i$, $i = 1, 2$]. The proof of Theorem 2 and Theorem 1 in the paragraph below proceeds as follows. We choose $\Sigma = \Sigma_0 \cup \Gamma$ and $\kappa_1, \kappa_2 : \Gamma \to \{1, 2, 3\}$ so that the spaces $X_i = (P, \varrho_{\kappa_i})$, $i = 1, 2$, from the Main Theorem have the required properties. The required equality

$$\operatorname{Clo}_m(X_1, \mathcal{K}_1) = \operatorname{Clo}_m(X_2, \mathcal{K}_2)$$

is shown by means of the equality of the sets of the corresponding Σ-terms in $H^{(m)}$ (i.e. we use the Main Lemma 3.6); the required non-isomorphism (or non-full-embeddability) of $\mathrm{Clo}_m(X_1, \mathcal{K}_1)$ and $\mathrm{Clo}_m(X_2, \mathcal{K}_2)$ is proved by the fact that for some $s \le m$ there are distinct numbers of classes of equivalent s-cells in $k_1(X_1^s, X_1)$ and $k_2(X_2^s, X_2)$.

6. Proof of Theorem 1 and Theorem 2

6.1. Proof of Theorem 1. Let c, u, s_1, s_2 in $\{0, 1, \dots\}$ be elements satisfying (*) (if some of them is ∞, the proof is easier and we leave it to the reader). We proceed as described in Remark 5.5. We choose

$$\Sigma = \Sigma_0 \cup \Gamma \text{ where } \operatorname{card} \Sigma_0 = 2^{\aleph_0},$$

the set Γ and the maps $\kappa_1, \kappa_2 : \Gamma \to \{1, 2, 3\}$ are chosen as stated below, depending on the mutual configurations of the numbers c, u, s_1, s_2. In all the cases, $X_1 = (P, \varrho_{\kappa_1})$, $X_2 = (P, \varrho_{\kappa_2})$ are the spaces as in the Main Theorem. We discuss the following cases:

1) $s_1 = s_2$:
   - 1,1) $c = u$: we choose $\Gamma = \{\sigma, \delta\}$, $\operatorname{ar}\sigma = s_1 + 1$, $\operatorname{ar}\delta = c + 1$, $\kappa_1(\sigma) = 2 = \kappa_2(\sigma)$, $\kappa_1(\delta) = 1$, $\kappa_2(\delta) = 3$;
   - 1,2) $c < u$: (*) implies $s_1 \le c$; we choose $\Gamma = \{\sigma, \gamma, \delta\}$, $\operatorname{ar}(\sigma) = s_1 + 1$, $\operatorname{ar}(\gamma) = c + 1$, $\operatorname{ar}(\delta) = u + 1$; $\kappa_1(\sigma) = 2 = \kappa_2(\sigma)$, $\kappa_1(\gamma) = 1$, $\kappa_2(\gamma) = 2$, $\kappa_1(\delta) = 2$, $\kappa_2(\delta) = 3$;
   - 1,3) $u < c$: (*) implies $s_1 \le u$; we choose Γ and κ as in the previous case;

2) $s_1 < s_2$:
   - 2,1) $c = u$: (*) implies $c \le s_1$; we choose $\Gamma = \{\sigma_1, \sigma_2, \gamma\}$, $\operatorname{ar}\sigma_i = s_i + 1$, $\operatorname{ar}\gamma = c + 1$ and $\kappa_i(\sigma_i) = 2$, $\kappa_2(\sigma_1) = 3$, $\kappa_1(\sigma_2)$ arbitrary, $\kappa_1(\gamma) = 1$, $\kappa_2(\gamma) = 3$;
   - 2,2) $c < u$: (*) implies $c = s_1$; if $u = s_2$, we choose $\Gamma = \{\sigma_1, \sigma_2\}$, $\operatorname{ar}\sigma_i = s_i + 1$ and $\kappa_i(\sigma_i) = 2$, $\kappa_2(\sigma_1) = 1$, $\kappa_1(\sigma_2) = 3$; if $u \ne s_2$, we choose $\Gamma = \{\sigma_1, \sigma_2, \delta\}$, $\operatorname{ar}\sigma_i = s_i + 1$, $\operatorname{ar}\delta = u + 1$ and $\kappa_i(\sigma_i) = 2$, $\kappa_2(\sigma_1) = 1$, $\kappa_1(\sigma_2) = 2$, $\kappa_1(\delta) = 3$, $\kappa_2(\delta) = 1$;
   - 2,3) $u < c$: (*) implies $u = s_1$; if $c = s_2$, we choose $\Gamma = \{\sigma_1, \sigma_2\}$, $\operatorname{ar}\sigma_i = s_i + 1$, $\kappa_i(\sigma_i) = 2$, $\kappa_1(\sigma_2) = 1$, $\kappa_2(\sigma_1) = 3$; if $c \ne s_2$, we choose $\Gamma = \{\sigma_1, \sigma_2, \gamma\}$, $\operatorname{ar}\sigma_i = s_i + 1$, $\operatorname{ar}\gamma = c + 1$ and $\kappa_i(\sigma_i) = 2$, $\kappa_1(\sigma_2) = 2$, $\kappa_2(\sigma_1) = 3$, $\kappa_1(\gamma) = 1$, $\kappa_2(\gamma) = 3$.

3) $s_2 < s_1$: this is analogous to the case 2).

We shall not count the sets of Σ-terms and the number of equivalent classes of cells in all the above cases. We show it only in the case 1,1), the rest of the long routine counting is left to the reader. Moreover, if we write $\mathrm{Clo}_m(X_1, \mathrm{Top}) \not\simeq \mathrm{Clo}_m(X_2, \mathrm{Top})$, this actually means, by 4.3, that neither $\mathrm{Clo}_m(X_1, \mathrm{Top})$ can be fully embedded into $\mathrm{Clo}(X_2, \mathrm{Top})$ nor $\mathrm{Clo}_m(X_2, \mathrm{Top})$ into $\mathrm{Clo}(X_1, \mathrm{Top})$ and analogously for $\mathrm{Clo}_m(X_1,\mathrm{Unif}) \not\simeq \mathrm{Clo}_m(X_2,\mathrm{Unif})$, $\mathrm{Clo}_m(X_i,\mathrm{Top}) \not\simeq \mathrm{Clo}_m(X_i,\mathrm{Unif})$. In the counting of 1,1), we discuss the following four possible cases:

a) $s_1 = s_2 = 0 = c = u$: then $\mathrm{Clo}_0(X_1,\mathrm{Top}) = \mathrm{Clo}_0(X_2,\mathrm{Top}) = \mathrm{Clo}_0(X_1,\mathrm{Unif}) = \mathrm{Clo}_0(X_2,\mathrm{Unif})$ because X_i^0 is a one-point space. Let us count the 1-cells. There are infinitely many 1-cells in $\mathrm{Clo}_1(X_1,\mathrm{Top})$ [namely all the maps $p_\delta \circ \ldots \circ p_\delta \circ p_\sigma$], two 1-cells in $\mathrm{Clo}_1(X_2,\mathrm{Top})$ [namely p_σ and p_δ], none in $\mathrm{Clo}_1(X_1,\mathrm{Unif})$ and infinitely many in $\mathrm{Clo}_1(X_2,\mathrm{Unif})$ [namely all the maps $p_\sigma \circ \ldots \circ p_\sigma \circ p_\delta$]. Hence $\mathrm{Clo}_1(X_1,\mathrm{Top}) \not\simeq \mathrm{Clo}_1(X_2,\mathrm{Top})$, $\mathrm{Clo}_1(X_1,\mathrm{Unif}) \not\simeq \mathrm{Clo}_1(X_2,\mathrm{Unif})$ and $\mathrm{Clo}_1(X_i,\mathrm{Top}) \not\simeq \mathrm{Clo}_1(X_i,\mathrm{Unif})$ for $i = 1,2$.

b) $s_1 = s_2 = 0 < c = u$: there is precisely one 1-cell in $\mathrm{Clo}(X_i,\mathrm{Top})$ [namely p_σ] and none in $\mathrm{Clo}(X_i,\mathrm{Unif})$, $i = 1,2$. We deduce that $\mathrm{Clo}_1(X_i,\mathrm{Top}) \not\simeq \mathrm{Clo}_1(X_i,\mathrm{Unif})$, $i = 1,2$. If $m \le c$, we have $\mathrm{Clo}_m(X_1,\mathrm{Top}) = \mathrm{Clo}_m(X_2,\mathrm{Top})$ and $\mathrm{Clo}_m(X_1,\mathrm{Unif}) = \mathrm{Clo}_m(X_2,\mathrm{Unif})$ because the sets of the corresponding Σ-terms in variables $x_0,\ldots,x_{m-1}$ without forbidden subterms are equal: the number $\mathrm{ar}\,\delta$ is too large, none term in variables $x_0,\ldots,x_{m-1}$ can contain a subterm $\delta(x_{\psi(0)},\ldots,x_{\psi(\mathrm{ar}\,\delta-1)})$ with *distinct variables* $x_{\psi(0)},\ldots,x_{\psi(\mathrm{ar}\,\delta-1)}$, hence $\kappa(\delta)$ does not influence these sets of terms. The situation changes for $m \ge c+1$. Let us count $(c+1)$-cells. There are infinitely many $(c+1)$-cells in $\mathrm{Clo}(X_2,\mathrm{Unif})$ [namely all the maps $p_\sigma \circ \ldots \circ p_\sigma \circ p_\delta$], one $(c+1)$-cell in $\mathrm{Clo}(X_2,\mathrm{Top})$ [namely p_δ] and none in $\mathrm{Clo}(X_1,\mathrm{Top})$ and in $\mathrm{Clo}(X_1,\mathrm{Unif})$. Hence $\mathrm{Clo}_{c+1}(X_1,\mathrm{Top}) \not\simeq \mathrm{Clo}_{c+1}(X_2,\mathrm{Top})$ and $\mathrm{Clo}_{u+1}(X_1,\mathrm{Unif}) \not\simeq \mathrm{Clo}_{u+1}(X_2,\mathrm{Unif})$.

c) $c = u = 0 < s_1 = s_2$: if $m \le s_1$, then there is no map $X_i^m \to X_i$ which is continuous but not uniformly continuous, hence $\mathrm{Clo}_m(X_i,\mathrm{Top}) = \mathrm{Clo}_m(X_i,\mathrm{Unif})$, $i = 1,2$. Let us count (s_1+1)-cells: there are no (s_1+1)-cells in $\mathrm{Clo}(X_i,\mathrm{Unif})$, $i = 1,2$, infinitely many (s_1+1)-cells in $\mathrm{Clo}(X_1,\mathrm{Top})$, namely $p_\delta \circ \ldots \circ p_\delta \circ p_\sigma$ and one (s_1+1)-cell in $\mathrm{Clo}(X_2,\mathrm{Top})$, namely p_σ. Hence $\mathrm{Clo}_{s_1+1}(X_i,\mathrm{Unif}) \not\simeq \mathrm{Clo}_{s_1+1}(X_i,\mathrm{Top})$. Let us count 1-cells: there is none 1-cell both in $\mathrm{Clo}(X_1,\mathrm{Unif})$ and in $\mathrm{Clo}(X_1,\mathrm{Top})$ but there is one 1-cell, namely p_δ both in $\mathrm{Clo}(X_2,\mathrm{Top})$ and in $\mathrm{Clo}(X_2,\mathrm{Unif})$. Hence $\mathrm{Clo}_1(X_1,\mathrm{Top}) \not\simeq \mathrm{Clo}_1(X_2,\mathrm{Top})$ and $\mathrm{Clo}_1(X_1,\mathrm{Unif}) \not\simeq \mathrm{Clo}_1(X_2,\mathrm{Unif})$.

d) $0 < c = d$ and $0 < s_1 = s_2$: we have that only p_σ and p_δ could be (representants of) cells in $\mathrm{Clo}(X_i,\mathrm{Top})$ and in $\mathrm{Clo}(X_i,\mathrm{Unif})$ and p_δ is a cell in $\mathrm{Clo}(X_2,\mathrm{Unif})$ and in $\mathrm{Clo}(X_2,\mathrm{Top})$ but neither in $\mathrm{Clo}(X_1,\mathrm{Top})$ nor in $\mathrm{Clo}(X_1,\mathrm{Unif})$ and p_σ is a cell in $\mathrm{Clo}(X_i,\mathrm{Top})$, $i = 1,2$, but not in $\mathrm{Clo}(X_i,\mathrm{Unif})$, $i = 1,2$. This gives the required statements.

6.2. Proof of Theorem 2. Let $m,n \in \{0,1,\ldots,\infty\}$, $m \le n$. We present the proof for $0 \le m < n < \infty$, the other cases are easier, the evident modifications are left to the reader. We choose

$$\Sigma = \Sigma_0 \cup \Sigma_{m+1} \cup \Sigma_{n+1}$$

where $\mathrm{card}\,\Sigma_0 = 2^{\aleph_0}$, $\Sigma_{m+1} = \{\sigma_i \mid i \in \text{ integers }\}$, $\Sigma_{n+1} = \{\gamma\}$ and

$$\kappa(\gamma) = 2\,, \quad \kappa(\sigma_i) = \begin{cases} 1 & \text{if } i \text{ is negative,} \\ 2 & \text{if } i = 0\,, \\ 3 & \text{if } i \text{ is positive.} \end{cases}$$

If $k \le m$, then $\mathrm{Clo}_k(X,\mathrm{Top}) = \mathrm{Clo}_k(X,\mathrm{Unif})$ because there are no forbidden terms in variables $x_0, \dots, x_{k-1}$ — the number of distinct variables is less than $\operatorname{ar}\sigma_i$ and $\operatorname{ar}\gamma$. But $\mathrm{Clo}_{m+1}(X, \mathrm{Top}) \neq \mathrm{Clo}_{m+1}(X, \mathrm{Unif})$ because the map $p_{\sigma_0} : X^{m+1} \to X$ is continuous but not uniformly continuous (because $\kappa(\sigma_0) = 2$). If $k \geq n+1$, then $\mathrm{Clo}_k(X, \mathrm{Top}) \not\cong \mathrm{Clo}_k(X, \mathrm{Unif})$ because $\mathrm{Clo}_k(X, \mathrm{Top})$ contains an $(n+1)$-cell p_γ but $\mathrm{Clo}_k(X, \mathrm{Unif})$ contains no $(n+1)$-cell. It remains to show that for k between $m+1$ and n, $\mathrm{Clo}_k(X, \mathrm{Top})$ is isomorphic to $\mathrm{Clo}_k(X, \mathrm{Unif})$. We describe the isomorphism by means of the sets of corresponding Σ-terms. For $0 \leq s \leq k$, the set of all continuous (or uniformly continuous) maps $X^s \to X$ correspond to Σ-terms in variables $x_0, \dots, x_{s-1}$ not containing forbidden subterms. Since $\operatorname{ar}\gamma > n$, γ cannot create forbidden terms in variables $x_0, \dots, x_{s-1}$ with $s \leq n$; the forbidden terms are $\sigma_i(x_{\psi(0)}, \dots, x_{\psi(m)})$ with $x_{\psi(0)}, \dots, x_{\psi(m)}$ distinct variables (i.e., $\psi : m+1 \to s$ one-to-one) and

$$i \leq 0 \text{ in Unif}, \quad i < 0 \text{ in Top}.$$

Let us denote by $\Sigma^{(s)}(\mathrm{Top})$ and $\Sigma^{(s)}(\mathrm{Unif})$ the sets of Σ-terms in variables $x_0, \dots,$ x_{s-1} not containing the forbidden subterms (in Top and in Unif). We define a bijection

$$b_s : \Sigma^{(s)}(\mathrm{Top}) \to \Sigma^{(s)}(\mathrm{Unif})$$

such that in any term $t \in \Sigma^{(s)}(\mathrm{Top})$ we replace any occurrence of any operational symbol σ_i by σ_{i+1}; clearly b_s is really a bijection and since the collection $\{b_s \mid s = 1, \dots, k\}$ preserves the substitution of terms, it really defines an isomorphism of $\mathrm{Clo}_k(X, \mathrm{Top})$ onto $\mathrm{Clo}_k(X, \mathrm{Unif})$.

7. The proof of the Main Theorem

7.1. In [12], super-extremally B-semirigid spaces are introduced as follows: Let $X = (P, t)$ be a Hausdorff space, $B \subseteq P$. We say that X is *super-extremally B-semirigid* if

(a) $(B, t/B)$ is a closed discrete subset of (P, t) and

(b) if P' is a set such that $P \setminus B \subseteq P' \subseteq P$ and t' is a Hausdorff topology on P' coarser than t/P' (i.e. every t'-open set is (t/P')-open) such that $B \cap P'$ is t'-closed and $t'/P \setminus B = t/P \setminus B$, then (P', t') is B-semirigid.

In [12], the following statement is given: for every cardinal number $\vartheta \geq 2^{\aleph_0}$ there exists a metrizable super-extremally B-semirigid space (P, t) such that $\operatorname{card} B = \operatorname{card}(P \setminus B) = \vartheta$. Moreover, the space is constructed as metric space (P, d) with $\operatorname{diam} \leq 1$ and $d(x, y) = 1$ whenever $x \neq y$, $x, y \in B$. We use this statement in our construction below.

7.2. But first, we need some easy observations. Let τ be a pseudometric on a set Q, let $B \subseteq Q$ be a closed subset of (Q, τ). Let a pseudometric u on B be given such that

$$u(a, b) \leq \tau(a, b) \text{ for all } a, b \in B.$$

For every $x, y \in Q$ set

$$d(x, y) = \min\left(\tau(x, y), \inf_{a, b \in B}(\tau(x, a) + u(a, b) + \tau(b, y))\right).$$

Then, clearly, d is a pseudometric on Q, $d \le \tau$ and d extends u, i.e. $d(a,b) = u(a,b)$ for all $a, b \in B$. Moreover,

for every $x, y \in Q$, $d(x,y) \ge \min(\tau(x,y), \tau(x,B) + \tau(y,B))$,
for every $x \in Q \setminus B$ and $y \in B$, $d(x,y) \ge \tau(x,B)$ and
$Q \setminus B$ is open in (Q,d) and both τ and d determine the same topology on it.

Let us denote d by $\tau * u$. Clearly, if $u_1 \le u_2 \le \tau/B$, then $\tau * u_1 \le \tau * u_2$.

7.3. Construction. Let Σ, Γ, Σ_0, $\mathbf{P} = (P, \{p_\sigma \mid \sigma \in \Sigma\})$, G_0, B, G_k, B_σ, $B_{\sigma,k}$ be as in 3.3. Let A be an infinite countable set disjoint with P, let $\{a_0, a_1, \ldots\}$ be a one-to-one sequence of its elements. If

$$\operatorname{card} \Sigma_0 \ge 2^{\aleph_0} \cdot \operatorname{card} \Gamma\,,$$

then $\operatorname{card} G_0 = \operatorname{card}(A \cup B)$, so that, by [12], we can find a metric τ on the set $P \cup A$ such that the metric space $Z = (P \cup A, \tau)$ is super-extremally $(A \cup B)$-semirigid and $\operatorname{diam} Z \le 1$, $\tau(x,y) = 1$ whenever $x, y \in A \cup B$, $x \ne y$.

Given a map

$$\kappa : \Gamma \to \{1,2,3\}\,,$$

we define two descending chains of pseudometrics,

$$u_\alpha \text{ on } B \quad \text{and} \quad \tau_\alpha \text{ on } P \cup A \quad (\alpha \in \mathrm{Ord})$$

as follows:
$u_0 = \tau/B$ and $\tau_0 = \tau$;
if α is a limit ordinal, then $u_\alpha = \inf_{\beta<\alpha} u_\beta$, $\tau_\alpha = \tau * u_\alpha$;
if $\alpha = \beta + 1$, then

$u_\alpha(x,y) = 1$ whenever $x \in B_{\sigma_1}$, $y \in B_{\sigma_2}$, $\sigma_1, \sigma_2 \in \Gamma$, $\sigma_1 \ne \sigma_2$, and for every $\sigma \in \Gamma$, u_α/B_σ is *the* pseudometric for which p_σ is an isometry of $(P^{\operatorname{ar}\sigma}, \tau_{\beta,\sigma,\kappa(\sigma)})$ onto $(B_\sigma, u_\alpha/B_\sigma)$ [i.e. we transfer the pseudometric $\tau_{\beta,\sigma,\kappa(\sigma)}$ from $P^{\operatorname{ar}\sigma}$ by p_σ onto B_σ] and $\tau_\alpha = \tau * u_\alpha$ again.

To finish this description, we have to say what are the pseudometrics $\tau_{\beta,\sigma,1}$, $\tau_{\beta,\sigma,2}$ and $\tau_{\beta,\sigma,3}$. Let us denote $\operatorname{ar}\sigma = n$. The pseudometric $\tau_{\beta,\sigma,3}$ is just the restriction τ_β^n/P^n of the pseudometric τ_β^n, i.e.

$$\begin{aligned} \tau_{\beta,\sigma,3}((x_0,\ldots,x_{n-1}),(y_0,\ldots,y_{n-1})) &= \\ \tau_\beta^n((x_0,\ldots,x_{n-1}),(y_0,\ldots,y_{n-1})) &= \max\nolimits_{i\in n} \tau_\beta(x_i,y_i)\,. \end{aligned}$$

The pseudometric $\tau_{\beta,\sigma,2}$ is defined by

$$\tau_{\beta,\sigma,2}(x,y) = \min(1, \tau_\beta^n(x,y) + |q(x) - q(y)|)$$

where q is the real valued function on $Z^n \setminus \{a\}$ defined by

$$q(x) = \frac{1}{\tau^n(x,a)}$$

where $a = (a_0, \ldots, a_{n-1})$ consists of the first n members of A.

The pseudometric $\tau_{\beta,\sigma,1}$ is defined by

$$\tau_{\beta,\sigma,1}(x,y) = \min(1, \tau_\beta^n(x,y) + |f(x) - f(y)|)$$

where f is the real valued function on P^n defined below:

a) if $n = 1$: we choose $b^{(1)} \in G_0$ and put

$$f(b^{(1)}) = 1\,, \quad f(x) = 0 \text{ for all } x \in P \setminus \{b^{(1)}\}\,;$$

b) if $n > 1$: we choose $b_0^{(n)}, \dots, b_{n-1}^{(n)} \in G_0$ distinct and for $b^{(n)} = (b_0^{(n)}, \dots, b_{n-1}^{(n)})$ we put $f(b^{(n)}) = 0$; for $x = (x_0, \dots, x_{n-1}) \in P^n \setminus \{b^{(n)}\}$ we put

$$f(x) = f_1(x) \cdot f_2(x)$$

where

$$f_1(x) = \frac{n \cdot \prod_{i=0}^{n-1} \tau(x_i, b_i^{(n)})}{\sum_{i=0}^{n-1} (\tau(x_i, b_i^{(n)})^n}$$

and

$$f_2(x) = \max\{0, \frac{1}{\varepsilon}(\varepsilon - \max_{i\in n}\{\tau(x_i, G_0), \tau(x_i, b_i^{(n)})\})\}$$

and ε denotes $\min_{i,j\in n, i\neq j}\{\tau(b_i^{(n)}, B), \tau(b_i^{(n)}, b_j^{(n)})\}$, i.e. $f(x) = 0$ if $x_i = b_i^{(n)}$ for at least one $i \in n$ but $\lim f(x) = 1$ if x approaches to $b^{(n)}$ ranging in $E^{(n)} = \{x \in P^n \mid \tau(x_i, b_i^{(n)}) = \tau(x_j, b_j^{(n)})$ for all $i, j \in n\}$ and f is uniformly continuous on every small subset of P^n.

By the transfinite induction, one can prove that really

$$u_0 \geq u_1 \geq \dots \quad \text{and} \quad \tau_0 \geq \tau_1 \geq \dots$$

(notice that $\operatorname{diam} Z \leq 1$ and $u_0(x,y) = 1$ whenever $x \neq y$ hence $u_1 \leq u_0$ so that $\tau_1 = \tau * u_1 \leq \tau_0$; the next steps are evident). Since there is only a set of distinct pseudometrics on $P \cup A$, the chain must stop. Hence there exists an ordinal α such that $\tau_\alpha = \tau_{\alpha+1} = \dots$. Let us denote by $\bar{\varrho}$ the resulting pseudometric on $P \cup A$ and by ϱ_κ its restriction to P.

7.4. Proposition. *The pseudometric $\bar{\varrho}$ is a metric.*

Proof. We construct a real symmetric function h on $(P \cup A) \times (P \cup A)$ such that $0 \leq h \leq 1$, $h(x,x) = 0$ and $h(x,y) > 0$ for $x, y \in P \cup A$, $x \neq y$, and we show that $\tau_\alpha(x,y) \geq h(x,y)$ for all $x, y \in P \cup A$ and all ordinals α.

We construct h inductively as follows:

1) for $x, y \in A \cup G_0$, we set $h(x,y) = \min(\tau(x,y), \tau(x,B) + \tau(y,B))$;

2) if h is defined for all $x, y \in A \cup G_k$, we extend the definition on $A \cup G_{k+1}$:

(a) for $x, y \in G_{k+1} \setminus G_k$, we define

$$h(x,y) = 1 \text{ if } x \in B_{\sigma_1}, y \in B_{\sigma_2}, \sigma_1, \sigma_2 \in \Gamma, \sigma_1 \neq \sigma_2,$$

$$h(x,y) = \max(h(x_0,y_0), \dots, h(x_{n-1},y_{n-1})) \text{ if } x, y \in B_\sigma, \operatorname{ar}\sigma = n \text{ and } (x_0, \dots, x_{n-1}) = p_\sigma^{-1}(x)\,, (y_0, \dots, y_{n-1}) = p_\sigma^{-1}(y)\,;$$

(b) for $x \in G_{k+1} \setminus G_k$, and $y \in G_k \cap B$, we define $h(x,y) = h(y,x)$ as in (a);
(c) for $x \in G_{k+1} \setminus G_k$, and $y \in A \cup G_0$, we define $h(x,y) = h(y,x) = \tau(y,B)$.

One can see easily that $h(x,y) > 0$ whenever $x \neq y$ and

(α) $h(x,y) = \tau(y,B)$ if $y \in A \cup G_0$, $x \in B$;
(β) $h(x,y) = 1$ if $x \in B_{\sigma_1}$, $y \in B_{\sigma_2}$, $\sigma_1, \sigma_2 \in \Gamma$, $\sigma_1 \neq \sigma_2$;
(γ) $h(x,y) = \max(h(x_0,y_0), \dots, h(x_{n-1},y_{n-1}))$ for $x, y \in B_\sigma$, $\sigma \in \Gamma$, $(x_0, \dots, x_{n-1}) = p_\sigma^{-1}(x)$, $(y_0, \dots, y_{n-1}) = p_\sigma^{-1}(y)$.

By (α), (β) and 1) above, we get $\tau \geq h$. Hence $\tau_\alpha \geq h$, by transfinite induction and 7.2. Consequently $\bar{\varrho} \geq h$.

7.5. Proposition. *All the spaces $(A \cup P, \tau_\alpha)$ are $(A \cup B)$-semirigid and the metric space (P, ϱ_κ) is B-semirigid.*

Proof. We started with the super-extremally $(A \cup B)$-semirigid space $(P \cup A, \tau)$. In our construction, every (pseudo)metric τ_α was given by the formula $\tau_\alpha = \tau * u_\alpha$ with $u_\alpha \leq \tau/B$, so that, by 7.2, τ_α determines on $(P \cup A) \setminus B$ (hence on $(P \setminus (A \cup B))$) the same topology as τ does. Moreover, since $\tau_\alpha \leq \tau$, the topology of $(P \cup A, \tau_\alpha)$ is coarser than the topology of $(P \cup A, \tau)$. Finally, the space $(P \cup A, \tau_\alpha)$, being a metric space, is Hausdorff. By the definition of the super-extremal semirigidity, $(P, \tau_\alpha/P)$ is B-semirigid (and $\varrho_\kappa = \tau_\alpha/P$ for some α) and all $(P \cup A, \tau_\alpha)$ are $(B \cup A)$-semirigid.

7.6. To finish the proof of the Main Theorem, it remains to show that the space $X = (P, \varrho_\kappa)$ satisfies (b) and (c) in the Main Theorem. The statement (b) is evident from the construction 7.3, every τ_α satisfies it. Thus, let us show that for every $\sigma \in \Gamma$, the map

$$p_\sigma : X^{\operatorname{ar}\sigma} \to X$$

is a suitable map of the type $\kappa(\sigma)$ (see 3.2). Let α be an ordinal number large enough so that ϱ_κ is a restriction of $\bar{\varrho} = \tau_\alpha = \tau_{\alpha+1} = \dots$. By the construction 7.3, the map $|p_\sigma| : P^{\operatorname{ar}\sigma} \to P$ carrying p_σ is an isometry of $(P^{\operatorname{ar}\sigma}, \tau_{\alpha,\sigma,\kappa(\sigma)})$ onto $(B_\sigma, \tau_\alpha/B_\sigma)$. Hence

$$p_\sigma : (P^{\operatorname{ar}\sigma}, (\tau_\alpha/P)^{\operatorname{ar}\sigma}) \xrightarrow{\text{identical map}} (P^{\operatorname{ar}\sigma}, \tau_{\alpha,\sigma,\kappa(\sigma)}) \xrightarrow{\text{isometry } |p_\sigma|} (B_\sigma, \tau_\alpha/B_\sigma).$$

Since $(\tau_\alpha/P)^{\operatorname{ar}\sigma} \leq \tau_{\alpha,\sigma,i}$ for $i = 1,2,3$, the map p_σ^{-1} is non-expanding hence uniformly continuous. Moreover, for $\kappa(\sigma) = 3$, p_σ is an isometry hence uniformly continuous. If $\kappa(\sigma) = 2$, then p_σ is continuous (because $a \notin P^{\operatorname{ar}\sigma}$) but not uniformly continuous. The later claim follows from the fact that $(P \cup A, \tau_\alpha)$ is $(B \cup A)$-semirigid (see 7.5) hence connected hence a is a limit point of $X^{\operatorname{ar}\sigma}$ so that the function q used in 7.3 is not uniformly continuous. The fact that $X^{\operatorname{ar}\sigma}$ is connected also implies that the function f used in 7.3 in the definition of $\tau_{\alpha,\sigma,1}$ is not continuous at $b^{(\operatorname{ar}\sigma)}$. Hence p_σ is not continuous whenever $\kappa(\sigma) = 1$. On the other hand p_σ is uniformly continuous on every small subset of $P^{\operatorname{ar}\sigma}$ because both the functions q and f are uniformly continuous on each small subset of $X^{\operatorname{ar}\sigma}$, see 7.3.

References

[1] J. Adámek, H. Herrlich, G. Strecker: *Abstract and Concrete Categories*, John Wiley & Sons, Inc., New York/Chichester/Brisbane/Toronto/Singapore 1990

[2] G. A. Grätzer: *Universal Algebra*, Springer Verlag 1979

[3] H. Herrlich: *On the concept of reflection in general topology*, Conference "Contribution to extension theory of topological structures", Berlin 1967

[4] H. Herrlich: *Topologische Reflexionen und Coreflexionen*, Springer Verlag, Berlin-Heidelberg-New York, Lect. N. in Math. (1968)

[5] W. F. Lawvere: *Functorial semantics of algebraic theories*, Proc. Nat. Acad. Sci. **50** (1963), 869–872

[6] S. MacLane: *Categories for the working mathematician*, Springer, Berlin-Heidelberg-New York 1972

[7] K. D. Magill Jr.: *A survey of semigroups of continuous selfmaps*, Semigroup Forum **11** (1975/76), 189–282

[8] R. McKenzie, G. McNulty and W. Taylor: *Algebras, lattices, varieties*, Volume 1, Brooks/Cole., Monterey, California 1987

[9] J. Sichler, V. Trnková: *Isomorphism and elementary equivalence of clone segments*, submitted

[10] W. Taylor: *The clone of a topological space*, Research and Exposition in Mathematics, vol. 13, Helderman Verlag 1986

[11] V. Trnková: *Semirigid spaces*, Trans. Amer. Math. Soc. **343** (1994), 305–325

[12] V. Trnková: *Continuous and uniformly continuous maps*, Topology and its Applications **63** (1995), 189–200

Simultaneous problems of clone segments in Top and in Unif

Věra Trnková
TRNKOVA@karlin.MFF.CUNI.CZ

Abstract. Continuous and uniformly continuous maps of finite powers of metric spaces are investigated. Functions on sets of metric spaces obtained by isomorphism of clone segments in Top and in Unif are fully characterized.

1992 Mathematics Subject Classification: 08A40, 54C05.

Key words: clones, products, continuous maps, uniformly continuous maps.

1 Introduction and the Main Result

1.1. *A clone* $\mathrm{Clo}\,(X, \mathrm{Top}\,)$ of continuous maps of a topological space X is the (concrete) category of all continuous maps of all finite powers ($X^0 = \{\emptyset\}$, $X^1 = X$, $X^2 = X \times X$, X^3, ...) of X. Its *n-segment* $\mathrm{Clo}_n(X, \mathrm{Top}\,)$ is its full subcategory generated by $X^0, \ldots, X^n$.

Clones and abstract clones (=finitary algebraic theories [4]) are extensively investigated in universal algebra, see e.g. [6]. Clones of topological spaces in the above sense are widely investigated in the monograph [8] which inspired also the present paper.

If X, Y are topological spaces, we investigate the value

$$\mathcal{V}(X, Y) = \sup\{n + 1 \mid \mathrm{Clo}_n(X, \mathrm{Top}\,) \simeq \mathrm{Clo}_n(Y, \mathrm{Top}\,)\}$$

where $\simeq$ denotes isomorphism of categories. Values $\mathcal{V}$ of these suprema are in the set $\{1, 2, \ldots, \infty\}$ and, by [9], every element of $\{1, 2, \ldots, \infty\}$ is equal to $\mathcal{V}(X, Y)$ for suitable metrizable spaces X, Y.

Let us investigate $\mathcal{V}$ for all pairs of spaces. Then, for every X, Y, Z, the following statements are fulfilled, evidently.

$$\begin{aligned} &(1) \quad \mathcal{V}(X, X) &&= \infty, \\ &(2) \quad \mathcal{V}(X, Y) &&= \mathcal{V}(Y, X), \\ &(3) \quad \mathcal{V}(X, Z) &&\geq \min\{\mathcal{V}(X, Y), \mathcal{V}(Y, Z)\}. \end{aligned}$$

These necessary conditions are also sufficient. In [7], the following theorem is proved:

Let V be a set, let $\mathcal{V} : V \times V \to \{1, 2, \ldots, \infty\}$ be a map which satisfies (1), (2), (3). Then there exists a collection $\{X_v \mid v \in V\}$ of metrizable spaces such that,

---

[0] Financial support of the Grant Agency of the Czech Republic under the grant no 201/93/0950 and of the Grant Agency of the Charles University under the grant GAUK 349 is gratefully acknowledged.

Eraldo Giuli (ed.), Categorical Topology, 269–278.

for every $v_1, v_2 \in V$,

$$\mathcal{V}(v_1, v_2) = \sup\{n+1 \mid \mathrm{Clo}_n(X_{v_1}, \mathrm{Top}) \simeq \mathrm{Clo}_n(X_{v_2}, \mathrm{Top})\}.$$

1.2. In [11], continuous and uniformly continuous maps of finite powers of metric spaces are investigated simultaneously. If X is a metric space, we take it also as the induced uniform space and the induced topological space. Besides of the clone $\mathrm{Clo}(X, \mathrm{Top})$ of the continuous maps of finite powers of the space X and its n-segments $\mathrm{Clo}_n(X, \mathrm{Top})$, we investigate also the clone $\mathrm{Clo}(X, \mathrm{Unif})$ of the uniformly continuous maps of the finite powers of X and its n-segments $\mathrm{Clo}_n(X, \mathrm{Unif})$, i.e. the categories of uniformly continuous maps of the spaces X^0, X^1, ..., X^n. Using the method developed in [11], we prove here the following strengthening of the above statement of [7].

Theorem. *Let V be a set, let $\mathcal{V}_c, \mathcal{V}_u : V \times V \to \{1, 2, \ldots, \infty\}$ be two maps, both satisfying* (1), (2), (3). *Then there exists a set P and a collection of metrics $\{\tau_v \mid v \in V\}$ on P such that, for every $v_1, v_2 \in V$,*

$$\begin{aligned}\mathcal{V}_c(v_1, v_2) &= \sup\{n+1 \mid \mathrm{Clo}_n((P, \tau_{v_1}), \mathrm{Top}) \simeq \mathrm{Clo}_n((P, \tau_{v_2}), \mathrm{Top})\} \text{ and}\\ \mathcal{V}_u(v_1, v_2) &= \sup\{n+1 \mid \mathrm{Clo}_n((P, \tau_{v_1}), \mathrm{Unif}) \simeq \mathrm{Clo}_n((P, \tau_{v_2}), \mathrm{Unif})\}.\end{aligned}$$

In fact, the method of [11] admits to get stronger results: if $n < \mathcal{V}_c(v_1, v_2)$, then the corresponding clone segments $\mathrm{Clo}_n((P, \tau_{v_1}), \mathrm{Top})$ and $\mathrm{Clo}_n((P, \tau_{v_2}), \mathrm{Top})$ are created by the same maps of $P^0, P, \ldots, P^n$ and if $n \geq \mathcal{V}_c(v_1, v_2)$, then neither $\mathrm{Clo}_n((P, \tau_{v_1}), \mathrm{Top})$ can be fully embedded into $\mathrm{Clo}_n((P, \tau_{v_2}), \mathrm{Top})$ nor $\mathrm{Clo}_n((P, \tau_{v_2}), \mathrm{Top})$ into $\mathrm{Clo}_n((P, \tau_{v_1}), \mathrm{Top})$; and analogously for $\mathcal{V}_u(v_1, v_2)$ and $\mathrm{Clo}_n((P, \tau_{v_1}), \mathrm{Unif})$, $\mathrm{Clo}_n((P, \tau_{v_2}), \mathrm{Unif})$.

The proof of the above theorem is presented in the next paragraphs 2 and 3. In 2, we recall briefly the general method developed in [11]; in 3, we apply this method to our case and prove our Theorem. In the last paragraph 4, we recall some results of [11] and present open problems.

1.3. Let us finish this paragraph with a corollary of the above Theorem. Let us recall that a metric space (V, d) is called *non-archimedean* (*=ultrametric*) if

$$d(v, v) = 0, \quad d(v, v') = d(v', v), \quad d(v, v'') \leq \max\{d(v, v'), d(v', v'')\}.$$

Clearly, the reciprocal values of our suprema $\mathcal{V}$ satisfy these axioms. By [3], every non-archimedean space is zero-dimensional and, conversely, every metrizable zero-dimensional space can be metrized by a non-archimedean metric. Inspecting the proof of [3], one can see that the constructed non-archimedean metric, metrizing a given metrizable zero-dimensional space, has its values only in the set $\{1, \frac{1}{2}, \frac{1}{3}, \ldots, \frac{1}{n}, \ldots, 0\}$. Hence we get the following consequence of our Theorem:

Let (V, t_1, t_2) be a bitopological space such that both the topologies t_1, t_2 are metrizable and zero-dimensional. Then there exists a collection $\{(P, \tau_v) \mid v \in V\}$ of metric spaces such that the reciprocal values of our suprema $\mathcal{V}_c$ and $\mathcal{V}_u$ metrize (V, t_1, t_2).

2 General method

2.1. In [11], a general method for constructing of metric spaces with some prescribed properties was developed. Our Theorem is obtained as an application of this method. We recall it briefly here.

We need some basic notions of universal algebra. They can be found in any monograph about universal algebra, see e.g. [2] or [6]. We describe our notation briefly, now.

Let $\Sigma = \bigcup_{n=0}^{\infty} \Sigma_n$ be a finitary signature of (mono-sorted) universal algebras, i.e. Σ_n is a set of n-ary operational symbols. If $\sigma \in \Sigma_n$, we write $\operatorname{ar}\sigma = n$. Let $\mathbf{P} = (P, \{p_\sigma \mid \sigma \in \Sigma\})$ be an initial Σ-algebra, i.e. an absolutely free Σ-algebra over the empty set of generators. As usual, if $\sigma \in \Sigma_0$, then the zero-ary operation p_σ is just an element of P. Denote

$$G_0 = \{p_\sigma \mid \sigma \in \Sigma_0\}, \quad B = P \setminus G_0 .$$

Then, for every $\sigma \in \Sigma_n$ with $n \geq 1$, $p_\sigma : P^n \to P$ is a one-to-one map. Let us denote $B_\sigma = p_\sigma(P^n)$. Denote $\Gamma = \bigcup_{n=1}^{\infty} \Sigma_n = \Sigma \setminus \Sigma_0$. Then, as it is well-known (see e.g. [2]),

$$B = \bigcup_{\sigma \in \Gamma} B_\sigma \quad \text{and} \quad B_{\sigma_1} \cap B_{\sigma_2} = \emptyset \text{ whenever } \sigma_1, \sigma_2 \in \Gamma,\ \sigma_1 \neq \sigma_2 .$$

2.2. In [11], the following Main Theorem is proved (we explain the notions which occur in it below):

Main Theorem. *Let $\Sigma = \bigcup_{n=0}^{\infty} \Sigma_n$ be a finitary signature of universal algebras, $\Gamma = \Sigma \setminus \Sigma_0$. Let $(P, \{p_\sigma \mid \sigma \in \Sigma\})$ be an initial Σ-algebra, $G_0 = \{p_\sigma \mid \sigma \in \Sigma_0\}$, $B = P \setminus G_0$. If*

$$\operatorname{card} \Sigma_0 \geq 2^{\aleph_0} + \operatorname{card} \Gamma ,$$

then, for every map

$$\kappa : \Gamma \to \{1, 2, 3\} ,$$

there exists a metric ϱ_κ on the set P such that the metric space $X = (P, \varrho_\kappa)$

(a) *is B-semirigid,*

(b) *if $\sigma_1, \sigma_2 \in \Gamma$, $\sigma_1 \neq \sigma_2$, then $\varrho_\kappa(B_{\sigma_1}, B_{\sigma_2}) = 1$ (where B_σ is as in 2.1) and*

(c) *for every $\sigma \in \Gamma$, the operation $p_\sigma : X^{\operatorname{ar}\sigma} \to X$ is a suitable map of type $\kappa(\sigma)$.*

2.3. Let us recall (see [9]) that a topological space X is called *B-semirigid* if B is its closed subset and every continuous selfmap $f : X \to X$ is either the identity or a constant or $f(X) \subseteq B$.

To explain all the notions which occur in the above Main Theorem, we have to recall the definition (see [11]) of a suitable map of type 1, 2, 3. For this reason, let us denote

$$\begin{aligned} P^n[i,B] &= \{(x_0,\dots,x_{n-1}) \in P^n \mid x_i \in B\} \text{ for } i \in n\,, \\ P^n[i,j] &= \{(x_0,\dots,x_{n-1}) \in P^n \mid x_i = x_j\} \text{ for } i,j \in n,\ i \neq j \text{ and}\,, \\ P^n[i,c] &= \{(x_0,\dots,x_{n-1}) \in P^n \mid x_i = c\} \text{ for } i \in n,\ c \in P\,, \end{aligned}$$

where P and $B \subseteq P$ are as above. Let $\mathcal{U}$ be a uniformity on P, $X = (P,\mathcal{U})$. Let $f : P^n \to P$ be a map. We say that

f is a suitable map of type 1 (or type 2 or type 3)

if it is one-to-one, its inverse $f(P^n) \to P^n$ is a uniformly continuous map of $(f(P^n),\mathcal{U}/f(P^n))$ onto X^n, the domain-restriction of f to any subset $P^n[i,B]$, $P^n[i,j]$, $P^n[i,c]$ for $i,j \in n$, $i \neq j$, $c \in P$, is uniformly continuous, but

type 1: f itself is not continuous;

type 2: f itself is continuous but not uniformly continuous;

type 3: f is uniformly continuous.

2.4. Let $\Sigma = \bigcup_{n=0}^{\infty} \Sigma_n$ be a finitary signature of universal algebras, $\Gamma = \Sigma \setminus \Sigma_0$, let $\mathbf{P} = (P, \{p_\sigma \mid \sigma \in \Sigma\})$ be an initial Σ-algebra, let G_0, B, B_σ be as in 2.1.

For every natural number m, the set $M^{(m)}$ of all maps $P^m \to P$ admits naturally a structure $\{\sigma^{(m)} \mid \sigma \in \Sigma\}$ of a Σ-algebra as follows:

for $\sigma \in \Sigma_0$, $\sigma^{(m)}$ is the constant map $P^m \to P$ with the value p_σ;
for $\sigma \in \Sigma_n$, $n \geq 1$, $\sigma^{(m)} : (P^m \to P)^n \to (P^m \to P)$ is defined by

$$\sigma^{(m)}(f_0,\dots,f_{n-1}) = p_\sigma \circ (f_0 \dot{\times} \dots \dot{\times} f_{n-1})$$

where the map $g = f_0 \dot{\times} \dots \dot{\times} f_{n-1} : P^m \to P^n$ is defined by

$$g(z) = (f_0(z),\dots,f_{n-1}(z))\,.$$

Hence the Σ-algebra $\mathbf{M} = (M^{(m)}, \{\sigma^{(m)} \mid \sigma \in \Sigma\})$ is just the direct power $\mathbf{P}^{P^m}$ of the initial algebra $\mathbf{P}$. Let

$$(H^{(m)}, \{\sigma^{(m)} \mid \sigma \in \Sigma\})$$

be the subalgebra of $\mathbf{M}$ (we denote the restrictions of the operations $\sigma^{(m)}$ to the subalgebra $H^{(m)}$ by $\sigma^{(m)}$ again) generated by the projections $\pi_i^{(m)} : P^m \to P$, $i \in m$ (i.e. $\pi_i^{(m)}(z_0,\dots,z_{m-1}) = z_i$). Then $H^{(m)} = \bigcup_{k=0}^{\infty} H_k^{(m)}$ where

$$\begin{aligned} H_0^{(m)} &= \{\pi_0^{(m)},\dots,\pi_{m-1}^{(m)}\} \cup \{\sigma^{(m)} \mid \sigma \in \Sigma_0\}\,, \\ H_{k+1}^{(m)} &= H_k^{(m)} \cup \textstyle\bigcup_{\sigma\in\Gamma}\{\sigma^{(m)}(f_0,\dots,f_{\operatorname{ar}\sigma-1}) \mid f_i \in H_k^{(m)} \text{ for all } i \in \operatorname{ar}\sigma\}\,. \end{aligned}$$

2.5. The standard construction of an absolutely free Σ-algebra over m variables $x_0, \ldots, x_{m-1}$ uses Σ-terms, introduced by the following inductive definition (see e.g. [2]):

every variable x_i, $i = 0, \ldots, m-1$, is a Σ-term and
every $\sigma \in \Sigma_0$ is a Σ-term;
if $\sigma \in \Sigma_n$, $n \geq 1$ and $t_0, \ldots, t_{n-1}$ are Σ-terms, then $\sigma(t_0, \ldots, t_{n-1})$ is a Σ-term;

(and there are no other Σ-terms than those obtained by finitely many applications of the above rules).

All the Σ-terms with the operations $\bar{\sigma}$, $\sigma \in \Sigma$, defined by

$$\begin{aligned} \bar{\sigma} &= \sigma \text{ if } \sigma \in \Sigma_0 \,, \\ \bar{\sigma}(t_0, \ldots, t_{n-1}) &= \text{ the term } \sigma(t_0, \ldots, t_{n-1}) \text{ for } \sigma \in \Sigma_n, n \geq 1 \end{aligned}$$

are well-known to form the absolutely free Σ-algebra over $x_0, \ldots, x_{m-1}$. Let λ be a map of the set of all Σ-terms in variables $x_0, \ldots, x_{m-1}$ into the set $H^{(m)}$ of maps $P^m \to P$ such that

$$\begin{aligned} &\lambda(x_i) = \pi_i^{(m)} \text{ for all } i \in m\,, \\ &\lambda(\sigma) = \sigma^{(m)} \text{ for all } \sigma \in \Sigma_0\,, \\ &\lambda(\sigma(t_0, \ldots, t_{n-1})) = p_\sigma \circ (f_0 \dot{\times} \ldots \dot{\times} f_{n-1}) \end{aligned}$$

whenever $\sigma \in \Sigma_n$, $n \geq 1$ and $\lambda(t_i) = f_i$ for $i = 0, \ldots, n-1$. It is well-known (and easy to see) that λ is a bijection onto $H^{(m)}$.

2.6. By means of the Main Theorem recalled here in 2.2, the following Main Lemma is proved in [11]:

Main Lemma. *Let $\Sigma = \bigcup_{n=0}^{\infty} \Sigma_n$ be a finitary signature, let $\mathbf{P} = (P, \{p_\sigma \mid \sigma \in \Sigma\})$, Γ, G_0, B, B_σ be as in 2.1. Let* $\operatorname{card} \Sigma_0 \geq 2^{\aleph_0} + \operatorname{card} \Gamma$. *Let $\kappa : \Gamma \to \{1,2,3\}$, ϱ_κ, $X = (P, \varrho_\kappa)$ be as in the Main Theorem. Then, for every natural number m, the set $C(X^m, X)$ (or $U(X^m, X)$) of all continuous (or uniformly continuous) maps of X^m into X is precisely the set of all maps f in $H^{(m)}$ for which the corresponding term $t = \lambda^{-1}(f)$ contains no subterm*

$\sigma(x_{\psi(0)}, \ldots, x_{\psi(n-1)})$ with $\sigma \in \Sigma_n$, $n \geq 1$ and $\kappa(\sigma) = 1$ (or $\kappa(\sigma) \in \{1,2\}$) where $\psi : n \to m$ is a one-to-one map (i.e. $x_{\psi(0)}, \ldots, x_{\psi(n-1)}$ are **distinct variables**).

Let us call the terms $\sigma(x_{\psi(0)}, \ldots, x_{\psi(n-1)})$ with $\sigma \in \Sigma_n$, $n \geq 1$ and $\psi : n \to m$ one-to-one

c-forbidden terms whenever $\kappa(\sigma) = 1$ and
u-forbidden terms whenever $\kappa(\sigma) \in \{1,2\}$.

2.7. By the above Main Lemma, we can, for every n, construct metric spaces X_1, X_2 such that

$$\operatorname{Clo}_n(X_1, \operatorname{Top}) = \operatorname{Clo}_n(X_2, \operatorname{Top}) \text{ but } \operatorname{Clo}_{n+1}(X_1, \operatorname{Top}) \neq \operatorname{Clo}_{n+1}(X_2, \operatorname{Top})\,.$$

In fact, we choose Σ with $\operatorname{card}\Sigma_0 \geq 2^{\aleph_0} + \operatorname{card}\Gamma$ and $\kappa_1, \kappa_2 : \Gamma \to \{1,2,3\}$ such that the corresponding sets of Σ-terms without c-forbidden subterms are or are not equal; analogously for clone segments in Unif and the u-forbidden subterms. However, the non-isomorphism $\not\simeq$ requires some further reasoning. This is outlined in 2.8–9 below.

2.8. Let Σ, $\Gamma = \Sigma \setminus \Sigma_0$, $\mathbf{P} = (P, \{p_\sigma \mid \sigma \in \Sigma\})$, G_0, B, $\kappa : \Gamma \to \{1,2,3\}$, ϱ_κ, $X = (P, \varrho_\kappa)$ be as above. Let $n \geq 1$. Let us call

representative continuous n-cell

every map $P^n \to P$ of the form

$$p_{\sigma_0} \circ \ldots \circ p_{\sigma_{k-1}} \circ p_\sigma$$

where $\sigma_0, \ldots, \sigma_{k-1} \in \Sigma_1$, $\sigma \in \Sigma_n$, all the maps p_{σ_i}, $i = 0, \ldots, k-1$, are not continuous, i.e. $\kappa(\sigma_i) = 1$, while p_σ is continuous, i.e. $\kappa(\sigma) \in \{2,3\}$. We admit also $k = 0$, i.e. the expression $p_{\sigma_0} \circ \ldots \circ p_{\sigma_{k-1}}$ could be missing. We mention explicitly that though p_{σ_i}, $i = 0, \ldots, k-1$, are not continuous, the map $p_{\sigma_0} \circ \ldots \circ p_{\sigma_{k-1}} \circ p_\sigma : X^n \to X$ itself is continuous because p_σ maps P^n into B and every p_{σ_i} restricted to B is uniformly continuous, see the definition of suitable maps of type 1 in 2.3. Moreover,

$$p_{\sigma_0} \circ \ldots \circ p_{\sigma_{k-1}} \circ p_\sigma \neq p_{\sigma'_0} \circ \ldots \circ p_{\sigma'_{l-1}} \circ p_{\sigma'}$$

whenever the words $\sigma_0 \ldots \sigma_{k-1}\sigma$ and $\sigma'_0 \ldots \sigma'_{l-1}\sigma'$ are distinct, this last fact follows from the properties of the operations on the initial algebra $\mathbf{P}$.

In the paragraphs 4 and 5 of [11], the following statement is proved:

Let Σ be a finitary type, $\Gamma = \Sigma \setminus \Sigma_0$, $\mathbf{P} = (P, \{p_\sigma \mid \sigma \in \Sigma\})$ the initial Σ-algebra, $\operatorname{card}\Sigma_0 \geq 2^{\aleph_0} + \operatorname{card}\Gamma$, $\kappa_1, \kappa_2 : \Gamma \to \{1,2,3\}$. Let ϱ_{κ_1}, ϱ_{κ_2} and $X_1 = (P, \varrho_{\kappa_1})$, $X_2 = (P, \varrho_{\kappa_2})$ be as in the Main Theorem in 2.2. Let $n \geq 1$. If the number of representative continuous n-cells $X_1^n \to X_1$ is different from the number of representative continuous n-cells $X_2^n \to X_2$, then, for every $k \geq n$, the clone segment $\mathrm{Clo}_k(X_1, \mathrm{Top})$ cannot be fully embedded into $\mathrm{Clo}(X_2, \mathrm{Top})$ and the clone segment $\mathrm{Clo}_k(X_2, \mathrm{Top})$ cannot be fully embedded into $\mathrm{Clo}(X_1, \mathrm{Top})$.

2.9. The situation concerning uniformly continuous maps is quite analogous. Let Σ, $\ldots$, $X = (P, \varrho_\kappa)$ be as above. Let $n \geq 1$. Let us call

representative uniformly continuous n-cell

every map $P^n \to P$ of the form

$$p_{\sigma_0} \circ \ldots \circ p_{\sigma_{k-1}} \circ p_\sigma$$

where $\sigma_0, \ldots, \sigma_{k-1} \in \Sigma_1$, $\sigma \in \Sigma_n$, all the maps p_{σ_i}, $i = 0, \ldots, k-1$, are not uniformly continuous, i.e. $\kappa(\sigma_i) \in \{1,2\}$, while p_σ is uniformly continuous, i.e. $\kappa(\sigma) = 3$.

In the paragraphs 4 and 5 of [11], it is also proved that, for the spaces $X_1 = (P, \varrho_{\kappa_1})$, $X_2 = (P, \varrho_{\kappa_2})$ as above, if the number of representative uniformly continuous n-cells $X_1^n \to X_1$ is different from the number of representative uniformly continuous n-cells $X_2^n \to X_2$, then, for every $k \geq n$, the clone segment $\mathrm{Clo}_k(X_1, \mathrm{Unif})$ cannot be fully embedded into $\mathrm{Clo}(X_2, \mathrm{Unif})$ and $\mathrm{Clo}_k(X_2, \mathrm{Unif})$ cannot be fully embedded into $\mathrm{Clo}(X_1, \mathrm{Unif})$.

In fact, stronger result is proved in [11] admitting also to compare the segments in $\mathrm{Clo}(X_1, \mathrm{Top})$ and $\mathrm{Clo}(X_1, \mathrm{Unif})$, but we shall not need it here.

In the next paragraph, we prove our Theorem by means of the method just described.

3 The Proof of the Theorem

3.1. Let a set V and maps $\mathcal{V}_c, \mathcal{V}_u : V \times V \to \{1, 2, \ldots, \infty\}$ satisfying (1), (2), (3) be given. By 2, it is sufficient to find a type Σ and maps $\kappa_v : \Gamma \to \{1, 2, 3\}$, $v \in V$, such that $\mathrm{card}\,\Sigma_0 \geq 2^{\aleph_0} + \mathrm{card}\,\Gamma$ and, for every v_1, v_2, the following statements are satisfied:

if $n < \mathcal{V}_c(v_1, v_2)$, then the set of all Σ-terms in variables $x_0, \ldots, x_{n-1}$ without c-forbidden subterms are the same for κ_{v_1} as for κ_{v_2} but the number of the representative continuous $\mathcal{V}_c(v_1, v_2)$-cells is distinct for κ_{v_1} and κ_{v_2} whenever $\mathcal{V}_c(v_1, v_2) < \infty$

and simultaneously

if $n < \mathcal{V}_u(v_1, v_2)$, then the set of all Σ-terms in variables $x_0, \ldots, x_{n-1}$ without u-forbidden subterms are the same for κ_{v_1} as for κ_{v_2} but the number of the representative uniformly continuous $\mathcal{V}_u(v_1, v_2)$-cells is distinct for κ_{v_1} and κ_{v_2} whenever $\mathcal{V}_u(v_1, v_2) < \infty$.

3.2. For this reason, it is sufficient to find a collection of pairs of sequences of cardinal numbers

$$\mathbf{C} = \{(\{\gamma_n^{(v)}\}_{n=1}^{\infty}, \{\delta_n^{(v)}\}_{n=1}^{\infty}) \mid v \in V\}$$

such that, for every $v_1, v_2 \in V$,

(α)	$\gamma_n^{(v_1)} \geq \delta_n^{(v_1)}$	for all $n = 1, 2, \ldots,$
	$\delta_1^{(v_1)} \geq \aleph_0$ and $\delta_n^{(v_1)} > \gamma_1^{(v_1)}$	for all $n = 2, 3, \ldots;$
(β)	$\gamma_n^{(v_1)} = \gamma_n^{(v_2)}$	for all $n < \mathcal{V}_c(v_1, v_2)$ but
	$\gamma_n^{(v_1)} \neq \gamma_n^{(v_2)}$	for $n = \mathcal{V}_c(v_1, v_2)$;
(γ)	$\delta_n^{(v_1)} = \delta_n^{(v_2)}$	for $n < \mathcal{V}_u(v_1, v_2)$ but
	$\delta_n^{(v_1)} \neq \delta_n^{(v_2)}$	for $n = \mathcal{V}_u(v_1, v_2)$.

Having such a collection, we construct Σ and $\kappa_v : \Gamma \to \{1, 2, 3\}$, $v \in V$, with the required properties as follows (as usual, each $\gamma_n^{(v)}$ is the set of all ordinals less than $\gamma_n^{(v)}$ and analogously for $\delta_n^{(v)}$; hence $\delta_n^{(v)} \subseteq \gamma_n^{(v)}$): the set Σ_n of all n-ary operational labels is just $\sup_{v\in V}\gamma_n^{(v)}$, $n = 1,2,\ldots$, and Σ_0 is a set with $\operatorname{card}\Sigma_0 \geq 2^{\aleph_0} + \operatorname{card}\bigcup_{n=1}^{\infty}\Sigma_n$. For every $v \in V$, $n \geq 1$, $\sigma \in \Sigma_n$, we put

$$\begin{aligned} \kappa_v(\sigma) &= 3 \quad \text{whenever } \sigma \in \delta_n^{(v)}\,; \\ \kappa_v(\sigma) &= 2 \quad \text{whenever } \sigma \in \gamma_n^{(v)} \setminus \delta_n^{(v)}\,; \\ \kappa_v(\sigma) &= 1 \quad \text{whenever } \sigma \in \Sigma_n \setminus \gamma_n^{(v)}\,. \end{aligned}$$

Then, for every $v \in V$, the number of the representative uniformly continuous n-cells is precisely $\delta_n^{(v)}$ and the number of the representative continuous n-cells is $\gamma_n^{(v)}$ (because $\delta_n^{(v)} > \gamma_1^{(v)}$ for $n = 2,3,\ldots$), so that, for the spaces X_1, X_2 on P with the metrics corresponding to κ_{v_1} and κ_{v_2}, $v_1, v_2 \in V$, we have

$$\begin{aligned} &\operatorname{Clo}_n(X_1,\mathrm{Top}) \not\simeq \operatorname{Clo}_n(X_2,\mathrm{Top}) \text{ whenever } n \geq \mathcal{V}_c(v_1,v_2) \text{ and} \\ &\operatorname{Clo}_n(X_1,\mathrm{Unif}) \not\simeq \operatorname{Clo}_n(X_2,\mathrm{Unif}) \text{ whenever } n \geq \mathcal{V}_u(v_1,v_2)\,. \end{aligned}$$

For $n < \mathcal{V}_c(v_1,v_2)$, the sets of the Σ-terms in variables $x_0,\ldots,x_{n-1}$ without c-forbidden subterms are the same for X_1 as for X_2: $\kappa_{v_1}(\sigma) \in \{2,3\}$ iff $\kappa_{v_2}(\sigma) \in \{2,3\}$ whenever $\operatorname{ar}(\sigma) = k \leq n$ and operations in $\bigcup_{k=n+1}^{\infty}\Sigma_k$ cannot create c-forbidden subterms *in variables* $x_0,\ldots,x_{n-1}$ because their arity k is too large to get term $\sigma(x_{i(0)},\ldots)$ with distinct variables $x_{i(0)},\ldots$; analogously for u-forbidden subterms whenever $n < \mathcal{V}_u(v_1,v_2)$.

3.3. It remains to construct a collection $\mathbf{C} = \{(\{\gamma_n^{(v)}\}_{n=1}^{\infty}, \{\delta_n^{(v)}\}_{n=1}^{\infty}) \mid v \in V\}$ with the properties (α), (β), (γ) in 3.2. Given $n \in \{1,2,\ldots\}$, we define binary relations C_n and D_n on V by

$$\begin{aligned} (v_1,v_2) \in C_n \quad &\text{iff } n < \mathcal{V}_c(v_1,v_2)\,, \\ (v_1,v_2) \in D_n \quad &\text{iff } n < \mathcal{V}_u(v_1,v_2)\,. \end{aligned}$$

The axioms (1), (2), (3) guarantee that all the relations C_n, D_n, $n = 1,2,\ldots$, are equivalences. Put $\mathcal{C}_n = V/C_n$, $\mathcal{D}_n = V/D_n$. Now, it is an easy exercise to find one-to-one maps of $\mathcal{C}_n$, $\mathcal{D}_n$, $n = 1,2,\ldots$, into the class of all cardinal numbers leading to a collection $\mathbf{C}$ with the required properties.

4 Concluding Remarks and Problems

4.1. In [11], the general method described here is applied to the investigation of the relation of continuous and uniformly continuous maps of finite powers of two metric spaces X_1 and X_2. Besides of our

$$\begin{aligned} c &= \sup\{n+1 \mid \operatorname{Clo}_n(X_1,\mathrm{Top}) \simeq \operatorname{Clo}_n(X_2,\mathrm{Top})\} \text{ and} \\ u &= \sup\{n+1 \mid \operatorname{Clo}_n(X_1,\mathrm{Unif}) \simeq \operatorname{Clo}_n(X_2,\mathrm{Unif})\} \end{aligned}$$

also

$$s_i = \sup\{n+1 \mid \operatorname{Clo}_n(X_i,\mathrm{Top}) \simeq \operatorname{Clo}_n(X_i,\mathrm{Unif})\}, \quad i = 1,2\,,$$

are investigated (the definition of c, u, s_1, s_2 is formally different in [11], but it plays no rôle). An easy proof is presented in [11] that the conditions below are necessary.

$$(*)\qquad \begin{cases} c \neq u & \Rightarrow \min\{s_1, s_2\} \leq \min\{c, u\}\,, \\ s_1 \neq s_2 & \Rightarrow \min\{c, u\} \leq \min\{s_1, s_2\}\,. \end{cases}$$

The proof that the conditions are also sufficient, given in [11], uses the above general method: for every c, u, s_1, s_2 satisfying $(*)$, a type Σ with $\operatorname{card}\Sigma_0 \geq 2^{\aleph_0} + \operatorname{card}\Gamma$, $\Gamma = \Sigma \setminus \Sigma_0$, and maps $\kappa_1, \kappa_2 : \Gamma \to \{1,2,3\}$ are chosen such that the clone segments in Top and in Unif of the metric spaces $X_1 = (P, \varrho_{\kappa_1})$ and $X_2 = (P, \varrho_{\kappa_2})$ realize the given quadruple c, u, s_1, s_2. A joint generalization of this result of [11] and the Theorem presented here would be to investigate not only the maps

$$\mathcal{V}_c, \mathcal{V}_u : V \times V \to \{1, 2, \ldots, \infty\}$$

satisfying the axioms (1), (2), (3) of 1.1, but also a map

$$s : V \to \{1, 2, \ldots, \infty\}$$

such that, for every $v_1, v_2 \in V$, the quadruple $c = \mathcal{V}_c(v_1, v_2)$, $u = \mathcal{V}_u(v_1, v_2)$, $s_i = s(v_i)$, $i = 1, 2$, satisfies $(*)$. Can every such triple of functions $\mathcal{V}_c$, $\mathcal{V}_u$, s be realized by clone segments in Top and in Unif in the above sense? This more complex problem has not been attacked.

4.2. In [11], the relation between isomorphism $\simeq$ and equality $=$ of clone segments is investigated. For every $n, m \in \{1, 2, \ldots, \infty\}$, $n \leq m$, a metric space X is constructed (by means of the general method: by means of a suitable choice of Σ and $\kappa : \Gamma \to \{1,2,3\}$) such that

$$\begin{aligned} \operatorname{Clo}_k(X, \mathrm{Top}) &= \operatorname{Clo}_k(X, \mathrm{Unif}) \quad \text{iff } k \leq n \text{ and} \\ \operatorname{Clo}_k(X, \mathrm{Top}) &\simeq \operatorname{Clo}_k(X, \mathrm{Unif}) \quad \text{iff } k \leq m\,. \end{aligned}$$

This offers a more complex problem than that in 4.1: to define functions $\mathcal{V}_c$, $\mathcal{V}_u$, s by means of the equality $=$ and other functions $\mathcal{V}'_c$, $\mathcal{V}'_u$, s' by means of $\simeq$. Which sixtuples of functions $\mathcal{V}_c$, $\mathcal{V}_u$, s, $\mathcal{V}'_c$, $\mathcal{V}'_u$, s' can be obtained from collections of spaces $\{X_v \mid v \in V\}$? Though this problem has not been attacked, the author believes that the general method of [11] described here in 2 could help.

References

1. J. Adámek, H. Herrlich, G. Strecker: *Abstract and Concrete Categories*, John Wiley & Sons, Inc., New York/Chichester/Brisbane/Toronto/Singapore 1990.
2. G. A. Grätzer: *Universal Algebra*, Springer Verlag 1979.
3. J. de Groot: *Non-archimedean metrics in topology*, Proc. Amer. Math. Soc. **7**(1956), 948–953.
4. W. F. Lawvere: *Functorial semantics of algebraic theories*, Proc. Nat. Acad. Sci. **50**(1963), 869–872.

5. S. MacLane: *Categories for the working mathematician*, Springer, Berlin-Heidelberg-New York 1972.
6. R. McKenzie, G. McNulty and W. Taylor: *Algebras, lattices, varieties*, Volume 1, Brooks/Cole., Monterey, California 1987.
7. J. Sichler, V. Trnková: *On elementary equivalence and isomorphism of clone segments*, Periodica Mathematica Hungarica **32**(1996).
8. W. Taylor: *The clone of a topological space*, Research and Exposition in Mathematics, vol. 13, Helderman Verlag 1986.
9. V. Trnková: *Semirigid spaces*, Trans. Amer. Math. Soc. **343**(1994), 305–325.
10. V. Trnková: *Continuous and uniformly continuous maps of powers of metric spaces*, Topology and its Applications **63**(1995), 189–200.
11. V. Trnková: *Clone segments in* Top *and in* Unif, this volume.

WORKSHOP ON CATEGORICAL TOPOLOGY
L'Aquila, Aug. 31 - Sept. 4

List of Participants:

ADÁMEK J., Braunshweigh, GERMANY
BENTLEY L., Toledo, U.S.A.
BRANDT R.D., Hannover, GERMANY
BRÜMMER G.C.L., Cape Town, SOUTH AFRICA
CAGLIARI F., Bologna, ITALY
CARBONI A., Genova, ITALY
CASTELLINI G.,Mayaguez, PUERTO RICO
CLEMENTINO M., Coimbra, PORTUGAL
CSÁSZÁR A., Budapest, HUNGARY
GÄHLER W., Potsdam, GERMANY
GAVIOLI N., L'Aquila, ITALY
GIULI E., L'Aquila, ITALY
HARDIE K.A.,Cape Town, SOUTH AFRICA
HERRLICH H., Bremen, GERMANY
HOLGATE D., Cape Town, SOUTH AFRICA
HUSEK M., Prague, CZECH REP.
KAMPS K.H., Hagen, GERMANY
KLEISLI H., Fribourg, SWITZERLAND
KOSLOWSKI J., Hannover, GERMANY
KÜNZI H.P., Berne, SWITZERLAND
LORD H., Upland, U.S.A.
LOWEN COLEBUNDERS E., Brussel, BELGIUM
LOWEN B.,Antwerp, BELGIUM
MANTOVANI S., Torino, ITALY
NEL H., Ottawa, CANADA
PEDICCHIO M.C., Trieste, ITALY
PIERANTONIO A., L'Aquila, ITALY

POPPE H., Rostock, GERMANY
PORST H.-E., Bremen, GERMANY
PREUSS G., Berlin, GERMANY
PULTR A., Prague, CZECH REP.
RICHTER G., Bielefeld, GERMANY
ROSICKÝ J., Brno, CZECH REP.
SCHRÖDER J., Phuthaditjhaba, SOUTH AFRICA
SCHWARZ F., Toledo, U.S.A.
SOBRAL M., Coimbra, PORTUGAL
SOUSA L., Viseu, PORTUGAL
STRAMACCIA L., Perugia, ITALY
THOLEN W., Toronto, CANADA
TRNKOVÁ V., Prague, CZECH REP.
TOAN NGUYEN, Bremen, GERMANY
TOZZI A., L'Aquila, ITALY